Praise for Dan Barber's

THE THIRD PLATE

THE NEW YORK TIMES BESTSELLER

PENGUIN BOOKS

THE THIRD PLATE

Dan Barber is the chef and co-owner of Blue Hill, a restaurant in Manhattan's West Village, and Blue Hill at Stone Barns, located within the nonprofit farm and education center Stone Barns Center for Food & Agriculture. He lives in New York City with his wife and daughter.

....

Praise for Dan Barber's *The Third Plate*

"Dan Barber's tales are engaging, funny, and delicious. . . . *The Third Plate* invites inevitable comparisons with Michael Pollan's *The Omnivore's Dilemma*, which Barber invokes more than once. And, indeed, its framework of a foodie seeking truth through visits with sages and personal experiments echoes Pollan's landmark tome. . . . But at the risk of heresy, I would call this *The Omnivore's Dilemma 2.0* . . . *The Third Plate* serves as a brilliant culinary manifesto with a message as obvious as it is overlooked. Promote, grow, and eat a diet that's in harmony with the earth, and the earth will reward you for it. It's an inspiring message that could truly help save our water, air, and land before it's too late."
—*Chicago Tribune*

"Not since Michael Pollan has such a powerful storyteller emerged to reform American food. . . . Barber is helping to write a recipe for the sustainable production of gratifying food."
—*The Washington Post*

"Compelling . . . *The Third Plate* reimagines American farm culture not as a romantic return to simpler times, but as a smart, modern version of it. . . . *The Third Plate* is fun to read, a lively mix of food history, environmental philosophy, and restaurant lore. . . . An important and exciting addition to the sustainability discussion."
—*The Wall Street Journal*

"There hasn't been a call-to-action book with the potential to change the way we eat since Michael Pollan's 2006 release, *The Omnivore's Dilemma*. Now there is. Dan Barber's *The Third Plate: Field Notes on the Future of Food* is a compelling global journey in search of a new understanding about how to build a more sustainable food system. . . . *The Third Plate* is an argument for good rather than an argument against bad. This recipe might at times be challenging, but what's served in the end is a dish for a better future. . . . Barber writes a food manifesto for the ages."
—*Pittsburgh Post-Gazette*

"Not long ago, Dan Barber, the chef and a co-owner of Blue Hill in Manhattan, and the animating force behind the Stone Barns Center for Food and Agriculture in Westchester County, came up with a dish he calls Rotation Risotto. It's a manifesto on a plate, a tricky play on the Italian classic that uses, instead of rice, a medley of lesser-known grains: rye, barley, buckwheat, and millet. . . . Taken together, they argue for a new way of thinking about the production and consumption of food, a 'whole farm' approach that Mr. Barber explores, eloquently and zestfully, in *The Third Plate: Field Notes on the Future of Food*."

—*The New York Times*

"When *The Omnivore's Dilemma*, Michael Pollan's now-classic 2006 work, questioned the logic of our nation's food system, 'local' and 'organic' weren't ubiquitous the way they are today. Embracing Pollan's iconoclasm, but applying it to the updated food landscape of 2014, *The Third Plate* reconsiders fundamental assumptions of the movement Pollan's book helped to spark. In four sections—'Soil,' 'Land,' 'Sea,' and 'Seed'—*The Third Plate* outlines how his pursuit of intense flavor repeatedly forced him to look beyond individual ingredients at a region's broader story—and demonstrates how land, communities, and taste benefit when ecology informs the way we source, cook, and eat."

—*The Atlantic*

"Barber's work is a deeply thoughtful and—offering a 'menu for 2050'—even visionary work for a sustainable food chain."   —*Publishers Weekly*

"In this bold and impassioned analysis, Barber insists that chefs have the power to transform American cuisine to achieve a sustainable and nutritious future."

—*Kirkus Reviews*

"Often, books about sustainability and America's food system are depressing and difficult to get through—but I should have realized I was in better hands. Somewhere in the first section, during a discussion about the structure of soil, I realized that I was completely engrossed. From that point on, it was a fight to read more slowly; I didn't want this book to end. . . . *The Third Plate* is impeccably researched, well written, and achieves its aim: to make us reconsider not only what we want to put on our plates, but how it got there in the first place."

—*Saveur*

"Dan Barber nails it . . . Here in this monumental time in the history of food in our country, Barber takes off the gloves and digs deep, providing a 'subterranean view' of the movement from (beneath) the ground up. Starting at the roots, he simplifies the often complex kaleidoscope of historical food regions, players, trends, ingredients, influences, and philosophies, and then paints a broad-stroke picture of neo-American cuisine using a fine-point brush. From the dynamic and unique perspective of a chef on a farm, who was initially on the hunt for better flavor, Barber carefully plots a Sherlock Holmes–style story line at times filled with clues, discoveries, and a-ha revelations. . . . To gain an educated, informed, and candid perspective on what the American dinner table should begin to look like as the peaceful era of farm-to-table dining dies on the vine, *The Third Plate: Field Notes on the Future of Food*, is a must-read."

—*Edible Magazine*

"In *The Third Plate*, Barber turns his chef's eye towards food: namely, the question of what we'll be eating in the future, taking into consideration changes in environment, supply and demand, and the way that we consume food. . . . It's a bit like *The Omnivore's Dilemma* as imagined by a chef curious about the future of food in his industry. . . . *The Third Plate* is a provocative look at how 'farm to table' cooking is sort of a beautiful fantasy, and suggests that to really eat and get food from the earth, we're going to have to change our ways."   —*Flavorwire*

"Dan Barber's new book, *The Third Plate*, is an eloquent and thoughtful look at the current state of our nation's food system and how it must evolve. Barber's wide range of experiences, both in and out of the kitchen, provide him with a rare perspective on this pressing issue. A must-read." —Vice President Al Gore

"In this compelling read Dan Barber asks questions that nobody else has raised about what it means to be a chef, the nature of taste, and what 'sustainable' really means. He challenges everything you think you know about food; it will change the way you eat. If I could give every cook just one book, this would be the one."

—Ruth Reichl, author of *Garlic and Sapphires* and *Tender at the Bone*

"Dan Barber is not only a great chef, he's also a fine writer. His vision of a new food system—based on diversity, complexity, and a reverence for nature—isn't utopian. It's essential."

—Eric Schlosser, author of *Fast Food Nation* and *Command and Control*

"I thought it would be impossible for Dan Barber to be as interesting on the page as he is on the plate. I was wrong."

—Malcolm Gladwell, author of *David and Goliath* and *The Tipping Point*

THE
THIRD
PLATE

*Field Notes
on the Future of Food*

DAN BARBER

PENGUIN BOOKS

PENGUIN BOOKS
Published by the Penguin Group
Penguin Group (USA) LLC
375 Hudson Street
New York, New York 10014

USA | Canada | UK | Ireland | Australia
New Zealand | India | South Africa | China
penguin.com
A Penguin Random House Company

First published in the United States of America by The Penguin Press,
a member of Penguin Group (USA) LLC, 2014
Published in Penguin Books 2015

Illustrations by John Burgoyne

Image on pages 42–43: The Land Institute

THE LIBRARY OF CONGRESS HAS CATALOGED THE HARDCOVER EDITION AS FOLLOWS:
Barber, Dan.
The third plate : field notes on the future of food / by Dan Barber.
pages cm
Includes bibliographical references and index.
ISBN 978-1-59420-407-4 (hc.)
ISBN 978-0-14-312715-4 (pbk.)
1. Natural foods—United States. 2. Seasonal cooking—United States.
3. Agriculture—United States. I. Title.
TX369.B3625 2014
641.3'02—dc23
2013039966

Printed in the United States of America
10

Designed by Marysarah Quinn

For Aria Beth Sloss

CONTENTS

A corncob, dried and slightly shriveled, arrived in the mail not long after we opened Blue Hill at Stone Barns. Along with the cob was a check for $1,000. The explanation arrived the same day, in an e-mail I received from Glenn Roberts, a rare-seeds collector and supplier of specialty grains. Since Blue Hill is part of the Stone Barns Center for Food and Agriculture, a multipurpose farm and education center, Glenn wanted my help persuading the vegetable farmer to plant the corn in the spring. He said the corn was a variety called New England Eight Row Flint.

There is evidence, Glenn told me, that Eight Row Flint corn dates back to the 1600s, when, for a time, it was considered a technical marvel. Not only did it consistently produce eight fat rows of kernels (four or five was the norm back then; modern cobs have eighteen to twenty rows), but it also had been carefully selected by generations of Native Americans for its distinctive flavor. By the late 1700s the corn was widely planted in western New England and the lower Hudson Valley, and later it was found as far as southern Italy. But a brutally cold winter in 1816 wiped out the New England crop. Seed reserves were exhausted to near extinction as most of the stockpiled corn went to feed people and livestock.

The cob Glenn had sent was from a line that had survived for two hundred years in Italy under the name Otto File ("eight rows"), which he hoped to restore to its place of origin. By planting the seed, he wrote, we would be growing "an important and threatened historic flavor of Italy while simultaneously repatriating one of New England's extinct foodways. Congratulations on your quest, Dan, and thank you for caring." Glenn added, in case I *didn't* care, that the Eight Row was "quite possibly the most flavorful polenta corn on the planet, and absolutely unavailable in the U.S." At harvest he promised another $1,000. He wanted nothing in return, other than a few cobs to save for seed.

If his offer sounds like a home run for Stone Barns, it was. Here was a chance to recapture a regional variety and to honor a Native American crop with historical significance. For me, it was a chance to cook with an ingredient no other restaurant could offer on its menu (catnip for any chef) and to try the superlative polenta for myself.

Yet I carried the corncob over to Jack Algiere, the vegetable farmer, with little enthusiasm. Jack is not a fan of growing corn, and, with only eight acres of field production on the farm, you can't blame him for dismissing a plant that demands so much real estate. Corn is needy in other ways, too. It's gluttonous, requiring, for example, large amounts of nitrogen to grow. From the perspective of a vegetable gardener, it's the biological equivalent of a McMansion.

In the early stages of planning Stone Barns Center, I told Jack about a farmer who was harvesting immature corn for our menu. It was a baby cob, just a few inches long, the kernels not yet visible. You ate the whole cob, which brought to mind the canned baby corn one finds in a mediocre vegetable stir-fry. Except these tiny cobs were actually tasty. I wanted to impress Jack with the novelty of the idea. He was not impressed.

"You mean your farmer grows the whole stalk and then picks the cobs when they're still little?" he said, his face suddenly scrunched up, as if he were absorbing a blow to the gut. "That's nuts." He bent over and nearly touched the ground with his right hand, then stood up on his toes and, with his left hand, reached up, high above my head, hiking his eyebrows to indicate just how tall a corn's stalk grows. "Only after all that growth will corn even *begin* to think about producing the cob. That big, thirsty, jolly green giant of a stalk—which *even* when it produces full-size corn has to be among the plant kingdom's most ridiculous uses of Mother Nature's energy—and what are *you* getting for all that growth? You're getting this." He flashed his pinky finger. "That's all you're getting." He rotated his hand so I could see his finger from all angles. "One tiny, pretty flavorless bite of corn."

—•—

One summer when I was fourteen years old, Blue Hill Farm, my family's farm in Massachusetts, grew only corn. No one can remember why. But it was the strangest summer. I think back to it now with the same sense of bewilderment I felt as a child encountering the sea of gold tassels where the grass had always been.

Before Blue Hill Farm became a corn farm for a summer, I helped make hay for winter storage from one of the eight pasture fields. We began in early August, loading bales onto a conveyor belt and methodically packing them, Lego-like, into the barn's stadium-size second floor. By Labor Day the room was filled nearly to bursting, its own kind of landscape.

Making hay meant first cutting the grass, which—for me, anyway—meant riding shotgun in a very large tractor for hours each day, crouching silently next to one of the farmers and studying the contours of the fields. And so, by way of no special talent, just repetition, I learned to anticipate the dips and curves in the fields, the muddy, washed-out places, the areas of thick shrubbery and thinned grasses—when to brace for a few minutes of a bumpy ride and when to duck under a protruding branch.

I internalized those bumps and curves the way my grandmother Ann Strauss internalized the bumps and curves of Blue Hill Road by driving it for thirty years. She always seemed to be going to town (to get her hair done) or coming back (from running errands). Sometimes my brother, David, and I were with her, and we used to laugh in the backseat, because Ann (never "Grandma," never "Grandmother") rounded the corners in her Chevy Impala at incredible speeds, maneuvering with the ease and fluency of a practiced finger moving over braille. Her head was often cranked to the left or to the right, antennae engaged, inspecting a neighbor's garden or a renovated screened-in porch. (She sometimes narrated the intrigue happening inside.) During these moments her body took over, autopiloting around corners without having to slow down, swerving slightly to avoid the ditch just beyond Bill Riegleman's home.

Often, on the last leg of the drive, Ann would tell us the story of how she came to buy the farm in the 1960s, a story she had told a thousand times before. Back then, the property was a dairy operation owned by the Hall brothers, whose family had farmed the land since the late 1800s.

"You know, I used to walk up this road every week for years; sometimes every day," she would say, as if telling the story for the first time. "I loved Blue Hill Farm more than any place in the world." At the top of Blue Hill Road was four hundred acres of open pasture. "But what a mess! I couldn't believe it, really. They had cows pasturing in the front yard. The house was run-down, and so dirty. They didn't have a front door—climbed in and out through the kitchen window, for heaven's sake. And you know what? I loved it. I loved the fields, I loved the backdrop of blue hills, I loved that I felt like a queen every time I came up here."

Whenever Ann saw the Hall brothers, she would let them know she wanted to buy the farm. "But they just laughed. 'Ms. Strauss,' they'd say, 'this farm's been in our family for three generations. We're never selling.' So I'd return the next week, and they'd say the same thing: 'Never selling.' This went on for many years, until one day I arrived at the farm and one of the brothers came running over, out of breath. 'Ms. Strauss, do you want to buy this farm?' Just like that! I couldn't believe it. He didn't even let me answer. 'This morning my brother and I got into the biggest fight. If we don't sell now, we're going to kill each other.' I said I was interested. For sure I would buy a piece of it. 'Ma'am,' he said, 'we're selling it now—the whole thing, or forget it. Right now.'

"So I said yes. I hadn't even been inside the farmhouse, and I didn't know where the property began and where it ended. But it didn't matter. What else was I going to say? I just knew this was the place."

The dairy part of Blue Hill Farm disappeared with the Hall brothers, but

Ann began pasturing beef cattle, because she wanted the fields to remain productive and because she enjoyed showing off the view to her friends; the image of cows dotting the iconic New England landscape is still fit for a coffee table book.

At the time, I didn't know about the importance of preserving that kind of view. I just enjoyed the tractor rides, the look back at the field lined with the long, curving windrows of just-cut grass, and then, as I got older, the hard work of baling and storing hay for the winter.

Which, as it happened, suddenly came to an end because of the summer of corn. The maize invasion meant the cows grazed at another farm, which meant the hours of fixing fences and lugging salt licks and watching the herd lie and chew cud before a rainstorm came to an end, too. And since you don't tend to a field of corn—in the same way you don't really tend to a houseplant— it meant the baler and the hay wagons, the farm interns, the red Ford F-150 pickup truck, the big iced tea jug, and all the sweaty work went with them.

To look out from the front porch at what had always been fields of grass transformed suddenly into amber fields of corn felt not quite right. Same home, new furniture. Endless rows of corn are one of those things that are beautiful to behold at a distance. They tremble in great waves with the slightest breeze, and you think of beauty and abundance. Up close it's a different story. For one thing, the abundance is relative. We can't eat feed corn— I tried to that summer. The enormous cobs line the stalks like loaded missiles, tasting nothing like the sweet stuff we chainsaw through in August. And there's little in the way of beauty. The long, straight rows take on a military-like discipline. They cut across bare soil, hard corners and creased edges replacing the natural contours of the field that I once knew so well.

—

I handed Jack the Eight Row Flint cob from Glenn and explained the situation, fearing that if the idea of growing corn offended him, the check for $1,000 might upset him even more. But I was wrong about both.

He loved the idea. "Look," Jack said to me—and in Jack's parlance, "Look" is a happy thing to hear. "Look" says: *I know I may have given you some differing opinion in the past, but there are exceptions to my rule, and this is one of them.* "This corn is the rare case of flavor driving genetics," he said, reminding me of the generations of farmers who had selected and grown Eight Row Flint for its superior flavor, not solely for its yield, as is the case with most modern varieties. "How often do you get to be a part of that in your lifetime?"

So far, so good. But Jack went a step further. He planted the Eight Row Flint like the Iroquois planted most of their corn—alongside dry beans and squash, a companion planting strategy called the Three Sisters. On the continuum of farming practices, Three Sisters is at the opposite end from how corn is typically grown, with its military-row monocultures and chemical-fed soil. The logic is to carefully bundle crops into relationships that benefit each other, the soil, and the farmer. The beans provide the corn with nitrogen (legumes draw nitrogen from the air into the soil); the corn stalk provides a natural trellis for the climbing beans (so Jack wouldn't need to stake the beans); and the squash, planted around the base of the corn and the beans, suppresses weeds and offers an additional vegetable to harvest in the late fall.

It was a masterful idea—mimicking the successful Native American strategy while taking out a small insurance policy on the Eight Row Flint. Even if the corn failed to germinate, Jack could still harvest the other crops, and in the meantime he'd show visitors to the Stone Barns Center a valuable historical farming technique. And yet I couldn't help but feel skeptical as I watched him plant the corn kernels and companion seeds into mounds of rich soil. I had nothing against honoring agricultural traditions, but I didn't need a sisterhood of beneficial relationships. I needed a polenta with phenomenal flavor.

As luck would have it (or maybe it was the sisterhood, after all), the Eight Row Flint had nearly perfect germination. Following the harvest in late September, Jack hung the corn upside down in a shed and waited for the moisture to evaporate. By late November, just in time for the long winter march of root vegetables, he triumphantly set a dried cob on my desk. It looked nearly too perfect, like a prop for an elementary-school production of the First Thanksgiving.

"Voilà!" he said, so pleased with himself he seemed to wriggle with the sheer joy of it. "They're ready to go. Tell me when you want them."

"Today!" I was feeding off Jack's energy. "We'll make polenta and then . . ." And then I realized something I hadn't considered: the corn needed to be ground. I didn't have a mill.

The truth is that I had never really considered the corncob behind the cornmeal. It hadn't crossed my mind once in twenty years of preparing polenta. Polenta was polenta. Of course I knew it came from corn, just as I knew bread came from wheat. Beyond the obvious, I had never needed to know more.

A week later, just before dinner service, our new tabletop grinder arrived. The engine whirred as it pulverized the kernels into a finely milled dust. I toasted the ground maize lightly and cooked it right away in water and salt. I'd like to say I cooked the Eight Row Flint the way Native Americans cooked it, stirring a clay pot all day with a wooden spoon over an open hearth. But the pot was carbonized steel, the spoon metal, and the hearth an induction cooktop that heats by magnetic force. It didn't matter. Before long the polenta was smooth and shiny. I continued stirring, which is when suddenly the pot began smelling like a steaming, well-buttered ear of corn. It wasn't just the best polenta of my life. It was polenta I hadn't imagined possible, so *corny* that breathing out after swallowing the first bite brought another rich shot of corn flavor. The taste didn't so much disappear as slowly, begrudgingly fade. It was an awakening. But the question for me was: Why? How had I assumed all those years that polenta smelled of nothing more than dried meal? It's really not too much to ask of polenta to actually taste like the corn. But back then, I couldn't have imagined the possibility until it happened. Jack's planting strategy, as artful as a sonnet, combined with the corn's impeccable genetics, changed how I thought about good food, and good cooking.

With remarkable, almost ironic regularity, I have found myself repeating this kind of experience. Different farm, different farmer, same narrative arc. I am reminded that truly flavorful food involves a recipe more complex than anything I can conceive in the kitchen. A bowl of polenta that warms your senses and lingers in your memory becomes as straightforward as a mound of corn and as complex as the system that makes it run. It speaks to something beyond the crop, the cook, or the farmer—to the entirety of the landscape, and how it fits together. It can best be expressed in places where good farming and delicious food are inseparable.

This book is about these stories.

If that sounds like a chronicle of a farm-to-table chef, it is—sort of.

Blue Hill has been defined by that term since Jonathan Gold, the head reviewer for *Gourmet* magazine, called us a farm-to-table restaurant just a few months after we opened Blue Hill in New York City in the spring of 2000. He visited our Greenwich Village restaurant on a night when asparagus was everywhere on the menu. It might have been because of the achingly short asparagus season, or because they were at the height of their flavor. Or because they had been grown locally and driven down to the city by Hudson Valley family farmers.

It was all of these things, but it was something much more straightforward, too. After returning from the farmers' market that morning and unloading a mountain of asparagus packed into the trunk of a yellow cab, I discovered another mountain of them already in the walk-in refrigerator—a week's worth, at least—and went into a rage about the disorganization in the kitchen. How could the market order have included asparagus when we were already overloaded? I had the cooks clean out the refrigerator and prep the cases of asparagus that were piled high and getting old. And I told them that they had to be used in every dish. I must have sounded serious, because they appeared in *every* dish. Halibut with leeks and asparagus, duck with artichokes and asparagus, chicken with mushrooms and asparagus. The asparagus soup that night even had the addition of roasted asparagus floating on its surface.

Instead of writing with puzzlement at the asparagus blitzkrieg, Jonathan Gold celebrated what he misinterpreted as intent. "What does it mean to be a farm-oriented restaurant in New York City?" he wrote as the opening line for the review, describing Blue Hill as a true representation of farm-to-table cooking. Farm-to-table is now a much abused descriptor, but back then the review pithily defined who we were, before we even knew who we were.

———

Farm-to-table has since gone from a fringe idea to a mainstream social movement. Its success comes with mounting evidence that our country's indomitable and abundant food system, for so long the envy of the world, is unstable, if not broken. Eroding soils, falling water tables for irrigation, collapsing fisheries, shrinking forests, and deteriorating grasslands represent only a handful of the environmental problems wrought by our food system— problems that will continue to multiply with rising temperatures.

Our health has suffered, too. Rising rates of food-borne illnesses, malnutrition, and diet-related diseases such as obesity and diabetes are traced, at least in part, to our mass production of food. The warnings are clear: because we eat in a way that undermines health and abuses natural resources (to say nothing of the economic and social implications), the conventional food system cannot be sustained.

Fixtures of agribusiness such as five-thousand-acre grain monocultures and bloated animal feedlots are no more the future of farming than eighteenth-century factories billowing black smoke are the future of manufacturing.

Though most of the food we eat still comes from agriculture that's mired in this mind-set—extract more, waste more—the pulse of common sense suggests this won't last. It will, in the words of the environmental writer Aldo Leopold, "die of its own too-much."

Farm-to-table—whose enthusiasts are called artisanal eaters and locavores—took root as the new food movement's answer to the conventional food system. It was also, undeniably, a reaction against a global food economy that erodes cultures and cuisines. It's about seasonality, locality, and direct relationships with your farmer. It's also about better-tasting food, which is why chefs have been so influential in broadening the movement. Most chefs support the farmers' market for the same reason that most doctors are drawn to prenatal care. As someone whose job it is to address the end result, how can you not care about the beginning? A growing number of chefs have joined the ranks of activists advancing the agenda of changing our food system.

———

The idea of chef as activist is a relatively new one.

It was the nouvelle cuisine chefs of the 1960s who, breaking with an onerous tradition of classic French cuisine, stepped out of the confines of the kitchen and launched modern gastronomy. They created new styles of cooking based on seasonal flavors, smaller portions, and artistic plating. In doing so, they established the authority of the chef, giving him a platform of influence that has only continued to expand.

Fifty years later, chefs are known for their ability to create fashions and shape markets. What appears on a menu in a white-tablecloth restaurant one day trickles down to the bistro the next, and eventually influences everyday food culture. After Wolfgang Puck reimagined pizza in the 1980s at his fine-dining restaurant Spago, in Los Angeles—smoked salmon instead of tomatoes; crème fraîche instead of cheese—gourmet pizza spread to every corner of America, eventually culminating in the supermarket frozen food aisle. We now have the power to quickly popularize certain products and ingredients—in some cases, as with certain fish, to the point of commercial extinction—and increasingly we do, with dizzying speed and effect. But we also possess the potential to get people to rethink their eating habits.

Which is where farm-to-table chefs have been most effective. Today the message has gone viral, highlighting the perils of our nation's diet and exposing the connections between how we eat and our heavy environmental footprint. We raise money for school lunch programs and nutrition education and shed light on the real costs of processed and packaged food. Michael Pollan's *The Omnivore's Dilemma* and Wendell Berry's *The Unsettling of America* are on our cookbook shelves, as much for reference as for inspiration. In Berry's words, we understand that eating "is inescapably an agricultural act, and that how we eat determines, to a considerable extent, how the world is used."

And yet, for all the movement's successes, and the accompanying shift in popular consciousness, the gains haven't changed, in any fundamental way, the political and economic forces shaping how most of the food in this country is grown or raised.

Nor, for that matter, have they changed the culture of American cooking. Americans have more opportunities to opt out of the conventional food chain than ever before (farmers' markets are ubiquitous; organic food is widely available) and more information about how to do it (innumerable cooking shows and easy access to a world of online recipes), but the food culture—the *way* we eat, which is different than *what* we eat—has remained largely unaffected.

How *do* we eat? Mostly with a heavy hand. For a long time, the prototypical American meal has featured a choice cut—like a seven-ounce steak or a boneless, skinless chicken breast or a fillet of salmon—and a small side of vegetables or grains. The architecture of this plate has shifted little throughout the years. It's become a distinctly American expectation of what's for dinner, seven days a week, every week of the year, protein-centric proof that our nation can produce staggering amounts of food.

And it persists even among the most forward-thinking farm-to-table advocates. That much became clear to me on a summer night just a year after we opened Blue Hill at Stone Barns. Standing in the kitchen a few minutes into the beginning of service and staring down at a collection of newly sauced entrées ready for the dining room, I experienced what I think ranks as a revelation. I started asking myself a series of questions that took a turn toward abstraction. Among them was: *Is a restaurant menu really sustainable?*

Chefs are often asked how their menus are created, especially how new dishes come into existence. Some of us are inspired by a favorite food from childhood, or we're drawn to rethinking classic preparations. A new kitchen tool may spark an idea, or a visit to the museum. As with anything creative, it's tough to pinpoint the origin, but whatever the process, the scaffolding for the idea forms first; assembling the ingredients comes later.

We forget that for most of human history, it happened the other way around. We foraged and then, out of sheer necessity, transformed what we found into something else—something more digestible and storable, with better nutrition and flavor. Farm-to-table restaurants promote their menus as having evolved in that order: forage first—maybe with a morning's stroll through the farmers' market—and create later. The promise of farm-to-table cooking is that menus take their shape from the constraints of local agriculture and celebrate them.

Blue Hill at Stone Barns was conceived with that promise of further shortening the food chain. David Rockefeller, grandson of patriarch John D.

Rockefeller, set out to preserve a memory—the place where he sipped warm milk from the lid of the milking jug. (The Normandy-style structures were built in the 1930s as part of the family's old eighty-acre dairy farm, twenty miles north of New York City.) He was also intent on making a tangible tribute to his late wife, Peggy, who raised breeding cattle on the farm and founded the American Farmland Trust to curb the loss of productive farmland.

Stone Barns Center for Food and Agriculture, along with the restaurant Blue Hill at Stone Barns, opened in the spring of 2004. Mr. Rockefeller donated the land and funded the renovation of the barns into an educational center, a place that he and his daughter Peggy Dulany envisioned would promote local agriculture with programs for children and adults. He also funded a working farm. Vegetables and fruits are managed by Jack, who oversees a 23,000-square-foot greenhouse and an eight-acre outdoor production field. The animals—pigs, sheep, chickens, geese, and honeybees—rotate around the more than twenty acres of pasture and woodlands, under the direction of Craig Haney, the livestock manager.

Take the harvests from the Stone Barns fields just outside the kitchen window, or from farms within a radius of a hundred or so miles, and incorporate them into the menu. How much more farm-to-table can you get?

But during that summer evening, the shortsightedness of the system—and perhaps the reason farm-to-table has failed to transform the way most of our food is grown in this country—suddenly seemed obvious. In just the first few minutes of a busy dinner service, we had already sold out of a new entrée of grass-fed lamb chops.

For much of that month, I had been preparing the waiters for the farm's first lamb—a Finn-Dorset breed fed only grass. The waiters learned about Craig's intensive pasture management, about how the sheep were moved twice a day onto the choicest grass, and how the chickens followed the sheep to help ensure even better grass for the next time around. It was among the more interesting things happening on the farm, if not the most delicious.

To honor the addition of lamb to the menu, we carefully sketched out a new dish, which included roasted zucchini and a minted puree made with the skins. I visited the farmers' market on an early-morning sweep to supplement whatever zucchini Jack promised to harvest.

That night, the waiters (convincing as waiters tend to be when they get their hands on a good story) succeeded in selling the lamb chops to each one of those first tables, sometimes to every diner at the table. There are sixteen individual chops per lamb. We had three animals, so forty-eight chops were ready for roasting, three to a plate. After months of work, years of grass management, a four-hour round-trip to the slaughterhouse, and a butcher breaking down the animals with the patience and skill of a surgeon, we had sold out in the time it takes to eat a hot dog.

Craig's lamb chops were replaced with grass-fed lamb chops from another farm. Diners, unaware of what they were missing, were happy. So where was the problem? A year into the life of Stone Barns, the farm's harvests were better than expected, the restaurant was busier than we'd anticipated, and our network of local farmers was expanding. With my sudden qualms about our tactics, I might have been accused of looking for the hole in the doughnut.

And yet, the night of the lamb-chop sellout, I began to think that the hole in our doughnut was the menu itself, or our Western conception of it, which still obeyed the conventions of a protein-centric diet. Sure, our meat was grass-fed (and our chicken free-range, and our fish line-caught) and our vegetables local and, for the most part, organic. But we were still trying to fit into an established system of eating, based on the hegemony of the choicest cuts. By cooking with grass-fed lamb and by supporting local farmers, we were opting out of the conventional food chain, shortening food miles, and working with more flavorful food. But we weren't addressing the larger problem. The larger problem, as I came to see it, is that farm-to-table allows, even celebrates, a kind of cherry-picking of ingredients that are often ecologically demanding and expensive to grow. Farm-to-table chefs may claim to base their cooking on whatever the farmer's picked that day (and I should know, since I do it often), but whatever the farmer has picked that day is really about an expectation of what will be purchased that day. Which is really about an expected way of eating. It forces farmers into growing crops like zucchini and tomatoes (requiring lots of real estate and soil nutrients) or into raising enough lambs to sell mostly just the chops, because if they don't, the chef, or even the enlightened shopper, will simply buy from another farmer.

Farm-to-table may sound right—it's direct and connected—but really the farmer ends up servicing the table, not the other way around. It makes good agriculture difficult to sustain.

We did away with the menus a year later. Instead diners were presented with a list of ingredients. Some vegetables, like peas, made multiple appearances throughout the meal. Others, like rare varieties of lettuce, became part of a shared course for the table. Lamb rack for a six-top; lamb brain and belly for a table of two. No obligations. No prescribed protein-to-vegetable ratios. We merely outlined the possibilities. The list was evidence that the farmers dictated the menu. I was thrilled.

And then, after several years of experimenting, I wasn't. My cooking did not amount to any radical paradigm shift. I was still sketching out ideas for dishes first and figuring out what farmers could supply us with later, checking off ingredients as if shopping at a grocery store.

Over time, I recognized that abandoning the menu wasn't enough. I wanted an organizing principle, a collection of dishes instead of a laundry list of ingredients, reflecting a whole system of agriculture—a cuisine, in other words.

The very best cuisines—French, Italian, Indian, and Chinese, among others—were built around this idea. In most cases, the limited offerings of peasant farming meant that grains or vegetables assumed center stage, with a smattering of meat, most often lesser cuts such as neck or shank. Classic dishes emerged—pot-au-feu in French cuisine, polenta in Italian, paella in Spanish—to take advantage of (read: make delicious) what the land could supply.

The melting pot of American cuisine did not evolve out of this philosophy. Despite the natural abundance—or, rather, as many historians argue, because of the abundance—we were never forced into a more enlightened way of eating. Colonial agriculture took root in the philosophy of extraction. Conquer and tame nature rather than work in concert with nature. The exploitative relationship was made possible by the availability of large quantities of enormously productive land.

Likewise, American cooking was characterized, from the beginning, by its immoderation—large amounts of meat and starch that grossly outweighed the small portions of fruits and vegetables. None of it was prepared with special care. In 1877, Juliet Corson, the head of the New York Cooking School, lamented the wastefulness of American cooks. "In no other land," she wrote, "is there such a profusion of food, and certainly in none is so much wasted from sheer ignorance, and spoiled by bad cooking." A real food culture—that *way* of eating—never evolved into something recognizable, and where it did, it was not preserved. Jean Anthelme Brillat-Savarin, the great gastronome who famously said, "Tell me what you eat: I will tell you who you are," would have found that difficult to do here.

With few ingrained food habits, Americans are among the least tradition-bound of food cultures, easily swayed by fashions and influences from other countries. That's been a blessing, in some ways: we are freer to try new tastes and invent new styles and methods of cooking. The curse is that, without a golden age in farming, and with a history that lacks a strong model for good eating, the values of true sustainability don't penetrate our food culture. To-day's chefs create and follow rules that are so flexible they're really more like traffic signals—there to be observed but just as easily ignored. Which is why it's difficult to imagine farm-to-table cooking shaping the kind of food sys-tem we want for the future.

What kind of cooking will?

—•—

In a roundabout way, I was confronted with that question not long ago. A food magazine asked a group of chefs, editors, and artists to imagine what we'll be eating in thirty-five years. The request was to sketch just one plate of food and make it illustrative of the future.

It brought out dystopian visions. Most predicted landscapes so denuded that we will be forced to eat down the food chain—all the way down, to in-sects, seaweed, and even pharmaceutical pills. I found myself sketching out something more hopeful. My one plate morphed into three, a triptych tracing the recent (and future) evolution of American dining.

The first plate was a seven-ounce corn-fed steak with a small side of vegetables (I chose steamed baby carrots)—in other words, the American expectation of dinner for much of the past half-century. It was never an enlightened or particularly appetizing construction, and at this point it's thankfully passé.

The second plate represented where we are now, infused with all the ide-als of the farm-to-table movement. The steak was grass-fed, the carrots were now a local heirloom variety grown in organic soil. Inasmuch as it reflected all of the progress American food has experienced in the past decade, the striking thing about the second plate was that it looked nearly identical to the first.

Finally, the third plate kept with the steak-dinner analogy—only this time, the proportions were reversed. In place of a hulking piece of protein, I imagined a carrot steak dominating the plate, with a sauce of braised second cuts of beef.

The point wasn't to suggest that we'll be reduced to eating meat only in sauces, or that vegetable steaks are the future of food. It was to predict that the future of cuisine will represent a paradigm shift, a new way of thinking about cooking and eating that defies Americans' ingrained expectations. I was looking toward a new cuisine, one that goes beyond raising awareness about the provenance of ingredients and—like all great cuisines—begins to reflect what the landscape can provide.

Since the best of them coevolved over thousands of years, tethered to deep cultural traditions and mores, how does one begin building a cuisine? In other words, how does the *Third Plate* go from imaginable to edible?

That question was not the starting point for this book; it is something that has evolved in the writing of it. I started, instead, with farmers, and with experiences like that Eight Row Flint polenta that challenged my assumptions as a chef and taught me, again and again, that truly delicious food is contingent on an entire system of agriculture.

To get a handle on what kind of cooking best supported this, I needed to uncover something more basic: What kind of farming is this? Local? Organic? Biodynamic? I learned that it goes beyond labels. It requires something broad to explain it. Lady Eve Balfour, one of the earliest organic farming pioneers, said that the best kind of farming could not be reduced to a set of rules. Her advice was prescient. She lived before organic agriculture became defined by just that—a set of rules—and before farming methods were used as marketing tools. The farming that produces the kind of food we really want to eat, she believed, depends "on the attitude of the farmer."

That's frustratingly vague advice, and yet I came to understand Lady Balfour's idea when I heard one farmer speak about the ultimate goal of good farming. "We need to grow nature," he said, and in doing so he revealed more than an insight. He was articulating an attitude, a worldview, and he might well have been speaking on behalf of all the farmers in this book.

To *grow nature* is to encourage more of it. That's not easy to do. More nature means less control. Less control requires a certain kind of faith, which is where the worldview comes into play. Do you see the natural world as needing modification and improvement, or do you see it as something to be observed and interpreted? Do you view humans as a small part of an unbelievably complicated and fragile system, or do you view us as the commanders? The farmers in this book are observers. They listen. They don't exert control.

It's hard to label these farmers, because it's hard to label an attitude, which was Lady Balfour's point. When King Lear asks the blind Gloucester how he sees the world, Shakespeare has him say, "I see it feelingly." The farmers in this book see their worlds feelingly.

~~

If the future of delicious food is in the hands of farmers who *grow nature* and abide by its instructions, we ought to become more literate about what that means. By and large, we tend to calculate sustainability based on the surface level of agriculture. We take what is measurable (increases in the use of pesticide and fertilizers, inhumane conditions in animal feedlots) and push alternatives (buy organic, choose grass-fed beef). These things are easy to quantify. They are things you can see.

The farmers in this book farm one level down. They don't think in terms of cultivating one thing. If your worldview is that everything is connected to something else, why would you? Instead, they *grow nature* by orchestrating a whole system of farming. And they produce a lot of things—delicious food, to be sure, but also things we can't easily measure or see. I learned this lesson many times, whether in wetlands or pastures. I was introduced to the kind of jam-packed diversity, both above and below ground, that I had read about but never really understood. It changed dramatically, and not only from farm to farm but from field to field. Each living community was vast, complex, and critical to the health of the whole system.

Had I been given a tour of Blue Hill Farm's fields during that summer of corn—or any monoculture field anywhere in the world, really—I would have discovered little to write about. Monocultures do that. They impoverish life and all its fantastic little ecosystems. They depopulate landscapes.

I confess I kept getting pulled into visiting the farms in this book because I was in pursuit of how an ingredient was grown or raised—whether it was flint corn, whole wheat, a fattened goose liver, or a fillet of fish. I went in search of answers to practical questions. A core finding of my experiences was that I was asking the wrong questions. Each time I tried to parse the specifics of how something was grown, I was instead pointed in the opposite direction: to the interactions and relationships among all the parts of the farm and then, with more time, to the interactions and relationships embedded in the culture and history of the place.

I was learning what John Muir, the American environmentalist, observed a century ago: "When we try to pick out anything by itself, we find it hitched to everything else in the Universe." Or rather, I was *relearning* what I'd discovered with that summer of corn, which supplanted the natural contours of the farm, stripped away the community of farmers, tractors, and hay bales—what Wendell Berry called the "culture" in agriculture—and left me without much of a summer.

Science teaches us that the answer to understanding the complexity of something is to break it into component parts. Like classical cooking, it insists that things need to be precisely measured and weighed. But interactions and relationships—what Muir called hitching, and we call ecology—cannot be measured or weighed. I found, for example, that the health of an aquaculture farm in southern Spain is connected to how we treat our soil, and that how we treat our soil determines, to a considerable extent, how we grow our grain, especially wheat, which is impossible to separate from how we choose our bread.

What we refer to as the beginning and end of the *food chain*—a field on a farm at one end, a plate of food at the other—isn't really a chain at all. The food chain is actually more like a set of Olympic rings. They all hang together. Which is how I came to understand that the right kind of cooking and the right kind of farming are one and the same. Our belief that we can create a sustainable diet for ourselves by cherry-picking great ingredients is wrong. Because it's too narrow-minded. We can't think about changing parts of our system. We need to think about redesigning the system.

A good place to start is with a new conception of a plate of food, a *Third Plate*—which is less a "plate," per se, than a different way of cooking, or assembling a dish, or writing a menu, or sourcing ingredients—or really all these things. It combines tastes not based on convention, but because they fit together to support the environment that produced them. The *Third Plate* goes beyond raising awareness about the importance of farmers and sustainable agriculture. It helps us recognize that what we eat is part of an integrated whole, a web of relationships, that cannot be reduced to single ingredients. It champions a whole class of integral, yet uncelebrated, crops and cuts of meat that is required to produce the most delicious food. Like all great cuisines, it is constantly in flux, evolving to reflect the best of what nature can offer.

And its realization will rely, at least in part, on chefs. They will play a leading role, similar to that of a musical conductor. The chef as conductor is an easy comparison: we stand at the front of the kitchen, cueing the orchestra, cajoling and negotiating, assembling disparate elements into something complete. I'm not the first to make the association. And yet there's a deeper, more interesting level of work related to the job of conducting, and it may inform the role of the chef for the future. This is the behind-the-scenes work, the preconcert study that investigates the history of the composition, its meaning and context. Once that's been determined, a narrative takes hold, and the job of the conductor is to interpret that story through the music. One could say that a cuisine is to a chef what a musical score is to a conductor. It offers the guidelines for the creation of something immediate—a concert, a meal—that will also ultimately be woven into the fabric of memory.

Today's food culture has given chefs a platform of influence, including the power, if not the luxury, to innovate. As arbiters of taste, we can help inspire a Third Plate, a new way of eating that puts it all together.

That's a tall order for any chef, not to mention eaters, but it's an intuitive one as well. Because, as the stories in this book suggest, it always takes the shape of delicious food. Truly great flavor—the kind that produces plain old jaw-dropping wonder—is a powerful lens into the natural world because taste breaks through the delicate things we can't see or perceive. Taste is a soothsayer, a truth teller. And it can be a guide in reimagining our food system, and our diets, from the ground up.

PART I
SOID

See What You're Looking At

O NE MORNING late in the spring of 1994, Klaas Martens finished spraying pesticides on his cornfield. He went to lift the sprayer to put it away, but he couldn't pick it up. Which was strange, because Klaas had been lifting the sprayer the same way for many years. He tried again, and again he could not lift it.

"My right arm just wouldn't work," he told me one night more than twenty years later as we sat around his kitchen table. "Less than an hour before, I could loft a bale of hay one-handed over my head."

"He could," his wife, Mary-Howell, said. "He absolutely could. It's my first memory of falling for Klaas. We had just started dating, and I remember coming to the farm to visit. I walked up to the barn and from the distance I could see Klaas towering over everyone, grabbing one-hundred-pound bags of grain feed and tossing them, literally, like they were feathers. I didn't know someone could be so strong."

Klaas, looking shy, reached across the table for the homemade butter, which Mary-Howell keeps in a white bucket that's frequently passed during meals. He gently dug his spoon in deep.

"I started having muscle spasms," he continued, dropping a mound of the butter on a slice of Mary-Howell's homemade bread. "Terrible spasms that went up and down the right side of my body."

"I remember standing there by the stove when he walked in the house with his big, protective Tyvek suit still on—we called it the 'zoot suit'—and green plastic gloves," Mary-Howell said. "I knew something was wrong."

"I think I said, 'Something is really wrong,'" Klaas said, softly.

"I thought something wasn't right weeks before," Mary-Howell said. "I was out in the yard on a beautiful June afternoon with our son—our daughter was on the way—and I smelled something I didn't like."

"It was 2,4-D," Klaas said, referring to the chemical herbicide commonly used to control weeds.

"It was, yes, absolutely it was 2,4-D, but I remember smelling it in a different way. It usually smelled like freshly cured leather. Now for some reason the smell had undertones of raw flesh."

Mary-Howell looked across the table at Klaas. "So he goes to an orthopedic surgeon. And get this: it was June—spraying time—here was a grain farmer, and the doctor thought nothing of the dead arm. He just sent Klaas home with a handful of muscle relaxers and painkillers." Mary-Howell got up to clear the dishes and turned around at the sink. "By this point I didn't need a doctor to tell me what was happening. My husband was being poisoned."

<hr />

Klaas, along with his sister, Hilke, and two younger brothers, Jan and Paul, was raised on a farm down the road.

Their father arrived at Ellis Island from Germany at the age of fourteen. This was 1927, and with him were his grandmother and six siblings. (His parents—Klaas's paternal grandparents—had already passed away.) Wary of mounting political turmoil, they had sold their farm and fled Europe in search of a new future in America. After World War I, Germans were not allowed to own land in the eastern United States, so the family moved to North Dakota and leased land to grow wheat. The crop failed, and in 1931, alarmed by the worsening conditions of the Dust Bowl, they moved again, to a dairy farm in Bainbridge, New York, where they finally prospered. But there were too many siblings and not enough land. In 1957, Klaas's father and his young wife broke away from the family and moved their dairy cattle to Penn Yan, a small town in the Finger Lakes region. Klaas was two years old at the time. They earned a profit the first year, then earned still more as new agricultural technology swept onto their farm. Improved grain varieties, including hybrid corn and a high-yielding wheat, and, more important, the widespread use of chemical fertilizers and pesticides, boosted yields beyond anything they could have imagined. As Klaas grew older and assumed more responsibility on the farm, he watched one record-shattering harvest after the next. "My father would stand at the grain bin and just scratch his head," he told me. The yields, which more than doubled in one year, fueled explosive profits. It was a magical time.

"Everything was happening so fast, we got drunk on yields," Klaas said. "It was like a drug addiction. The first year, there was an incredible response from the chemicals, but we didn't notice that it kept taking more and more chemicals to get the same yields."

Weed resistance soon followed, which called for more chemicals, which led to even more stubborn weeds. Within a few years, Klaas was applying combinations of different herbicides to get rid of weeds they had never seen and didn't know existed. Klaas's father wasn't happy, and though profits remained strong (thanks to Klaas's increasingly creative chemical cocktails), he was skeptical that yields could be sustained. Convinced that his children's future was at risk, he encouraged each of his sons to experiment with new sources of income on the farm. Klaas saw it as an opportunity to pursue his "lifelong hankering" to grow bread wheat.

"Up until that point, we focused only on animal feed," he said. "I really wanted to grow grain that fed people *directly*."

He didn't feel good about the inefficiencies of feeding grain to animals. Which is a fair concern—it takes roughly up to thirteen pounds of grain to produce one pound of beef, for example. Klaas didn't want to farm in a way that put more pressure on limited resources. And he wanted to grow wheat that Mary-Howell, by now his wife, could use to bake her bread. Still, to his brothers, this seemed an absurd venture. At the time, farmers in the Northeast grew only pastry wheat, which was easier to cultivate.

Undeterred, Klaas grew a test plot. And it worked. The harvest was exceptional. But then he confronted a larger problem: Who was going to buy bread wheat from upstate New York? No distributors handled it locally, because no one else was growing it locally. But Klaas found a Mennonite neighbor who offered to buy it.

"They were thrilled to have local wheat for their bread," he said. The neighbor talked to Klaas about the wheat's flavor and baking quality. "It reinforced the idea that I wanted to grow grain for people, not for cows."

Klaas's father died in 1981, and, after several years of disagreements and difficult harvests, Klaas and his brothers decided to divide the land. The process was not easy. Klaas credits the story of Abraham and Lot for giving him the wisdom to make a deal. The Bible tells of Abraham traveling to Canaan with his nephew, Lot. With limited grazing land, it soon became clear that there wasn't enough for both of them to raise their respective herds. Abraham, fearing conflict, proposed that they settle separately. As the eldest, he had the right to choose whatever land he wanted, but instead he allowed Lot to make the first choice. Lot chose the beautiful, rich valley, while Abraham was left with the rugged hills and little drinking water.

After hearing the story one Sunday in church, Klaas stopped demanding what he felt was a fair and equitable distribution of the land. He took less than his third, including fields that were in poorer condition and not as productive. It was the same land—now thriving—that Klaas and Mary-Howell's house stands on today.

—

For dessert, Mary-Howell served rhubarb cake, which was so light and airy, I was sure it had been made with cake flour. It wasn't, Mary-Howell told me, in a voice that suggested she wasn't surprised I had guessed wrong. In fact, the cake was made with whole wheat flour—a variety called Frederick, grown on the farm and hand-milled in her kitchen that morning. She pointed to a tiny old mill no larger than a toaster oven, next to the microwave.

Each bite of cake brought a whiff of wheat. Just as the sweetness of the sugar and the vegetative tang of the rhubarb made the cake delicious, the wheat itself was unmistakably present, and it made a prosaic dessert richly textured and interesting.

"We were scared," Mary-Howell said, continuing with the story of Klaas's numb arm. "I mean, here we were, completely on our own, just the two of us, and Klaas can't use his right arm."

I asked when they stopped using chemicals. "That day," they said.

While interest in organic agriculture had already begun to grow by the early 1990s, organic grain farming was still essentially unknown. "We literally had no one to turn to," Mary-Howell said. "There were tons of successful organic vegetable farmers. And there were some good organic dairies, to be sure. The organic market was booming. But grains? We didn't know one farmer."

"Which is when people started planning our auction," Klaas said. Mary-Howell laughed and accused him of exaggerating. "Oh yes, indeed," he insisted. "There was talk at the coffee shop in Dresden. I heard that firsthand. And then there was old man Ted Spence . . ."

"Oh, God, Teddy . . ." Mary-Howell shook her head.

"He came over here one day, drove right up to the house," Klaas said, pointing to just outside their front door. "He rolls down his window and yells, 'Klaas, your dad would be disgusted with the way you're farming.'"

"He said that," Mary-Howell confirmed. "He absolutely said that."

By luck, a few weeks later, a local farm paper carried an advertisement from a large mill that wanted to pay for certified organic bread wheat. Sitting at the kitchen table, Klaas and Mary-Howell couldn't believe the coincidence.

"It was like a hand from God was reaching down to us," Klaas said. "We jumped at the chance."

THE FERTILE DOZEN

I first met Klaas in 2005, at a gathering hosted by Jody Scheckter, the contro-versial former world champion race-car driver known for his erratic driving and for what was perhaps the worst accident in Formula One history. Having turned his attention to organic farming, Jody created Laverstoke, a two-thousand-acre farm in Hampshire, England, that he determined would be the best in the world. Jody being Jody, he really meant the best. So he reached out to Eliot Coleman.

Eliot, a widely revered organic vegetable farmer and author from Maine, is a Gandhi-like figure for the sustainable agriculture movement. He did not invent organic farming, of course, just as Gandhi did not invent the doctrine of nonviolent resistance, but countless small farmers and gardening enthusi-asts have absorbed the philosophy through his teachings. I was given a copy of Eliot's back-to-the-earth guidebook, *The New Organic Grower*, in college, and in my early twenties I took it with me when I went to California to ap-prentice in a bread bakery.

Jody commissioned Eliot to identify the twelve most important farmers in the world—half from the United States, half from Europe—and bring them, at Jody's expense, to England for a three-day discussion on how best to use his land. Eliot framed the event as a once-in-a-lifetime summit of the world's greatest agricultural minds. He called the group "the Fertile Dozen."

Eliot, who by this point had become a friend (and later would be a trusted adviser during the creation of the farm at Stone Barns Center), called me a few weeks before the meeting to ask if I'd be interested in preparing the final dinner. It wasn't so much a question as a foregone conclusion.

I spent the day at Laverstoke shuttling between the kitchen and a corner of the large room where the twelve men sat around an old English table (King Arthur's Round Table came to mind) explaining their farming methods and philosophies. They were brilliant, engaging, passionate, and inspiring in a way that you know will stay with you for a lifetime.

There was Joel Salatin, in the days before he was made famous by Michael Pollan in his book *The Omnivore's Dilemma*, speaking about energy exchange and pasture-based farming; Willem Kips, from Denmark, who married traditional biodynamic farming with modern technology for high yields; Frank Morton, an Oregon seed breeder who quietly revolutionized American salad with his new varieties of greens; Thomas Harttung, whose pioneering community-supported agriculture program today supplies organic vegetables to more than 45,000 homes in Denmark and Sweden; Fons Verbeek, of the Netherlands, who spoke about animal-vegetable relationships; Joan Dye Gussow, the nutritionist and innovative organic gardener considered by many to be the founding voice for the local-food movement; and Amigo Bob Cantisano, a California organic farmer, adviser, and creator of the Ecological Farming Conference, with a résumé almost as impressive as his salty-gray Tom Selleck mustache. One after another, without pretension or exaggeration, these farmers described their unique contributions to farming.

No one spoke directly about how their work translated into crops with more flavor, because it was simply understood. I got hungry just sitting there.

And then Klaas Martens rose to tell his story. Standing six foot three, with his John Deere baseball cap askew and his overalls hiked alarmingly high, he looked more Gomer Pyle than agricultural statesman. I decided to get back to the kitchen, but as I turned to leave, Klaas offered the group a simple question: "When do you start raising a child?" Just like that. It was an oddball opening to a talk about his life's work, but Klaas's humble, practical tone drew everyone's attention. I stayed for the answer.

Klaas said he'd come to the question through his interest in the Mennonite community, a group he had known over the years and greatly respected. He explained that Mennonites forbid the use of rubber tires on their farm tractors. The Fertile Dozen shook their heads in near unison. Klaas smiled, acknowledging the severity of the decree—steel-tired tractors inch along, slow as oxen.

He said one day he got up the nerve to ask a Mennonite bishop why rubber tires were forbidden. The bishop answered Klaas's question with a question: "When do you start raising a child?" According to the bishop, Klaas told us, child rearing begins not at birth, or even conception, but one hundred years before a child is born, "because that's when you start building the environment they're going to live in."

Mennonites, he went on, believe that if you look at the history of tractors with rubber tires, you see failure within a generation. Rubber tires enable easy movement, and easy movement means that, inevitably, the farm will grow, which means more profit. More profit, in turn, leads to the acquisition of even more land, which usually means less crop diversity, more large machinery, and so on. Pretty soon the farmer becomes less intimate with his farm. It's that lack of intimacy that leads to ignorance, and eventually to loss.

Around the table, heads nodded in silent recognition: Klaas had just described the problem with American agriculture.

CHAPTER 2

IF FEELING HUMBLED in the face of nature is what you're after, skip the Grand Canyon and stand in a large field of wheat. Or stand in any grain field next to dozens of other, contiguous grain fields. The wide, ripe expanse doesn't just surround you, it envelops you. It makes you feel small. I once heard the environmental lawyer and activist Robert Kennedy Jr. speak of an epiphany he had. God talks to human beings through many vectors, he said, but nowhere with such clarity, texture, grace, and joy as through a growing field of wheat.

A few years after meeting Klaas at Laverstoke, I stood in the middle of one of his wheat fields in Penn Yan, New York, and saw what Kennedy meant. I had never been to Penn Yan—didn't even know it existed until I met Klaas—and though it's only forty-five minutes from downtown Ithaca and the hubbub of Cornell University, it feels more like central Kansas than upstate New York.

The scene reminded me of a painting I once saw in grade school. A crew of seamen, sailing at a time when conventional wisdom had it that the world was flat, quaked with fear and knelt in prayer as their ship slowly approached the edge of the horizon. Their expressions of despair would be appropriate if you found yourself about to fall off the face of the earth, but I had trouble sympathizing. To my adolescent mind, the men looked a little silly, their fear exaggerated.

And yet from the vantage of that wheat field, I thought maybe those men had been on to something. The idea that the world is *not* flat seemed, at that moment, sort of radical. I raked my gaze back and forth, enormity and abundance in every direction. The rain had just cleared, and the air was still thick with odor and color. To the east, beyond Klaas's fields, I could see his neighbor's fields—a figure of a man on a tractor was no larger than a grasshopper—and, beyond this, his neighbor's neighbor's fields, until eventually the grass just dropped off into a kind of oblivion.

Klaas leaned over, broke off a stalk of emmer wheat, and brought it to his mouth for a taste. He chewed thoughtfully, separating the wheat kernel from its chaff and rolling it around in his mouth. Klaas's features sometimes seem to have outgrown his frame. His hands flap around like empty ski gloves when he speaks, and his shoulders are so wide you're tempted to inspect the back of his jacket to make sure he didn't leave the coat hanger in. He embodies a particular brand of solidness—the German immigrant farmer who plowed our country's midsection with nothing more than grit and determination. And yet Klaas is an irrepressibly cheerful man, generous and humble.

I asked Klaas why he found it important to grow wheat. He paused to examine another stalk. "The nice thing about wheat is how it's tied to Western civilization, to the cradle of civilization. The history of wheat is the history of a sociable crop."

He was right. For centuries, wheat was a community builder, a grain whose benefits were reaped only through cooperation and effective social organization—farmers grew it, millers ground it, and bakers turned it into sustenance and pleasure. In his book *Seeds, Sex & Civilization,* Peter Thompson says all three of the world's great grains—wheat, corn, and rice—provided the foundations for civilization. But, he wrote, "whereas the foundations provided by maize and rice were sufficient to build walls," wheat's inherently communal qualities "provided the keystones of arches to support the edifices of urban civilizations."

The story of wheat is the story of who we are.

Klaas broke off a kernel and held it in his big hand. "This is probably what someone was threshing when Ruth showed up," he said, adding that emmer was one of the first domesticated crops. He shook his head. "It humbles me just holding it."

God may or may not communicate through wheat, but for sure *we* communicate by carpeting so much of our landscape with grain. The middle of the wheat field in Penn Yan was insignificant—a mere nursery compared with the Corn Belt of the Midwest, or the plowed-up prairie of the Plains. Today nearly 60 percent of American cropland is in grain production—corn, wheat, and rice, mostly. Wheat—which, worldwide, covers more acreage than any other crop—is planted on fifty-six million acres in the United States. Vegetables and fruits, by comparison—what most everyone, including chefs, fixate on—occupy just 5 percent of our cropland.

Why haven't we talked more about wheat? While we've been obsessed with record corn harvests—as impressive and record-breaking as they are—wheat still blankets much of our country's midsection. It also constitutes a large percentage of our diet—more than 130 pounds per person, every year. That's more than beef, lamb, veal, and pork put together. It's more than poultry and fish, too. If you don't count corn sweeteners, we eat more wheat than every other cereal combined.

But rarely do we consider how it's grown. If we want to improve the condition of our food system and create a food tradition that thoughtfully ties together the disparate parts, focusing only on fruits and vegetables is like planning a new house but designing only the doors and windows. It misses the big picture.

Klaas acknowledged the disconnect. "I see people go to all the trouble to visit the farmers' market and really take the time to pick out the best peach, or stand in line for a grass-fed steak that's treated the way a cow ought to be treated," he said. "And then on their way home they buy packaged bread in the store." He removed his cap and ran his hand over a mop of matted-down hair. "That's bread made with wheat that's adulterated and dead, even more than the fruits and vegetables they successfully avoided purchasing a half hour before. And I mean dead, like a rotten tomato, which you would never eat."

He turned to me. "So how is this possible? How do we get to the point that we willingly, even happily, eat the equivalent of a rotten tomato?" He paused, looking out at his fields as a gentle breeze made the wheat sway in unison. "It happens," he said, "because we've lost the taste of grain."

―――

My office sits in the corner of the kitchen at Blue Hill at Stone Barns. The drafting chair at my desk faces out so I can observe the cooks, catch mistakes, and sometimes even head off small disasters.

One night, not long after my fateful conversation with Klaas, I was sitting at my desk and watching the kitchen wind down for the end of service. It's a scene I've witnessed a thousand times before, the cooks slowing to the rhythm of the late orders. But for some reason, on that night I noticed something I hadn't been conscious of before. Wheat was everywhere.

In one corner, a waiter cleaned the bread station for the evening, saving the unused loaves for the pigs' dinner. Over by the stove, Duncan the fish cook sprinkled the last order of trout with flour before roasting it. Across from him, the meat cook wrapped a loin of pork with an herb dough. An intern organized trays of fresh ravioli and thick-cut spaghetti. And there was Alex, the pastry chef, serving his white-chocolate-and-cardamom cake with dried fruit strudel. Trays of after-dinner cookies and small pastries flew past on the way to the dining room.

Suddenly Jake, the pastry sous chef, came into view hauling a fifty-pound bag of all-purpose flour, which he heaved into the flour bin just outside my office. It was his second fill of the day. A white flurry hovered in the air, as in a just-shaken snow globe. As it drifted toward the window of my office and fell away, I was reminded of standing with Klaas and watching his fields stretch to the end of the horizon. Back then I'd been struck by how much the story of agriculture is really about grain. The kitchen scene that night had me realizing that the story of our menu is really about grain, too, particularly wheat.

When Klaas complained of his neighbors' visiting the farmers' market for fruits and vegetables, only to then carelessly purchase bread at a supermarket, he might as well have been complaining about me. As the owner of a farm-to-table restaurant—actually a restaurant *in the middle* of a farm—I've gone on and on (and on and on) about local fruits and vegetables with no more apologies for repetition than a peanut vendor in a ballpark. Like most chefs, I can name the heirloom variety of this or that tomato, or the breed of cattle with the most flavorful grass-fed steaks. We root around obsessively for all these things because they taste better, and because we know the people, and the practices, that produced them. The soft, white dust dumped into the container bin twice a day was the most generic thing in our kitchen, but I knew more about the construction of our stove than how the flour had been farmed.

I wanted to learn the taste of wheat (or relearn it), and to do that, I needed to learn its history. What could account for its odd duality—the all-purpose little grain that is everywhere on my menu but about which I knew close to nothing?

CHAPTER 3

IN THE MYTH OF PYGMALION, a sculptor falls in love with his female statue and helps bring her to life. The story of wheat is the anti-Pygmalion: in our ten-thousand-year effort to sculpt a more perfect grain, we've succeeded mostly in making it more dead.

Can something be *more* dead? Technically, no. And yet as I began to dig into the story of wheat in the United States, I learned that it suffered exactly that: several stages of degradation and death. Who's responsible for killing wheat? It's no mystery—what makes the story of American wheat so interesting, and so tragic, is just how obvious it all was. Culinary historian Karen Hess once called it "the conjugation of seemingly unrelated events." Everyone and no one killed wheat. It was the perfect murder.

It began innocently enough. Domesticated wheat wasn't even here when Columbus arrived, as opposed to corn, which flourished. The Spanish were the first to bring wheat to the New World, and other European immigrants did the same when they settled the colonies. It failed miserably at first, but with great effort on the part of the early settlers, it eventually took hold. Long before wheat became synonymous with the Midwest, the East Coast was America's breadbasket. Gristmills dotted the countryside—one for every seven hundred Americans in 1840. Once ground, flour had a shelf life of only about one week, and if you wanted a loaf of bread, you baked your own. That meant bringing your wheat to the mill or milling it yourself.

With the help of farmers, wheat adapted itself to specific regions. But it thrived especially in the milder climate of the mid-Atlantic—Pennsylvania, Maryland, and New York. As of 1845, wheat was grown in every county in New York, including four acres in Manhattan. Wheat had distinctive characteristics, flavors, and baking qualities, not just from state to state (Massachusetts "Red Lammas" versus Maine "Banner wheat") but from farm to farm, and from year to year. Diversity flourished. Farmers tasted the raw kernels in the field to assess their protein content and when it was time for harvest. Women adjusted recipes according to the condition of the flour. These were good times for wheat. After all, what more could a grass seed want than to find itself thriving in a new world?

The opening of the Erie Canal, in 1825, completed the link between the Eastern Seaboard and the Midwest, establishing new trade routes and creating a milling hub around Rochester, New York—soon to be known as the Flour City. Railroads soon followed, which coincided nicely with our nation's longing for cheaper and less crowded farmland. And wheat went along for the ride. This was nothing sinister, just inevitable. But something significant happened here that served as a harbinger of times ahead: for the first time in America, wheat started to be grown far away from where it was consumed.

The roller mill appeared in the late 1800s, just in time to expand the divide between the wheat field and the table. It was a technological breakthrough that revolutionized the wheat industry just as the cotton gin had done for the cotton industry a century earlier. Until its widespread use, people used stone mills. Stone mills, like the one we use at Blue Hill, work like molars, crushing the kernels between two large stones. They are effective, but slow and tedious, and they do little to separate the kernel into its component parts, a key development in the drive to industrialize flour.

A few years ago, Klaas's wife, Mary-Howell, showed me a picture of a wheat kernel in cross section. It looked like an ultrasound image of a six- or seven-week-old human gestational sac, which isn't a bad comparison; a wheat kernel is a seed, after all. The grain's embryo, or "germ," is surrounded by the starchy endosperm—the stuff of refined white flour—which stores food for the germ. Surrounding the endosperm is the seed coat, or bran, which protects the germ until moisture and heat levels indicate it's time to germinate. (Later that same day, I returned to the field with Klaas and saw, in a bizarre neonatal vision, the wheat as a phalanx of plant stalks holding their embryos up high in the air, as if they were torches.)

Whereas stone mills had crushed the tiny germ, releasing oils that would turn the flour rancid within days, roller mills separated the germ and bran from the endosperm. This new ability to isolate the endosperm allowed for the production of shelf-stable white flour, able to be stored and transported long distances. Overnight, flour became a commodity.

It's hard to fathom that merely removing a temperamental little germ could revolutionize a staple grain. But that's just what happened. The settling of the Great Plains and the advent of roller-mill technology meant that white flour was suddenly cheaper and more readily available. Small wheat farms, including those in the former grain belt of New York, couldn't compete. Farmers chewing kernels in the field and gristmills dotting the landscape became the stuff of folklore. The homogenization of the U.S. wheat industry had begun.

The whiter flour became, the greater the demand. To be fair, that's been the history of wheat for thousands of years. But for all its efficiency, steel couldn't match the old-school grindstone in two key respects. In fully removing the germ—that vital, living element of wheat—and the bran, the roller mill not only killed wheat but also sacrificed nearly all of its nutrition. While the bran and the germ represent less than 20 percent of a wheat kernel's total weight, together they comprise 80 percent of its fiber and other nutrients. And studies show that the nutritional benefits of whole grains can be gained only when all the edible parts of the grain—bran, germ, and endosperm—are consumed together. But that's exactly what was lost in the new milling process.

There was another cost as well, just as devastating. Stone-milled flour retained a golden hue from the crushed germ's oil and was fragrant with bits of nutty bran. The roller mills might have finally achieved a truly white flour, but the dead, chalky powder no longer tasted of wheat—or really of anything at all. We didn't just kill wheat. We killed the flavor.

THE PRAIRIE

Our nation's prairie became collateral damage along the way.

What did I know of the prairie before I developed an interest in wheat? Nothing, really. I doubt that I would have been able to locate it on a map. I definitely didn't know that at one point, not that long ago, our country was more than 40 percent open prairie, a lush expanse of grassland that extended from Missouri to Montana and straight down to Texas. And even if I had known these things, I couldn't have said why it mattered to a chef.

Then I met Wes Jackson. Wes is the folksy and eloquent cofounder of the Land Institute, in Salina, Kansas, where he leads research into how to breed grain crops—wheat in particular—so that they can be planted once and harvested year after year. Domesticated wheat—the wheat we eat—is an annual crop, which means that every year new seed is sown.

If it were to instead become perennial, like wheat grows in the wild—if it could be "native to its place," as Wes likes to say—agriculture's worst offenses, like plowing and the need for chemical fertilizers, could be avoided.

In 2009, Wes and I attended a food conference in California as part of a panel about the future of food. When asked by the moderator to describe his work, Wes simply said, "I'm solving the ten-thousand-year-old problem of agriculture." To his mind, agriculture's problem is not mega-farms or feedlots or chemical fertilizers. The problem is agriculture itself.

On the walk back to the hotel that evening, I asked him about the possibility of his perennial wheat appearing anytime soon, a question I later learned annoys Wes, because he hears it so often. But he only cranked up his slow prairie drawl and said, not immodestly, "If you're working on a problem you can solve in your own lifetime, you're not thinking big enough." He said he wanted to show me what he meant.

I followed him to his room, where he handed me a cardboard shipping tube. "You are the first to see this," he said. I must have had a look of *Why me?* because he added, "We shipped them here the day they arrived. I knew I wouldn't sleep tonight if I didn't have a nice, long look-see." I started uncorking the tube. He stopped me. "Go ahead and roll it out, but do it in the hallway. It won't fit in the room."

I unfurled the photographic banner onto the hallway carpet. It was twenty-two feet long and reached down the corridor, past the doorways of two other rooms. Wes bent down and evened out the crinkles. On the left was a life-size profile of perennial prairie wheat, showing the plant both above and below the soil. Aboveground, the stalks, leaves, and seed head took up less than half the photograph. Belowground, the wheat's root system was at least eight feet long—a Rapunzel-like tangle of thick fibers anchored deep in the soil.

I stepped back. The roots merged into what looked like the trunk of a sequoia tree, only growing down instead of up. "That's nature investing— digging into the soil, seeking nutrients and moisture," Wes said as I studied what once had been the underbelly of the prairie.

To the right of this, a photo showed another patch of wheat, above and below ground. But this was modern wheat, the kind that's planted each year and, as Wes reminded me, "occupies sixty million acres of real estate in this country alone." Aboveground, the wheat was a much shorter copy of its

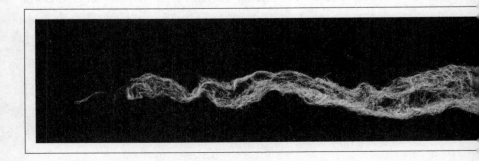

perennial cousin. But belowground, the roots were wispy, thin hairs, barely an arm's length in depth. Compared with the perennial, they looked laughably anemic, needle threads next to those dreadlocks. Such are the roots that blanket the prairie and fill those bags of white flour dumped into the bin in front of my office. I was looking at the roots of my cuisine.

"Those wimpy little things," Wes said, smiling. "There's your problem right there."

—

Until the 1800s, almost everyone who visited the Great Plains thought the problem was the prairie itself. The massive land area was called the Great American Desert, which, from the perspective of people accustomed to things like trees, is a forgivable first impression. But also a mistaken one.

In fact, there was plenty of aboveground diversity in the prairie. Add to the grasses the surrounding two hundred or so broadleaf flowering plants, the forbs, shrubs, and sedges, and what you had was a kaleidoscope of natural variety—a richly purposeful system in which grass and plant depended on one another to thrive.

And yet, the true wealth of any prairie exists in the soil, where the majority of the biomass resides (unlike, say, a rainforest ecology, where the richness, or biomass, is mostly above the surface). Wes likes to remind his audiences that the soil's richness results from a lucky geological break. A few million years ago, glaciers formed in the continent's far north. Frozen rivers

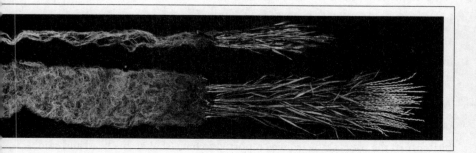

stripped northern Canada to hard rock and dumped ancient dirt on top of the already rich soil of this country's midsection. As fierce prairie winds distributed the dirt, it was the grasses that clung to it, holding it long enough to consolidate the mass into soil. The rich root systems absorbed nutrients from the soil and knit the soil together.

For the prairie, this was the greatest insurance policy against erosion and extreme weather fluctuations. The weather in the Plains was—and still is—unpredictable, fierce, and destructive—desertification on the one hand, flash floods on the other. The root systems' ability to store energy and nutrients ensured that the prairie grass could always grow back quickly. And the grass, in turn, kept the rich soil in place as millions of bison fertilized it over thousands of years, depositing more nutrients into the soil's natural fertility bank.

We've been drawing from the account ever since the first settlers tried to dig in with their plows, an effort that, from above (or, more to the point, from below), must have appeared comical. The root systems were so dense, the plows snapped and clogged. Looking at the entangled roots of just one small patch of perennial wheat made it easy to see why. One square yard of prairie turf can contain twenty-five miles of these massively thick roots; the coal-black topsoil can run to a depth of a dozen feet. (Wes reminded me, with glee, that the average topsoil on the East Coast is closer to six inches.)

In 1837, an Illinois blacksmith named John Deere solved the problem by inventing a cast-steel plow that could cut through the deep roots and rip up the grass for planting. Like the roller mill, the steel plow arrived at a fortuitous moment—just at the time when thousands of "sodbusters" were crashing deep into the Plains. President Abraham Lincoln sweetened the deal in 1862 by signing the Homestead Act, which promised 160 acres of free land to anyone who could claim and cultivate it for five years.

Biologist Janine Benyus, in her book *Biomimicry: Innovation Inspired by Nature*, describes the misplaced heroism of the settlers who worked to replace perennial prairie grass with annual wheat: "A Sioux Indian watching a sodbuster turn the roots skyward was reported to have shaken his head and said,

'Wrong side up.' Mistaking wisdom for backwardness, the settlers laughed as they retold the story, ignoring the warning shots that fired with each popping root." The more you learn about the destruction of the prairie, the more difficult it becomes to see a modern wheat field as a thing of beauty, in the same way it is hard to see beauty in a clear-cut forest.

The new wheat didn't exactly thrive on the Great Plains at first. Varieties grown in the East did poorly with less rain and extreme variations in temperature. Disease was common. So were low yields and outright failures. It wasn't until the 1870s, when hard winter wheat, the drought-resistant "Turkey Red" introduced by Mennonite immigrants, replaced the traditional soft wheat, that it took hold. Hard wheat suited the new steel roller mills as well, making the now assembly-line-like refining process even more efficient.

Wes's banner in the hallway blocked a couple on the way to their room. "Good evening, folks," Wes said cheerfully. "We're making an analysis of our nation's depleted capital. Care to join us?" The couple smiled uncomfortably and shuffled alongside the two root systems.

Wes pointed to the annual wheat. "Of course, this wheat won out. Sixty million acres of puny roots that we need to fertilize because it can't feed itself. Puny roots that leak nitrogen, that cause erosion and dead zones the size of New Jersey." Wes smiled beatifically, gums and all. "This wheat won out, but what you're looking at is the failure of success."

By the early 1900s, westward expansion amounted to a twenty-million-acre experiment.

And the wheat kept growing. When Europe ran out of wheat during World War I, the American government stepped in, guaranteeing wheat prices. The Enlarged Homestead Act of 1909 followed, doubling the amount of free land to 320 acres per settler, and the wave of settlement became a tsunami. In 1917, a record forty-five million acres of wheat was harvested; by

1919, it was seventy-five million acres. Much of the gain came from plowing marginal land—areas of North Dakota and the southern Plains where soils were thinner and there was less water for irrigation—but for the time being, it didn't matter.

Historian Donald Worster argues that by the time the war effort ended, the Midwest's grain economy had become inseparable from the industrial economy. "The War integrated the plains farmers more thoroughly than ever before into the national economy—into its networks of banks, railroads, mills, implement manufacturers, energy companies—and, moreover, integrated them into an international market system." The grasslands were remade. There was no turning back.

So the plows kept plowing until the rain suddenly stopped. The soil, naked, anchorless, and now dry, turned to dust and, in 1930, started to blow. Dust coated everything, consuming surfaces, bed linens, and attic floors (which routinely collapsed under the accumulation). It buried fence posts, cars, and tractors in enormous drifts. And this was only the light stuff. The heavier soil clumps ripped fences apart and whacked down telephone poles as they blew across the landscape. During the worst of these storms, visibility was zero, and vegetables and fruits died from the storms' electrical charge. The drought ushered in a biblical infestation of insects, which devoured any wheat that survived, and a plague of jackrabbits emerging from their habitats in search of food.

Klaas remembers his aunts telling him about the Dust Bowl. The storms were so severe that the family would set the dinner table with the plates upside down. By the time they served dinner, the table linen would be imprinted with rings of dust. The family lived with the hardship until the farm itself went under.

⌒

Over the course of the next decade, our country's midsection heaved hundreds of thousands of years' worth of incomparably rich soil into the air.

Some regions lost more than 75 percent of their topsoil. The decade came to be known as the Dirty Thirties, and it marks one of the worst environmental disasters in our history. In his book *The Worst Hard Time: The Untold Story of Those Who Survived the Great American Dust Bowl*, Timothy Egan describes one of the dust storms:

> A cloud ten thousand feet high from ground to top appeared. . . . The sky lost its customary white, and it turned brownish then gray. . . . Nobody knew what to call it. It was not a rain cloud. . . . It was not a twister. It was thick like coarse animal hair; it was alive. People close to it described a feeling of being in a blizzard—a black blizzard, they called it—with an edge like steel wool.

One of the largest of the storms hit in the spring of 1935—Black Sunday. It didn't die in the prairie but moved east, gathering strength as it went.

The following Friday, a scientist named Hugh Bennett stood on the floor of the U.S. Senate, arguing for the creation of a permanent Soil Conservation Service. Even though photos of Black Sunday had appeared in newspapers around the country the same morning, most senators believed they had already done enough for the people of the prairie. Just as Bennett was wrapping up his plea, an aide appeared at the podium and whispered in his ear. "*Keep it up,*" he said. "*It's coming.*" Bennett kept talking. A few minutes later, he stopped talking. The chamber turned dark. A giant copper dust cloud blew through Washington for an hour.

"This, gentlemen, is what I'm talking about." Bennett said, pointing to the windows. "There goes Oklahoma." Eight days later, Congress signed the Soil Conservation Act into law. Some call the incident the beginning of the environmental movement in America.

The white mushroom cloud, the one that billowed up from the flour bin in the restaurant's kitchen and slowly drifted toward my office window, could be thought of as the modern manifestation of the Dust Bowl, with all-purpose flour now playing the part of prairie topsoil. Which is to say the degradation of the prairie is still reaching us like it reached Hugh Bennett a century ago, as vital a topic now as it was the day he stood in front of the Senate and argued for reform.

THE MODERN PRAIRIE

Writers have spared no ink in making the case that the Dust Bowl era is a parable of man's hubris. In his essay "The Native Grasses and What They Mean," Wendell Berry writes, "As we felled and burned the forests, so we burned, plowed, and overgrazed the prairies. We came with visions, but not with sight. We did not see or understand where we were or what was there, but destroyed what was there for the sake of what we desired."

This blindness began with our nation's earliest European settlers, many of whom weren't themselves landowners in Europe and had little experience farming. If you have a hankering, as I do, for the old days of our young republic, when farming was what farming should be—small, family-owned, well managed and manicured, a platonic paradigm of sustainable agriculture—think again. Today's industrial food chain might denude landscapes and impoverish soils, but our forefathers did much of the same. They just had a lot less horsepower.

Even George Washington criticized the exploitative methods of "slovenly" farmers who, spoiled by the abundance of fertile land and natural resources, "have disregarded every means of improving our opened grounds."

Colonial farmland was quickly run down. Forests were cleared for coveted virgin land. In his book *Larding the Lean Earth*, historian Steven Stoll identifies the detrimental precedent that came to define American farming:

> In a common pattern, farmers who had occupied land for only 20 or
> 30 years reduced the fertile nutrients in their soils until they could
> no more than subsist. Either that, or they saw yields fall below what

they expected from a good settlers' country and decided to seek fresh acres elsewhere. Forests cut and exported as potash, wheat cropped year after year, topsoils washed—arable land in the old states of the Union had presented the scares of fierce extraction by 1820.

This attitude only intensified as we pursued Western expansion. As Alexis de Tocqueville noted in his famous study of the country, farmers approached farming with the attitude of capitalists rather than conservationists. "Almost all the farmers of the United States combine some trade with agriculture; most of them make agriculture itself a trade," he wrote in *Democracy in America*. "It seldom happens that an American farmer settles for good upon the land which he occupies." Americans arrived on the prairie to settle the West on their own terms. We set out to conquer rather than to adapt—unable, or just unwilling, to adjust our sight to the needs of the new ecology. There was so much abundant and enormously productive land available that vigilant soil management became an Old World idea.

I had understood this, to some degree, for years, but what I hadn't understood until that evening when Wes and I studied root systems in a hotel hallway was what that blindness had wrought. We didn't just replace the deep root system of perennials with puny annuals. We replaced the prairie's ecosystem, one of the most diverse in the world, with 56 million acres of monoculture. Today, almost all the hard wheat grown in the prairie comes from just two varieties, which, in the words of writer Richard Manning, is "a spanning of the scale of genetic possibilities from A to B."

Look out on a field in the middle of Kansas or North Dakota and what you see are grain fields so uniform they look like tabletops, the prairie manifestation of a desecrated grave. Wheat, as Klaas describes it—as a social crop, as a community builder, as the story of who we are—no longer really exists. At least not in the way we're farming it. Or eating it.

Not long after my evening with Wes, I consulted a map of the United

States. I was looking for the breadbasket states—the "Wheat Belt," as I'd heard it called countless times before, without knowing the coordinates. On the map, it appeared as a thick strip, a belt of land running from North Dakota all the way down to Texas, passing through South Dakota, Nebraska, Kansas, and Oklahoma. Wheat is the primary crop of these six states.

By chance, I came across another map in my search, a census of population changes in the first decade of the twenty-first century. I put it next to the map of the Wheat Belt, and the comparison was as stark and alarming as Wes's banner. The census map showed how the same stretch of six states has become shockingly depopulated. The Wheat Belt is emptying out, even as the rest of America grows denser. It is a relentless decline in numbers that began in the Dust Bowl years ago and has never really ended. What makes this population drop so remarkable is that the phenomenon seems almost wholly confined to these Wheat Belt states, indeed that while much of America continues to grow, the former heart of the country's grain production is today in demographic free fall. In Kansas alone, six thousand towns have vanished in the past eighty years. In many parts, the population is like those wispy annual roots—sparser today than at the end of the nineteenth century, when the census deemed them "frontier."

One explanation for the population declines can be traced back to advances in farming technology and the subsequent consolidation of farmland. New tractors and other farm machinery did more work in less time. Take, for example, the combine, introduced in the 1830s. Until then, farmers spent hours harvesting, threshing (separating the edible part of the grain from the surrounding chaff), and cleaning their grain to prepare it for milling. True to its name, the combine consolidated these functions into a single machine, mechanizing the harvest and, in keeping with the Mennonites' predictions, enabling fewer farmers to farm even more land. Between 1950 and 1975, the number of farms in the country declined by half, as did the number of people on farms. And the average size of farms nearly doubled, from 216 acres in 1950 to 416 acres in 1974. Nowhere were these trends more apparent than in the Wheat Belt.

But underlying all of this was a lack of biological diversity. Verlyn Klinkenborg, an author and editor who often writes about agricultural issues, argues that biological complexity has direct implications for social and cultural robustness. In other words, the Wheat Belt's cultural decline is a reflection of its denuded landscape—the product of "what nature has made of us and we have made of nature."

However unwittingly, chefs and bakers have played a part in that decline. Profiting from mountains of cheap flour, we've bought into the system. We're complicit in the depopulation of the prairie, just as we have blood on our hands for the death of wheat.

CHAPTER 4

L ATE ONE SWELTERING June morning a year and a half after my initial trip to his farm, I found myself back in Klaas's fields. Klaas, intent on schooling me in the diversity of grass, told me to focus on a small circle of ground, no more than a few feet in diameter. We inched around the perimeter as he provided a grass-by-grass playlist.

"Here's the wild garlic, and the yellow rocket, and that right there is"—Klaas squatted low for a gopher's-eye view, brushing the other grasses aside—"yup, wild radish, right underneath." I followed him, head down. The intensity of his obsession for the smallest detail—a divot in the ground, a grass that looked out of place—seemed almost comical, considering the vastness that surrounded us.

"Okay . . ." Klaas said slowly, stopping at a particularly lush spot. "Oat grass. And there's fleabane, bladder campion, foxtail, dandelion, red clover, chamomile, quack grass—couch grass in England."

"Weeds?" I asked.

"Grasses and legumes and forbs, and, yes, weeds. But 'weed' is an arbitrary word. When I was in Agronomy 101, the definition of a weed was anything that grows where you don't want it to grow. How preposterous is that?"

Aldo Leopold asked the same question in his 1943 essay "What Is a Weed?" and answered it with a warning about blacklisting grasses for no other reason than what we mistakenly perceive to be their value.

"If I lose crop yields to weeds, I'm the one doing something wrong,"

Klaas continued. "They're not doing damage to the pasture. The opposite, actually; they tell me when *I'm* doing damage." He pointed. "Orchard grass. I love orchard grass. Very pleasing to look at," he said, taking his own advice. "And, hey! Hairy vetch—you know vetch, right? Great cover crop. And here's sow thistle. I think that's sow thistle—yep, sure, that's thistle, for sure."

The year Klaas stopped using chemicals, he began reading old farming books to learn how to eradicate weeds naturally. "I discovered that either they didn't write a lot of books about weed control before 1945 or someone had thrown them all away," he said. After a long search, he came across a book in a Cornell University library written by a German agricultural researcher named Bernard Rademacher. Rademacher was the leading authority on weeds in the 1930s, and he participated in some of the first research on chemical herbicides.

"Rademacher wrote something that turned our thinking upside down. He argued that successful weed control is all about promoting the growth of the crop: 'Vigorous plant stands are the best means for eradicating weeds.'" Klaas gave me a goofy smile. "*Vigorous plant stands are the best means for eradicating weeds*—I read that to Mary-Howell and we just looked at each other and said: 'Duh! Focus on the best plants! How come we didn't think of that?'"

Klaas realized that the best way to ensure a healthy plant was through healthy soil. Attend to the plant's needs and it will take care of itself.

———

I first learned about plant health from Eliot Coleman, who writes about the subject often. He was my Rademacher, but instead of writing about plant health and weeds, Eliot spoke of plant health and pest management.

Eliot claims that a healthy plant, living in healthy soil, doesn't *need* pest eradication, because pests don't attack healthy plants. It's a simple idea, but powerful. Feed the soil, and the tiny creatures that live in it, with care and attention, and the pests will almost always be incapable of inflicting damage. In that sense, the idea of "Mother Earth" is turned around. *We* actually have to mother the earth, by feeding the soil properly. If we don't, plants get sick. Or, as Eliot prefers to say, stressed. The sick part comes later.

The easiest way to understand plant stress is to think about what happens to our own bodies when we're overworked or sleep-deprived. Our immunity weakens, and we're more vulnerable to colds or, over time, disease. A small aphid attack is the plant equivalent of a cold; a flea beetle infestation is like a disease. As the plant's health deteriorates, pests overtake its natural defenses.* Applying a pesticide spray, as chemical farmers do, works to eradicate the pest but not the root cause. Take medicine for flu symptoms and you can function—your friends might not even notice—but you're not healthy.

Eliot once illustrated this point with the story of a flea beetle attack he witnessed early in his career at an organic garden in Maine. In a large field of cabbage, he noticed a lone rutabaga plant that had somehow been mixed in by mistake. The cabbage plants were "healthy and vigorous," but the rutabaga was swarming with flea beetles. A few days later, it was dead. Upon closer inspection, the farmer realized that the rutabaga had become so overgrown it had detached from its roots:

> Whatever the cause of its demise, the flea beetles knew the plant was under stress three days before there were any visible signs. When the rutabaga died the flea beetles did not move to the cabbages. Because the cabbages were not stressed. Obviously insects cannot maintain a population on unstressed plants because such plants do not provide

---

* How does an insect detect a sick plant? "Sick plants smell different," Eliot told me. In fact, there's some science behind the idea. Dr. Philip Callahan, a former entomologist who is also an electronics and radio expert, discovered that tiny hairs on the antennae of insects pick up variances in wavelengths emitted from plants. There are odor wavelengths, or "broadcasts," depending on the plant's temperature. In his book *Tuning in to Nature*, Callahan showed that stressed plants produce vastly different odor molecules from those of healthy plants, emitting wavelengths an insect can readily "see."

conditions conducive to insect nutrition and multiplication. The cabbages remained pest free. The organic farmer would look for the cause. The chemical farmer would treat the symptom and just spray.

Treating causes instead of symptoms is as elegant, but not as simple, as it sounds. To address the cause, you need to look for underlying problems—which means you need to have a certain kind of worldview.

It helps if your worldview includes the belief that nature knows best. A plant suffering from an infestation of pests is not a shortcoming of nature; it's a plant you're not mothering well. Either the nutrient balance in the soil is wrong or your crops aren't being rotated properly or the variety cultivated is wrong for the area—or any one of dozens of other possibilities. Your job is to figure it out. Since the chemical farmer has the option of spraying the problem away, he tends not to bother.

BUILDING FENCES

My grandmother Ann never spoke about weeds or plant health. I'm sure she knew nothing about the subject. But every spring, she had coffee with Mr. Mitchell, the farmer who owned the cattle that grazed Blue Hill Farm, and Mr. Mitchell always said the same thing. "Ms. Strauss, my cows don't eat much of anything until they get to your farm. I don't know what it is about your grass, but they fatten right up." When Ann heard "they fatten right up," she looked like a child in front of a mountain of ice cream. She was proud of her grass, even if she had nothing to do with its health or virtue.

Mr. Mitchell's praise worked just as well for an impressionable kid like me. I knew Blue Hill Farm was the most special place in the world, but now there was proof—bragging rights that meant a lot (who could argue with a fattened cow?), even if I never bragged (because who would care?).

Then a change happened on the farm that, at the time, I hardly noticed.

The cows began to congregate at the fence line, necks down, to eat the perimeter grass. They stretched their necks under the fence to reach grasses that apparently were worth the strain. Then they reached over the barbed wire for a few bites. Pretty soon the cows ignored the fence altogether, pushing their way through and roaming the property. They tended to find their way to the gardens and then fan out all over the lawn. By August this was happening two or three times a week. I would wake to Ann snapping up her window shade and hear, "Oh, for heaven's sake." (If they were already in her beloved flower garden, it was "Oh, God damn it.") The cows appeared gratified by their extra effort. Ann was furious.

By late morning, Mr. Mitchell's sons, Robert and Dale, would arrive to usher the cows back into the pasture. We'd spend the next hour "fixing fence," as the brothers called it—walking the line to find the weak places the cows might exploit. It felt like meaningful work, and it placated my grandmother, but by the end of the summer its futility had become clear. Unless we built the Berlin Wall, there was no way to ensure the cows would remain in the pasture.

Many years later, long after my grandmother died, I read the work of André Voisin, a French biochemist who studied the link between soil, animal, and human health. Voisin tells the story of French farmwomen in the eighteenth century who fed their cattle with roadside grass because they had no land of their own. These cows ended up yielding large amounts of milk, much more than other people's pastured cows did. So farmers purchased the women's calves for large sums, only to realize that once on pasture, the cows produced less milk. The farmers thought they were buying superior animals, but in fact it was the free, roadside grass that made the difference. Its diversity met the cows' nutritional needs better than any poorly managed pasture.

Perhaps the cows at Blue Hill Farm had been telling us something similar. Mr. Mitchell's praise of the land might not, for all I know, have been genuine—a small lie to soften my grandmother and persuade her to let the cattle graze for free. Or maybe the health of the pasture had diminished over time. Either way, the cows were telling us *something*, but instead of wondering what it was, we went ahead and fixed fence. We secured the borders, or tried to, because handling the symptoms of the problem (a faulty fence line) was a lot easier than dealing with its root cause (a hungry cow).

THE LANGUAGE OF THE SOIL

Klaas took Rademacher's wisdom of weed control and made it his own. He moved away from his lifelong mission to destroy weeds—a Sisyphean pursuit for any farmer—and took up a suite of interrelated strategies to strengthen the plant.

"You can't compartmentalize farming," Klaas told me. "Soil fertility here, crop rotations over there, weed control somewhere else—it doesn't work over time. All decisions should connect to plant vigor." If plant vigor is properly supported, weeds can't compete. Which was Klaas's point. Is a weed really a weed if it doesn't compete with the plant?

Klaas and I walked to another field, this one growing spelt, an ancient species of wheat that has become one of the more profitable crops on Klaas's farm. I wanted to learn about the mechanics of growing wheat like spelt, but Klaas was still focused on weeds.

"At a certain point you start to notice patterns—and when I see patterns, either in a repetition of certain grasses or the absence of others, I see it as the soil talking to me," he said. "I know it sounds corny, but the soil has a language, and its language is at least partly expressed through what weeds are growing and not growing, what's strong and what's weak. The trick is to learn the language of the soil. To learn what the weeds are telling us."

If you pay attention to which weeds proliferate, the soil will tell you what it needs. The presence of chicory or wild carrot or the lovely Queen Anne's lace means the soil is low in fertility, a classic problem that arises when you harvest crops without returning nutrients to the soil. Milkweed is a sign that the soil lacks zinc; wild garlic means low sulfur. Foxtail grows most often in soils where water is poorly filtered, and thistle thrives when soil is too compact, so there's not enough room and air for proper germination.

In Klaas's telling, soil communicates pretty clearly, perhaps even more clearly than people. It doesn't mince words, doesn't get passive-aggressive when hungry or annoyed. *What* it needs is expressed directly. *How much* of what it needs depends on the type of soil and where it's located. Factoring in these variables, Klaas can take corrective steps for the next planting to satisfy the soil's demands.

That's not to suggest that soil is *needy*. Soils can just as easily be overwhelmed by too much fertilization. I learned this lesson from Eliot Coleman as it relates to pests. Soils spoiled with an excess of nutrients become too rich. They lose balance. And pests eventually attack the weakness.

"The crops start to resemble a guy on the street who's had too much to drink," Eliot told me once. "He slowly walks toward you, and you know something isn't quite right, but from where you're standing you can't put your finger on the problem. That's what your crop starts to look like before an insect attack—not quite right."

Klaas approved of the analogy. "Among the hardest lessons to learn in farming is that too much of a good thing isn't good," he said. Because farmers see how fertilization brings growth, they're often motivated to overfertilize as insurance. The effect can be destructive. Galinsoga, a weed feared by every small farmer, grows out of soil thirsty for carbon—which it needs to sober up after the inebriating effects of too much fertilizer. It's how the soil fights to balance the excess nutrients.

I told Klaas about a farmer who once told me that the sight of galinsoga is a declaration of war, that he would mow all his crops, take down the peas and the carrots—everything—just to get rid of the stuff. Klaas said he must have been applying too much fertilizer and guessed—correctly—that his farm was not organic.

"On a small farm," he said, "you can deal with the problem of weeds by constantly weeding, or you can become a nozzlehead and just spray the weed or the pest away."

On a grain farm the size of Klaas's, weeding by hand is impossible. So weed suppression (without spraying the problem away) means listening to the soil and maintaining plant health. If your farm becomes so large you can't identify the different grasses and what they are telling you about the health of the soil, then you're not really farming, at least not at the right scale. Limiting yourself to what you can see, either by choice or, like the Mennonites, through the decree of steel tires, isn't such an antiquated idea.

A CASE OF VELVETLEAF

So how do you grow great wheat?

I think that's what I was learning. But I couldn't be sure, because by late afternoon that same June day, Klaas and I had covered only two fields, and all the talk had been about wild grasses (formerly known to me as weeds). I was also learning that while I traffic, for the most part, in simple sums, farmers like Klaas deal in calculus. Klaas had spent the day not so much answering questions as making connections, a habit that made him slightly less frustrating to consult than the Oracle of Delphi.

I finally got my answer when we walked into a field growing soybeans, another profitable crop for the farm. Klaas reached for a broad-leaved green plant that seemed to be everywhere and, to my rookie eye, looked very nearly as vibrant as the soybeans around it. Velvetleaf, he told me. He turned over one of its leaves and smiled broadly. The underside was blanketed with tiny flies. There must have been several hundred, maybe even several thousand, on that one leaf. He turned over another infected leaf, and then another. Walking down the row, he continued turning leaves over for me to inspect, like a magician showing his cards.

"I wanted you to see what I consider to be my greatest success," he said.

That the velvetleaf is a noxious weed (yes, absolutely a weed) and the tiny flies are a troublesome pest named whitefly not only didn't trouble Klaas—it thrilled him. He was in that moment as satisfied and contented as I would ever know him to be.

Why the exuberance in the middle of a field riddled with pests and weeds? Klaas found inspiration in another soil scientist. "When we went organic, I started reading Dr. William Albrecht. "You know how sometimes we read something and immediately recognize we're not going to think quite the same way again? That's what it was like to read Albrecht. He said, 'See what you're looking at.' I love that quote. Think about that. It requires close observation without prejudice. How often do we see something without really *seeing*?"

Klaas looked down at the velvetleaf blanketed with flies and held it in his hands. "A field of velvetleaf with an infestation of whiteflies can actually be a good thing. It can be the greatest success, actually, but you'll only understand it if you're able to see what you're looking at."

⸺

Albrecht, the longtime chairman of the Department of Soils at the University of Missouri, was born on a farm in central Illinois in 1888. After spending some time as a Latin professor, Albrecht studied biology and agricultural science at the University of Illinois as farmers plowed up the surrounding prairie soil. He earned a medical degree but then abandoned formalized medicine because (like Eliot Coleman) he found treating disease much less interesting, and a lot less effective, than exploring its causes. Albrecht went to the cause.

He began with a simple observation, which more or less informed his life's work. He saw cows straining their necks to eat grass on the other side of the fence, much like the cows I had once watched at Blue Hill Farm. Why, he asked himself, would a cow risk entanglement with barbed wire when acres of pasture were free for the taking? The question (which I never bothered to ask about *our* cows as a boy) led Albrecht to discover that cows—supposedly "dumb beasts," passive and indiscriminate about their diet—actually spend their days in one long, exhaustive search for more nutritious pasture. Cows determine their next bite by swiping their facial hairs against the tips of the grass. The hairs act like antennae, sensitive to the grass's richness. A quick calculation is made: Does this plant hold enough nutrition to be worth expending the energy required to take a bite? In many cases, Albrecht noticed, the cows didn't bother.

He had a hunch that these discerning diners made choices based on minerals potentially available in certain grasses. Chemistry, in other words, could explain a cow's preference. Albrecht watched as cows walked past what was commonly considered "good grass" to eat seventeen different types of weeds that had been fertilized with calcium limestone, magnesium and phosphate. No wonder the cows deemed the other grass a waste of their energy.

"The cow is not classifying forage crops by variety name, nor by tonnage yield per acre nor by luscious green growth," he wrote, but she is more adept than any biochemist at assessing their true value. Albrecht came to the humbling conclusion that we can't really identify a healthy pasture just by looking at it—we have to really *see* it, and that requires a deeper understanding.

"Albrecht embraced chemistry for the answers," Klaas told me. "He made the direct connection between soil health and pests and weeds by determining which nutrients were lacking, or, more to his brilliance, which nutrients were out of balance in the soil. He always said, 'Feed the soil and let the soil feed the plants.' And he saw the whole thing as a measurable, repeatable process."

Either cows are smarter than people or they're just better eaters.

OPERATION VELVETLEAF
ERADICATION

"The field was a total disaster back then," Klaas said, still holding the velvetleaf in his hand. He and Mary-Howell had rented the adjoining farm in 1994, after the previous farmer abandoned it out of frustration. The farmer had "misused" the plow, Klaas explained. There were several places where the topsoil was so whisper thin, you could see the subsoil.

"It was a crime lab of past sins," Klaas said. "He said this field was worthless. He told me it would never produce anything of value. And when I got in here to really look around, I started to believe him. Weeds were exploding. The velvetleaf was this tall—" Klaas reached his hand high above his head. "Some grew to twelve feet. They looked like trees, with root systems like trees. They were so strong, I couldn't pull them out with my hands."

That first summer, he decided to just walk the field and get to know it, mimicking Albrecht's patience for detail. Klaas saw a field depleted of fertility and set out to restore it by planting a series of different crops.

For more than three decades, the previous farmer had primarily grown corn. "I remember his corn started to go downhill," Klaas said. The crop got smaller, and Klaas suspected it was also losing sugar content. "Albrecht always said compromised fertility shows up first in the quality of what's harvested," he said.

"You can taste decline?" I asked.

"Yes, you can. Before yield declines, before weed and pest pressures, you can taste it. The doom before the doom." Considering that the farmer had been growing feed corn, I thought Klaas might be exaggerating its need to be delicious. Then again, if we're going to eat the beef, why wouldn't we want the cow to be provided with the best feed possible? This was Albrecht's point.

Klaas's first move was to plant spelt. But it wasn't the spelt itself he was after, because in the mid-'90s a large market for spelt didn't exist. He wanted spelt in the ground so it could pave the way for clover. Spelt is a perfect soil builder, because its extensive root systems reach deep into the earth, creating air pockets that allow the soil to breathe (aeration) and beginning what Klaas referred to as "the cleanse." And spelt's stalk, or straw, which Klaas plows back into the soil, provides a hearty boost of carbon.

Clover was planted in the early spring, interspersed with the spelt that was just beginning to peek up out of the ground. Clover does the miraculous work of "fixing" nitrogen: grabbing it from the atmosphere, where nitrogen is abundant, and storing it in its roots. It also provides the soil with sugars, proteins, and minerals, and it furthers the soil's aeration by attracting earthworms. All this had been missing from the previous farmer's chemical regime. Synthetic fertilizers supplied the corn with nitrogen for fast growth, but the soil itself was ignored, never paid back for the harvests. According to Klaas, it was "like burning your house to keep it warm each year."

Nitrogen from clover is better than nitrogen from chemicals, Klaas told me later, because all the carbon keeps it in a stable form. "Without carbon, and without microbial activity in the soil, you can't hold on to the nitrogen," he explained. "It leaches like crazy."

He turned and pointed to Seneca Lake, less than a mile from where we were standing. It was picturesque and serene, but Klaas pointed out something I couldn't have seen on my own. "Seneca used to be so clear it looked like a tall glass of water. But the nitrogen leaching from soils like this, over so many years, has darkened it. It's polluting the great lake because the soils of Penn Yan can't hold on to the fertility."

As the weather started to warm that spring, both the clover and the spelt grew in earnest. Since the spelt had a head start, it dominated the clover and was soon ready for harvest. (Had the clover dominated the spelt, we would be calling clover a noxious weed, which is why Klaas sees *weed* as such an arbitrary term.) With the spelt no longer competing with it for sunlight and nutrients, the clover exploded.

"At this point, there were choices to be made," Klaas said. "If we had cows, I'd have run them on the clover. Have you ever seen a cow on good clover grass? It's a beautiful thing. It's rocket fuel for ruminants. Even better, there's no loss of minerals, because the manure returns them to the soil." I pictured a herd of his cattle grazing energetically through the field, but then I remembered that Klaas didn't have cows, or any animals at all, on his farm.

"Mary-Howell doesn't want to have to deal with animals," he explained, adding quickly, "I don't blame her, of course."

But Klaas and Mary-Howell view soil organisms as a kind of livestock in and of themselves. "What I say is, we have livestock—we have a lot of livestock on the farm—they're just very small," Mary-Howell once told me. "But we have to tend them with just as much intention and care as if they were larger livestock. Which means we have to feed them."

Klaas plowed the clover into the ground, along with what was left of the spelt stalks, which he kept instead of selling as straw. Nitrogen from the clover infused the soil, and the straw provided the expected hit of carbon—all the while supplying a rich diet for the soil's "very small" livestock. The sick land started to heal.

———

Klaas turned again to Albrecht and took a soil test. "That's when it got really interesting," he said.

By "interesting," Klaas meant complicated for those without a degree in chemistry. Soil tests, one of Albrecht's many contributions to the field, measure mineral levels in the soil. That means macronutrients—nitrogen, phosphorus, potassium, calcium, magnesium, sulfur—as well as micronutrients (otherwise known as trace elements) such as copper, iron, and manganese, which are needed only in small quantities. All are key ingredients in healthy plants—and tasty foods.

The results of the soil test were surprising. The nitrogen levels had returned so completely that Klaas decided to plant corn. "I planned on wait-ing," he explained, "but the soil test gave me a green light, and since we're running a business here, and organic corn is always at a premium, I figured I'd go for it." Even though the velvetleaf still persisted, the corn yield was good, with a sugar content that, while lower than those of corn harvests from his other fields, was still respectable. The gamble had been worth it.

So what was next? Most farmers would follow a successful corn harvest with soybeans or, if feeling lucky, another planting of corn. Plant the most profitable crop, in other words, if the soil allows it.

"I went with yellow mustard," Klaas said, and then he leaned his head back and smiled mischievously. My expression didn't change, which I could tell confused him. Had I known about the improbability of planting yellow mustard, I would have said, "Holy shit, Klaas. You planted a weed in your already weed-infested field?!" That's what his Penn Yan neighbors said. Why cultivate a weed, they wondered, and on top of that lose all potential revenue on the harvest?

The answer came from the Albrecht soil test, which had indicated a defi-ciency in sulfur, one of the macronutrients that's essential for the develop-ment of vitamins in plants and plays an important role in root growth. Actually, the soil test merely proved what Klaas had already noticed: the field was filled with yellow flowers. Older farmers had always told Klaas that yellow-flowered weeds thrive in sulfur-depleted fields.

"Here's a classic case of 'see what you're looking at,'" Klaas said. "It wasn't until the soil test confirmed a deficiency in sulfur that I could see the yellow flowers for what they were—a virtual advocacy group for the addi-tion of sulfur."

Eliot Coleman has compared a good organic farmer to a skilled rock climber. Both "are interested in solutions, each one simpler and more elegant than the last." On first glance, the analogy seems a little asymmetrical: one inelegant move by the rock climber can be catastrophic, while a bad rotation decision by the organic farmer might only amount to a few yellow flowers in his field. But as the mistakes accumulate, the results for the farmer can be just as disastrous, including a much less delicious harvest.

And herein lies the point: if the farmer is made to believe that his rotation decisions don't really matter, that their equivalent "next step" up the rock face isn't critical, then he is not motivated to forgo a profitable harvest to plant yellow mustard.

Klaas turned the mustard into the soil, allowing the sulfur to take full effect. But he knew he needed to again restore nitrogen levels. "I could have planted soybeans for the nitrogen at this point—they're a legume, after all, so they fix nitrogen. But to be honest," he said, leaning in very close to me as if he were afraid of offending the surrounding crop, "soy is a little lazy. It's a profitable plant, but it's lazy. I got more out of what I went with: kidney beans. My kidneys get top dollar for canning. Not huge money, but who cares, right? I'm banking the nitrogen that the beans are fixing into the soil for a future crop."

After the kidney beans, the obvious choice was to go back to corn. "I went with wheat instead," he said, turning away too late to hide another grin. "I chose wheat because I don't like to go too long before I plant a crop that actually feeds people. Going back to corn was totally possible—the nitrogen was there—but at the end of the day, I want people eating what I grow."

The fact that a farmer can make more money feeding animals than feeding people is a problem with the marketplace, Klaas said, reiterating a point he had made to me many times before. "If too much of the farm is cultivated in corn, I'm encouraging a diet where herbivores [grass eaters like cows and sheep] are fed grain." He paused to consider the implications. "It gives me indigestion just thinking about it."

But then the question was, what kind of wheat? Modern wheats are hard on the soil—extractive rather than restorative, like spelt—and Klaas is judicious about planting them. "I obviously get more money from the newer varieties, but, again, I pay for it," he said. "The full accounting includes what's drawn from the soil bank." Klaas chose emmer, an ancient wheat once grown throughout the American Northeast. Emmer is known for its large root systems, providing optimum yields without large fertility requirements.

Almost as soon as the emmer had been harvested, Klaas noticed a change in the velvetleaf. "It was still there in big numbers, but it wasn't as tall. It didn't look as healthy," he said. "The environment was changing."

A group of agricultural scientists, having heard about Klaas's experience with the velvetleaf, came to his farm to study it. "They visited every week and eventually pinned the declining health of the velvetleaf on a fungus, anthracnose," Klaas said. "But anthracnose had never killed anything on an organic farm. They were convinced, I think, because they were fungal guys. They knew fungus, and to them the velvetleaf was dying because of a fungus attack." He laughed. "They turned out to be right—the velvetleaf had anthracnose—but they were wrong, too. That's not what was killing the velvetleaf. They completely missed the whiteflies." Klaas turned the leaf over once more so I could again see for myself. It seemed impossible that anyone could have missed them.

"It's true. It wasn't until I came out here one day that I saw the velvetleaf for what it had become—a diminished, sick plant—not a healthy plant that happened to have a fungal virus. That's when I could see it. Just like Albrecht said: *See what you're looking at.* The whiteflies suddenly appeared for me. They're nature's cleanup crew, attacking the least fit species."

I looked at the soybean plants (Klaas's most recent rotation) and realized they were growing a mere three inches from the velvetleaf, yet they remained completely untouched by the whiteflies. The soil conditions had changed so drastically that the soy had become the dominant species, impervious to attack, while the velvetleaf, once twelve feet tall with a mile of hardened roots, had shrunk. It was literally suffocating, a weed shriveling in on itself.

CHAPTER 5

WHY SHOULD A CHEF CARE about how farmers manage weeds and pests? Spraying the problem away has damaging environmental implications, so there's that, of course. And because we prepare food, chefs are closer to farming than, say, lawyers and accountants are. But everyone eats, which means that chefs presumably shouldn't be more outraged about bad soil management than lawyers or accountants are. Poor soil health, with its resulting weed and pest problems and the chemicals needed to solve them, affects chefs no more than it affects everyone else.

Or does it? The more I learned from Klaas, the more I saw the error in that way of thinking. These kinds of questions matter quite a bit more to chefs, because how soil is managed, and how a farmer negotiates weeds and pests, is the single best predictor of how food will taste.

Fruits, vegetables, animals—and, for that matter, grains—grown in poor soils make it harder for chefs to cook delicious food. *Truly* delicious, like the Eight Row Flint corn polenta I tasted as if tasting polenta for the first time. If soil is compromised, there can be no such thing as great food. Which is unfortunate, since the short history of soil in this country, like that of wheat, is another story of degradation and death.

A VERY SHORT HISTORY OF SOIL

Soil is alive, and not just in the metaphysical sense. It inhales and exhales, procreates, digests, and constantly changes temperature. When growing properly, soil organisms breathe as we do—taking in oxygen and releasing carbon dioxide into the air. In that sense, an open field of pasture grass is not unlike a packed stadium of football fans during a tense fourth quarter.

And like us, soil contains a lot of tiny living creatures—a complex community of bacteria, microbes, fungi, worms, grubs, bugs, and slugs. I remember Klaas dropping to his knees during my first visit to his farm and digging to fill his palm with dirt. "Right here, there are more organisms in this soil than the entire population of Penn Yan!" he said. "That's a lot of life to feed." I raised my eyebrows in a futile attempt to look impressed. But Klaas was merely being modest; the number of creatures in soil is really much greater. One teaspoon of the good black stuff has been said to contain more than a million living organisms, though scientists now consider even this figure far too conservative—it's well over a billion. Soil is literally teeming with life.

Klaas's handful of soil most likely contained ten thousand different species of microbes—not individual organisms, but *species*—all of which aggressively modify their habitats to suit themselves. And yet they are also so interrelated, so connected to one another and the surrounding ecosystem, that studying them individually under a microscope has until very recently been next to impossible. They simply cannot survive long enough without their neighbors.

Soil has a personality, too. It manipulates its environment to get what it needs (think of weeds), and, according to Klaas, it talks to you—if you learn to speak the language. In his book *The Tree*, Colin Tudge describes living tissue, with its complex design and endless capacity for self-renewal, as "not a thing but a performance." The same is true of soil. Now that I was beginning to see that food's flavor depends on soil health, and that soil health depends on a thriving population of organisms, I wanted to know, if only in the crudest way, how the performance works.

———

Soil covers itself. On any bare patch, a green carpet of plants and weeds sprouts immediately, protecting the earth from the elements. Whispers of grass, poking their way out of sidewalk cracks, are evidence that even urban soil, sheathed in concrete, hankers for protection.

As the plants grow up, the roots grow down, until eventually the roots (and dead grass, if cut and left on the surface) decompose into a sheet of humus. This nearly magical process—recommissioning plant material into organic matter—is accomplished by soil organisms, from earthworms and insects down to bacteria, the proletariat of the soil. The humus is then turned into salts—not the *salt* we sprinkle on food, but nitrates and phosphates, which help the plants grow. Add animals into a system, and their manure can help bring about the same outcome, but more quickly—in just a few months rather than several years. There is faster decay and faster growth.

That's a rough explanation of a complex system, but the core of the idea is that soil sustains itself (and actually improves itself) by working in a circle. Live roots become dead roots. Dead roots become food for soil organisms. What's not eaten either nourishes new grass or becomes humus, a kind of long-term bank account that provides for the future needs of plants.

What happens when farming is introduced? We upset the balance. By harvesting crops, we extract and export (and eventually eat) soil fertility, which requires that an equal or greater amount be returned to the soil. Sir Albert Howard, the British scientist and father of organic agriculture, called this the Law of Return, the word *law* suggesting—rightly, it turns out—that it's nonnegotiable. Unless fertility is restored, soils suffer. Restoring fertility is the key to soil health, which means it's key for flavor, as well.

There are three parts to soil fertility, and Klaas described them to me by using the analogy of a successful company. The first is the profit part—we cash in on this in the harvest. Then there is the working capital—the engine of any business, which in the case of soil fertility are things like manures and composts that feed the soil directly. Finally, there's the reserve, the bankable funds that feed the company's productivity over the long run. Humus serves this purpose; as Albrecht said, it's what gives any soil its "constitution." Without all three in good working order, a company is likely to go bankrupt.

Farmers have pretty much always understood soil fertility—even if they couldn't explain it—and when they broke the law of return, either because they didn't have animal manure or because they didn't know how to apply it, they simply moved on to virgin land. Virgin land brims with fertility, so one needn't be concerned with depletion—until, of course, it *is* depleted—as colonial Americans quickly learned.

In their book *Empires of Food: Feast, Famine, and the Rise and Fall of Civilizations*, Evan Fraser and Andrew Rimas argue that, historically, food empires such as ancient Rome and Greece and medieval Europe built their success on the same system of careless banking. They grew food and transported it long distances to feed a growing population. They cashed in on the fertility without paying back the bank. This worked for a while, but ultimately the soils stopped producing.

A CHEMICAL APPROACH

Among the most important ingredients for fertility is nitrogen. Plants demand it; without it, they cannot grow. Nitrogen can be restored to the soil in two ways. The first is through leguminous plants like beans and peas (or, in Klaas's example, clover), which "fix" nitrogen from the air.

The other is through manure, which contains nitrogen (in the form of either ammonium or organic material) as well as other valuable nutrients. Historically, it was held in such high esteem for its value to the farm that, as late as the 1900s, a French girl from the countryside had her dowry measured by the amount of manure produced on her family's farm. But this method of fertilization has a couple of drawbacks. Not all of manure's nitrogen is available to the soil. And it's a time-consuming process: animals graze slowly, and as long as they're doing so, the land isn't available for growing food.

Farmers found their solution in 1840 with the publication of *Chemistry in Its Application to Agriculture and Physiology*, by the German chemist Justus von Liebig. Rather than recycle nutrients, Liebig suggested that farmers could simply add certain chemical amendments to the soil. He reduced soil fertility to just three nutrients indispensible for plant growth: nitrogen, phosphorus, and potassium—N-P-K, in shorthand.

It seems crazy to think that soil's rich biological fertility could be ignored in favor of just three chemical elements. But from a farmer's point of view, the logic is tantalizing. If it's the minerals in manure that provide fertility, why not just add in the minerals and forget the manure altogether? Age-old and laborious farming techniques no longer seemed so important when a one-way transfer of nutrients proved not only possible but also highly efficient.

David Montgomery, in his book *Dirt: The Erosion of Civilizations*, argues that Liebig's discovery was a pivotal point in humans' understanding of the universe, in that it showed us how to manipulate nature:

> Now a farmer just had to mix the right chemicals into the dirt, add seeds, and stand back to watch the crops grow. Faith in the power of chemicals to catalyze plant growth replaced agricultural husbandry and made both crop rotations and the idea of adapting agricultural methods to the land seem quaint . . . large-scale agro-chemistry became conventional farming.

If the death of wheat is impossible to pin on any single thing—a "conjugation of seemingly unrelated events"—the demise of soil is more transparent.

There was motive (farmers working to produce as much as possible), justification (increasingly exhausted soils), and now the means (science). Liebig's findings opened the door for a few simple drivers of plant growth to replace the natural complexity of healthy soils.

Liebig's N-P-K model might have revolutionized farmers' thinking, but it did not do it overnight. Initially, at least, buying minerals was prohibitively expensive.

Another German chemist named Fritz Haber would help change that. In 1909, Haber succeeded in tapping into the atmosphere's reservoir of nitrogen gas and turning it into molecules useful to living things. Like legumes, his new process "fixed" nitrogen, but in a handy and highly concentrated chemical form that farmers could simply feed to their soil. This Haber-Bosch process (Carl Bosch gets credit for upsizing the invention, in 1913, to factory scale) produces liquid ammonia, the raw material for making nitrogen fertilizer. By the time World War II was over, some of the munitions factories that had produced so abundantly for the war effort had been converted, in some cases literally overnight, to producing chemical fertilizer. (Ammonium nitrate is also the key ingredient in explosives.) Suddenly our attention turned from winning the world war to winning the war with nature.

If there is such a thing as a smoking gun in the murder of soil, this was probably it. Natural limits on crop growing became irrelevant. As long as there was nitrogen (the air has a limitless supply) and the energy to run ammonia factories (which, thanks to the growth of the petroleum industry, there was), farmers would never again need to run animals on farmland or rotate their crops. Specialization was suddenly not only possible but practical.

Writer and journalist Michael Pollan, who's often described the pitfalls of an industrialized food chain, sees this relentless drive to monoculture as the "original sin" of agriculture. Monocultures breed more monocultures: once you've determined what it means to be efficient, and you have the technology to do it, why wouldn't you, for instance, remove cows and haying from your farm and just grow corn?

Which is exactly what happened. In 1900, diversification (at least on some level) was inherent to agriculture; 98 percent of farms had chickens, 82 percent grew corn, and 80 percent raised milk cows and pigs. Less than a hundred years later, only 4 percent of farms had chickens, 25 percent grew corn, 8 percent had milk cows, and 10 percent raised pigs. And, in many cases, the farms producing these commodities did so exclusively.

Armed with synthetic fertilizers and new plant varieties (bred to soak up more and more nitrogen), grain farmers saw incredible results. Wheat yields at least doubled between 1900 and the 1960s. And corn showed even more staggering increases. Today we're growing corn on fewer acres yet harvesting four times as many bushels (10.8 billion in 2012, versus 2.7 billion in 1900).

Along with the new monocultures, meat production was transformed as well. Cows no longer needed to roam the fields and supply manure for soil fertility, so there was no reason for them to leave the barn. The farmer brought the field to them. As the system became more refined, and the animals' living conditions more confined, farmers gained far greater control over protein production. Feed mills, feedlots, slaughterhouses—the entire animal food chain became industrialized.

———

Not coincidentally, flavorful food dies a little, too, at this point. What Pollan called the original sin of monoculture also helped pave the way for the original sin of food preparation: large-scale food processing. The specialization of farming, and the reduction in crop prices, allowed the food-processing industry to take hold. Using technology developed to feed the military during World War II, the industry created processed food that saved time and liberated women from cooking.

The rise of the American processed-food industry is nearly always examined through the lens of convenience, and of course there's truth to the claim. But at the heart of these changes was Haber's invention, which, by freeing the farmer from the constraints of nature, allowed the industrialization of the food industry to flourish. Some see Haber's invention as the savior of mankind. Today about three billion people depend on synthetic nitrogen to grow the crops they eat, and in coming years the number will likely only grow. Others argue that Haber's science has saturated the planet with excess nitrogen, creating a profound chemical dependency in agriculture and, in the process, precipitating many of the most vexing environmental problems we face today: soil erosion, global warming, pollution of streams and rivers, and the fouling of the world's oceans.

For good or ill, it's hard to think of many scientific discoveries that have had a more profound effect on the world. It's also hard to think of a discovery that's been more disastrous for the flavor of food.

SPEAKING FOR THE SOIL

Not everyone bought into the degradation of soil life. The problems with the chemical approach to farming (what we now call conventional agriculture) were apparent almost from the beginning. As early as 1924, Rudolf Steiner, the Austrian philosopher and father of biodynamic agriculture, was warning farmers against a reductive approach to growing food, because he believed it harmed subtle ecological relationships that were not fully understood.

But the opposition might have found its most influential spokesperson in an English botanist named Sir Albert Howard. Like William Albrecht, Howard saw soil health as a kind of universal string theory, a belowground answer to everything possible aboveground. Howard's prescription was to begin "treating the whole problem of health in soil, plant, animal and man as one great subject." He studied soil science but also botany, plant science, animal science, medicine, and, late in his career, economics, because he felt all of these disciplines were ultimately connected.

Disenchanted with the narrowness of university research (he found the work of "learning more and more about less and less" stifling), Howard decided to put his learning into practice. At the age of thirty-two, he was sent to India as a development worker to teach Indian peasants some of the modern ways of growing food. He remained for twenty-five years and ultimately determined that it was the Indian peasants who had taught *him*. Nature, he came to understand through close field observation, was the "supreme farmer." Weeds and pests were his "professors of agriculture."

His book *An Agricultural Testament*, published in 1940, later emerged from this work and became the bible for the organic movement. If healthy soils were to last—if they were to be "sustainable," as we now say—they would need constant feeding. Howard advocated on behalf of the soil's tiny underground livestock. When fed well, he argued, they do their work to support the fertility of land and lead to a lot of good things, including more flavorful food. According to Howard, vegetables raised on a diet of N-P-K are "tough, leathery and fibrous: they also lack taste." But vegetables grown with humus are "tender, brittle, and possess abundant flavour."

Howard's virtue was neither pious nor self-serving. He was calm and steady in his faith in nature, and his writing is so matter-of-fact (this from a lab scientist, remember) and effortlessly contemporary that his books read more like thoughtful journals than the organic manifestos they would become.

"The maintenance of soil fertility is the real basis of health and of resistance to disease," he wrote in *An Agricultural Testament*. Howard saw farmers heading in the other direction. He saw the chemical trend as, at best, shortsighted and, at worst, a folly that would result in the collapse of soil's productive capacity. Artificial manures, he believed, "lead inevitably to artificial nutrition, artificial food, artificial animals and finally to artificial men and women."

Healthy soil brings vigorous plants, stronger and smarter people, cultural empowerment, and the wealth of a nation. Bad soil, in short, threatens civilization. We cannot have good food—healthy, sustainable, or delicious—without soil filled with life.

CHAPTER 6

BACK WHEN THE STONE BARNS CENTER opened for business in the spring of 2004, the soils growing the vegetables for Blue Hill were already filled with life. That wasn't just luck. Ten years after I first read Eliot Coleman's advice on growing healthy plants, he was hired to consult on the center's creation. Eliot's charge was to identify the best land on the property for cultivating vegetables.

The first time he came to visit the site was on a crisp fall afternoon in 2002. As the daylight faded in the cold, slow burn of late November, Eliot seemed anxious. He had decided on a relatively flat, healthy field running the length of the property. But then we came to the bottom of another long field, sloping upward from the largest of the stone barns (Blue Hill's future dining room), where the dairy cows had been milked in the 1930s and '40s. Eliot stopped and scanned the six-acre expanse.

"The cows probably pastured here," he said softly, almost to himself. When I turned to look, Eliot had dropped his bag and started running. He traversed the old pasture, slaloming back and forth, dodging rocks and thistle, all along rotating his head to study the position of the waning sun. He ran past what would later become home to rows of tomatoes, cucumbers, fava beans, parsnips, and eventually the Eight Row Flint corn, until he arrived at the highest point, where he raised his finger to the sun. Then he was off again, to the northeast corner of the field, where he stopped, hands on hips, and scanned the land intently. Even in his early sixties (and still today), there was something hopeful and blithe in his pursuit. Eliot is like a wild horse—curious, observant, sly, and energized by an intuitive connection to nature. I watched him in awe.

"Hey, fucking cool," he said on his return, his dirty blond hair gleaming in the dull light. His eyes were wide and nearly pulsating; he looked to be breathing through them. He had a fistful of soil, and he turned his hand up so I could have a look.

"Black enough for you?" he asked. "*This* is the field. Forget the other one. It's making me hungry just thinking about what you're going to grow here." I asked if he had changed his mind based on the field's position in relation to the sun.

"The sun? Oh, hell, no. I was looking at the sun because it's so damn beautiful right now." He squinted his eyes, admiring the last of it setting in the distance. "No, I wanted to make sure this had been pasture for the cows."

Eliot explained that the field closest to the milking barns was usually the one most grazed by the cows (why walk the cattle farther than necessary at 5 a.m.?) and therefore the most mineralized by manure. In this case, he was right. The field, we later confirmed, had once been grazed by the Rockefeller family's dairy cattle.

"I bet there's a deep layering of organic matter," Eliot said. "It's going to grow vigorous, absolutely delicious plants."

Having mapped out the land, Eliot was assigned his second task: help find a farmer. Here he had the good sense to consult Amigo Bob Cantisano, the legendary and widely influential sage of California organic agriculture. I first met Amigo at Laverstoke—he had been among Eliot's Fertile Dozen. Mutton-chopped and dreadlocked, he has a formidable silvery-black mustache, making him look more Pancho Villa–like than plain old Bob-like. (Which may explain why he's simply called "Amigo"—the name his girlfriend's brother gave him in high school.)

Amigo recommended Jack Algiere, a young farmer he had grown to admire after consulting with him on an organic olive farm. "I've worked with so many farmers over time," he told me. "Sometimes I'll work with a guy who could have been a rocket scientist if he had wanted—you know, they've really just got it going on up here," he said, drilling his index finger into the side of his skull. "Jack's one of those."

16.9

Jack turned out to be as gifted as Amigo Bob promised, and as curious as Eliot about how to grow great-tasting food.

One day, during an especially cold stretch of winter in 2006, a few years after Blue Hill at Stone Barns opened, Jack came running into the kitchen, smiling big. Jack has curly hair and—especially back when his beard was still full and flowing—the look of a man who works closely with nature. You might say (although he wouldn't) that he's sort of a cross between Paul Bunyan and a young Bob Dylan.

On this particular day, he held two big handfuls of bunched carrots, their green tops waving in the air like pom-poms. It's hard not to be taken by Jack's electric good cheer in moments like these—showing off a new variety, or a perfectly ripe vegetable. You'd think such displays would happen often in a kitchen that's connected to a working farm, but the truth is that we tend to ignore one another, the farmers and the cooks, precisely because we're so close. The morning harvest arrives, it gets organized in the receiving room and stored in the coolers, and by the end of dinner service it's gone.

"We're sort of like a marriage," Jack once said. "We need to do one of those date nights every week just so we can actually talk."

Jack placed the carrots on a cutting board and took a step back, allowing us to admire his work. The last time he'd displayed his wares like this, it was an exotic variety of ginger, and before that an "extra dwarf" bok choy that fit into my palm. But carrots? They were always growing—in the field during the spring and summer and in the greenhouse most of the winter and spring. They were usually good carrots, sometimes exceptionally good, but did they deserve such swagger?

"Sixteen-point-nine, pal," he finally said. "Sixteen-point-freakin'-nine."

"Sixteen-point-nine?" I repeated, not understanding.

"Brix," Jack said, removing a small handheld refractometer from his pocket as evidence. Refractometers, which look like high-tech spyglasses, are popular tools for measuring the Brix, or amount of sugar present, in a fruit or vegetable. They've been used for years to verify levels of sweetness in grapes, helping winemakers determine ideal harvest times.

But Brix also indicates the presence of healthy oils and amino acids, proteins, and—this is key—minerals, those ingredients that Albrecht recognized were so critical for flavor. A 16.9 reading means the carrots were 16.9 percent sugar—and bursting with minerals. It's an extraordinarily high number, which Jack made sure I understood, even as the cooks, being cooks, drifted away to get back to work.

"Off-the-charts high," Jack said, watching me take a bite. He wasn't kidding. The variety, mokum, had been shown to reach a Brix of 12, a fact Jack discovered before his visit to the kitchen. So the astonishingly delicious mokum carrots I tasted that day were, in fact, off the charts.

ON JACK ALGIERE

Jack grew up on a farm tucked away at the end of a mile-long driveway along the Pawcatuck River in southern Rhode Island. In the mornings, his mother would open the kitchen door to shoo her son into the forest and fields, not to return until dinnertime. His carefree explorations gave him a passion for nature that the biologist E. O. Wilson, who enjoyed the same kind of childhood, calls "biophilia"—an "innate tendency to focus on life and lifelike processes."

By the time Jack graduated from high school, he had decided to farm for a living. That summer, he worked in a large greenhouse close to home, growing vegetables, shrubs, trees, and a variety of flowers, all started from tissue cultures rather than seed.

"That's how greenhouses generally work," he explained to me. "You grow five thousand plants that are carbon copies of one another." Genetic uniformity like this—a super-monoculture, essentially—under a closed roof is extremely susceptible to plant disease, which is why organic farming in greenhouses is so rare.

Jack saw the failings of the system up close. "One morning I went to water the geraniums," he told me. "I noticed a tiny black spot on one of the stems, which means you're about to be in major doo-doo." The black spot was made by a type of mold. And because of the crop's uniformity, Jack knew that none of the plants would have a natural defense against it. By noon that day the black spots had covered all the stems.

Jack consulted with the farm's owner, Bud Smith. "Bud was traveling at the time, but he had me go to this closet where we kept the full arsenal of chemicals and load up the strongest fungicide they had. I remember he said: 'Jack, don't fuck with this stuff.' If Bud was saying that, this had to be toxic as all hell." Bud had him wear a special protective suit and instructed him to seal the openings with duct tape.

"I caught a glimpse of myself as I walked out to the greenhouse. Like a space monster," Jack said. And then he sprayed—four entire houses filled with geraniums. "By the time I had finished spraying the first half of that first house, I was weeping. And I wept again inside when we realized the chemical had failed," Jack said. "The geraniums that survived looked really weird and ended up going to a local cemetery's urns, if you can believe the irony. But the memory of spraying stayed with me. It's the closest I've felt to being in the middle of a battlefield during a totally senseless war. I walked out of there and said to myself, *I don't want to be doing this.* And do you know what? I never did it again."

He intended to quit but was torn by his respect for Bud's great talent as a greenhouse grower. So Jack went to Bud's office to talk to him. "I walked in and saw Bud sitting at his desk, looking up at me," Jack said. "He knew. He absolutely knew what had happened. It was the first time I'd ever seen an adult look vulnerable. So instead of quitting, I blurted out, 'Isn't there another way?' I realized at that moment that Bud didn't want to be spraying chemicals—he hated it, actually, and so does every farmer farming conventionally. All Bud said was, 'What can I do? The customer wants geraniums, a lot of them, and they'll only pay so much.'"

Jack convinced Bud to let him take over a few of the greenhouses and convert them to organic. "I made a million and a half mistakes," he said, "but converting those greenhouses totally changed my life. Without the license to try—to see that it could really work, organically—I probably would have quit farming altogether."

Realizing he needed to learn more, Jack studied horticulture at the University of Rhode Island. During his second year of studies, he had another epiphany. "The departments, the professors—they were all there acting as enablers, keeping the chemical industry going because someone somewhere had determined it was worth keeping alive," he said. "It was as if I was in school to learn how better to kill the geraniums instead of how to prevent the fungus."

Then he discovered the library's collection of agriculture books, which included works by Sir Albert Howard and Rudolf Steiner. "I read them and it clicked," he said. "I mean, it all just came together. Steiner was crying out to farmers in the mid-1920s, basically saying, 'Don't be fooled by chemicals!' He really spoke to me, because that's exactly what I was trying to say to Bud: *Don't be fooled by all of this*. I didn't know enough at the time to know that I was right, but when I read Steiner I suddenly took on a kind of confidence I had never experienced before."

When Amigo Bob contacted him about the job at Stone Barns, Jack and his wife, Shannon, were happily farming someone else's land in Connecticut. He took the job mainly for the chance to bring together the ideas of Howard

and Steiner on a farm of his own design, one that connected the soil to everything around it—the flora and fauna, as he likes to say, but also the culture of
the place.

The project didn't start off well. "Literally just before Stone Barns is to
open, on my first day of work, 9 a.m., I drive up to the gate at the entrance to
the farm and I find myself behind a large herbicide-spraying truck," Jack
told me. "Spray the Problems Away Inc. or something like that. We sit there,
both of us waiting for the gate to open, and I'm thinking, *Who's this guy?* So I
honk my horn and get out of the car. 'Excuse me,' I say to the guy. 'What are
you doing here today?' He looks at me all confused and says, 'I'm here to
spray,' just like that. I'm thinking, wait a second, this is supposed to be a freakin' organic farm, right? The gate opens and we both drive in, the chemical
truck and me, and I'm banging the dashboard, screaming, '*This can't be fucking happening.*' I had just left a great job, moved with my wife to a place
where we didn't know a soul, and the first minute of the first day on a farm
I'm about to spend—oh, I don't know, the next forty years—is going to get
showered with an herbicide.

"We pull up to the offices, and I tell the guy to hang out a second. There
are like three hundred people running around in construction hats, but no
one to deal with this. Finally I get to the head of the construction company
and ask about the chemical guy with the big truck. He consults his work
sheet for the day. Sure enough, right there on the schedule is a 9 a.m. appointment to spray Rodeo in the pond between the future greenhouse and the
future outdoor vegetable field. Supposedly there was an outbreak of phrag-
mites, an invasive grass, and since the Stone Barns Center hadn't officially
formed yet, these construction guys were just doing what they needed to do.
It was like I was walking into the Wild West.

"I said, forget it. This is going to be an organic farm. If the guy sprays
this stuff, game over. The head of construction looks at me, sympathetic,
even though he has absolutely no idea what Rodeo is or how toxic it could be
for the farm, or even that I'm barely ten minutes into the first hour of the first

day of my new job. I got ready to send the chemical guy home when the construction head discovers that the Rodeo spray, and a whole schedule of spraying for the next month, had already been contracted for. Thirty-five thousand dollars. Done, papers signed. I left and called James [Ford, the founding executive director of the Center], and he and I hatched a plan right there. The chemical company could keep their money, but they wouldn't spray. And that's what happened: $35,000 to do essentially nothing. I keep thinking, even to this day, what if I had showed up late for work?"

SOIL AND FLAVOR

The 16.9 mokums lasted only a few services, but the small harvest left a big impression. Which is why, the following week, early on a frigid January morning, I stood with Jack in front of a row of future 16.9's incubating in the rich soil of the greenhouse. He had offered to explain to me in detail how the carrots had come to be so delicious.

The 23,000-square-foot greenhouse was calm and quiet, save for the soft hum of the overhead fans. Jack wore a look of pride as he surveyed the rich black soil that spanned the building. The soil came from the excavation of the Stone Barns parking lot, which partly explains Jack's fondness. After the construction crews unearthed the virgin soil, Jack rescued it from the dumpster. Then he created a recipe for the highest-quality compost, mixing it in to build up the soil's organic matter. He applies a wheelbarrow of his compost to every new row of vegetables.

I was familiar with the power of compost (what I understood of it) and impressed by the quality of Jack's personal blend. So I had a sense of how, after several years of building fertility, the soil could now nurture a carrot with a Brix of 16.9. But how exactly?

Jack pointed to the soil. "There's a war going on in there!"

"War" seemed a funny way to put it. The process had always struck me as extremely cooperative: Leaves and needles and grasses eventually die, forming a brown carpet of carbon on top of the soil. Herbivores (such as cows) and birds (chickens) periodically disturb the surface, allowing soil organisms (worms) to reach up and pull this organic material deeper into the dirt, where it—along with other material such as dead roots—is broken into nutrients available to the plants.

Jack went on to explain this war, which is when my understanding of the soil organisms' shared objective—the notion that everyone works together for the betterment of the soil community—became more complicated. There is a whole class system. First-level consumers (microbes), the most abundant and minuscule members of the community, break down large fragments of organic material into smaller residues; secondary consumers (protozoa, for example) feed on the primary consumers or their waste; and then third-level consumers (like centipedes, ants, and beetles) eat the secondaries. The more Jack explained it, the more it started to sound like a fraught, complex community. Organisms within each level may attack a fellow comrade (say, a fungi feeding on a nematode—or vice versa), or any of the tiny eaters can, and often do, turn on their own kind.

All of this subterranean life, Jack explained, is forced to interact— "cooperatively, yes, but also violently and relentlessly to maintain the living system."

To call this war may be a little extreme. When I ran Jack's analogy by soil scientist Fred Magdoff, he likened the process to a system of checks and balances. "To me there is real beauty in how it works," he said. "When there is sufficient and varied food for the organisms, they do what comes naturally, 'making a living' by feeding on the food sources that evolution provided. Sure organisms eat one another, but is that war? We eat carrots, but are we declaring war on carrots? What you have is a thriving, complex community of organisms."

Which is precisely what we want for better-tasting food. The result of all this activity—combat or cooperation, you choose—is that insoluble molecules are broken down and rearranged into forms accessible to plants. It's a process analogous to coffee making. Imagine, a farmer once told me, the difference in taste between a cup of coffee dripped through whole beans and one dripped through beans that have been broken apart into micro-granules.

Some of these microscopic nutrients combine to form phytonutrients, chemical compounds that are the building blocks of taste. "Like let's say calcium," Jack said. "Taste doesn't come just from calcium—not directly, at least. It comes from a more complex molecule that gets eaten, taken apart, and put back together in a different way. The plant takes this, and all the other molecules, and catalyzes them into phytonutrients. Taste doesn't come from the elemental compounds. It comes from the synthesis."

Phytonutrients—like amino acids, esters, and flavonoids—are key to the flavor of the mokum carrot, or whatever vegetable, grain, or fruit you're growing, Jack said. He crouched low to the ground to smooth out an uneven patch of soil. "And, not unimportantly—actually, *most* importantly," he continued, "phytonutrients are vital to building the immune systems of plants. They are part of the building blocks for vigor."

When insecticides and fungicides are used, they usurp the plant's natural defenses, which means the plant produces fewer phytonutrients. Studies show that organic fruits and vegetables typically contain between 10 and 50 percent more antioxidants and other defense-related compounds than are found in conventional produce.

Some scientists suggest it's one of the reasons organic food tastes better than conventional food. As soil biologist Elaine Ingham explained to me, "Phytonutrients are the building blocks for all of the flavor compounds. A lot of those flavor compounds are quite complex, and it takes quite a bit of energy and requires quite a diversity of building blocks in order to make them. So you have to have a plant with really good nutrition for those flavors to be expressed. It's not all that simple to have something that tastes really good. It's a lot easier to get something that has sweetness to it, but those really subtle complex flavors? You really have to have a healthy plant to have that."

I thought of Klaas and the velvetleaf—his soybean crop's resistance was not only a sign of healthy soil but a promise of great flavor as well.

"That's just it," Jack said when I mentioned Klaas's work. "The development of flavor, and the health of the plant, are the same freakin' thing. You don't get one without the other. If I treat the soil's microorganisms right, if they have everything they need to prosper, they'll do the work for me. At that point you just need to put it on the plate, basically."

As we left the greenhouse, Jack acknowledged that the precise mechanics of flavor creation are still mysterious. He realized this many years ago, after experimenting with brining olives. At first he chose distilled vinegar, which, when used as a brine, produced a predictable olive—delicious, but uniform in flavor. "Then I used a live vinegar," he said, "and after six months to a year, with all the fungi and bacteria in there, some olives would turn out sweet like fruit, some smoky, some had a roasted flavor almost. It was wild! The same thing is true for soil. You have different things going on, catalyzing new flavors, reaching the full potential and expression of the plant. It's the action that's important. But who really knows what the hell is going on in there?"

The admission took me by surprise, if only because Jack always seemed to know *exactly what was going on in there*. But eventually I realized he had it just right. I thought of Sir Albert Howard, who, writing in 1940, could not have named the full roster of microorganisms. Nor would he have known a phytonutrient if he saw one. Nor could he have described the chemistry behind well-composted soils—even though he was a chemist, and the father of compost. He didn't need to. I suppose that, like Jack, Howard was fine with not knowing. Where there is a bit of mystery, respect—even awe—fills the void.

A little ignorance keeps us from wrongly thinking it's possible to manipulate the conditions for every harvest. It's humbling to not know the *how*, and in the end it's probably a lot healthier. In the words of ecologist Frank Egler, "Nature is not more complex than we think, but more complex than we can think."

If a great-tasting carrot is tied to the abundance of soil organisms, a bad-tasting carrot comes from the absence of soil life. Which is the big distinction between organic and chemical agriculture. The nutrients in compost are part of a system of living things. They are constantly absorbed and rereleased as one organism feeds on another, so they're continuously available as plants need them. The supply to the plant comes in smaller quantities than it does with fertilizer, but it comes in a steady stream. It's slow release, versus one heavy shot of chemicals. The disparity is enormous.

To administer the heavy shot, soil is bypassed. Synthetic fertilizer, in soluble form, is fed directly to the plant's root. "It's a fast system," Jack said. "Whoosh! Water and nutrients are just flushing through. You can get your crops to bulk up and grow very quickly."

This is one of the reasons conventional salad lettuce—iceberg lettuce from the Salinas Valley of California, for example—often tastes of virtually nothing. It's almost all water, and the nitrates saturate the water, leaving no room for the uptake of minerals.

Thomas Harttung, another of the Fertile Dozen farmers at Laverstoke and founder of the largest organic farming group in Europe, has compared it to cooking: "Imagine a wonderfully balanced Italian main course full of herbs and other fresh ingredients. You then drop the salt bowl into it—rendering it totally inedible. The other taste notes 'die.'" Industrially produced grains, vegetables, and fruits taste of almost nothing because the nitrates have crowded out the minerals.

To bypass the network of living things is to deprive the plant's roots of the full periodic table of the elements the soil provides. But it also deprives the soil organisms of their food source. When Klaas said the number of organisms in his fistful of dirt was greater than the population of Penn Yan, he added, "That's a lot of community life to feed." He meant it as an obligation. "What kind of soil life are we going to promote in our fields, and what kind of flavor are we going to get in our mouths, if we feed soil life garbage?"

Why limit the hand that feeds you? As Eliot Coleman once said, "The idea that we could ever substitute a few soluble elements for a whole living system is like thinking an intravenous needle could administer a delicious meal."

A SUBTERRANEAN VIEW

Late one afternoon the following November, Jack finished his carrot tutorial by excavating a three-foot ditch in the vegetable field next to the fall crop of mokums. We climbed into the trench to examine a cross-sectioned wall of black dirt. It reminded me of the glass-enclosed ant farms I studied in seventh-grade biology. But in the dim light, this soil looked both exposed and secretive. Jack, my subterranean escort, pointed with a small stick to the exposed earth, hoping to illustrate once more how flavor starts in the soil.

"You should see this, because everyone talks about the chemistry of soil, or the biology," Jack said, running his hand along the wall, "but without the right physical structure, say goodbye to chemistry and biology. Nothing works."

The root systems created what appeared to be small highways and back roads, allowing organisms the freedom to move around. It brought to mind the interior of a well-made loaf of bread—moist, textured, and filled with irregular bubbles. The miles of white, wispy root hairs clenching the dirt in Jack's trench looked like the strands of gluten in bread that allow it to expand in the oven. Unhealthy soil, by comparison, resembles cake mix—dry and packed down, with no spaces for air to circulate or organisms to maneuver. (No wonder Klaas advocates for rotations of spelt; its large, deeply penetrating root systems create space for the community to thrive.)

Pointing again with the stick, Jack circled the narrow areas around the roots, the rhizosphere. It's the soil's most competitive environment, where organisms thrive in densities up to one hundred times greater than in other parts of the soil. The roots, sensing nutrients in the area, drill into the soil to take advantage of the rich possibilities for nutrition. In healthy conditions, organisms called mycorrhizal fungi bind to the root tissue, forming an unbeatable partnership that allows the root to reach even deeper into the earth, extracting what the soil has to offer.

"You can look at plants that produce mycorrhizal fungi like you look at oil companies," Jack said. "These companies invest the heavy costs of searching for oil if they believe it's a region rich with resources. The roots work like that. It's an incentive economy." He said plants will spend as much as 30 percent of their energy to build these fungal root extensions in order to tap into the tiniest spaces in the soil and get the nutrients there.

It turns out that the mechanism is a prerequisite for great wine. I learned this from Randall Grahm, the iconoclastic winemaker of Bonny Doon Vineyard, in Santa Cruz, California. "Mycorrhizae are microbial demiurges—they bring minerals into the plants," he told me. "What does that taste like? Persistence. The best wines are powerfully persistent. You breathe out your nose and you taste the wine over again, or you leave the bottle open for a week and the wine still tastes alive. Persistence doesn't fade, and it doesn't oxidize. That's from the minerals."

Jack got his finger into a nook of soil to show where the minerals are retrieved. "Here's where they suck out the phosphorus, or the copper or zinc, and all that comes up into the root with some stored water from the soil." He shook his head. "Brilliant, right? But you see what I'm saying? It's not just chemistry or biology down here. It all works if the physical structure is welcoming to the organisms and the fungi. At the end of the day, the plant's just looking for a good dinner, but he's got to be able to get to the table."

Peeking my head out of the ditch, I saw the final minutes of golden light bathing the upper stretches of the vegetable field. I remembered the fall afternoon eight years earlier (nearly to the day, and the hour) when Eliot

Coleman galloped back and forth across the land and sensed—correctly, as it turned out—a deep layer of organic matter below our feet.

"In a healthy system," Jack said, waving his hand to indicate all the vegetation above us, "everything you're looking at has a corresponding weight of roots and organisms belowground. *Everything.*"

A corresponding weight? It seemed almost impossible to imagine. As ecologist David Wolfe says, human beings are "subterranean-impaired." We're unable to see what's underneath us. It took a visit to the control room (and, as I stood in the ditch, a nematode's view of the underworld) to change how I looked at a landscape: what we see aboveground—the plants, trees, wildflowers, shrubs, and grasses—is mirrored by root systems belowground. I suddenly thought back to Wes Jackson's image of perennial wheat, with its iceberg-like proportions. It's not green and lush and filled with sunlight, nor does it inspire painters or poets or picnics, but nature's underworld is at least half of the story.

0.0: THE INDUSTRIAL CARROT

Back in the kitchen, Jack brought out his refractometer to test another batch of mokums. They scored well again, with Brix readings between 12 and 14. Someone pulled a case of stock carrots from the refrigerator. Grown in Mexico, these workaday carrots are large, uniform, and fast-growing, which makes them cheap fodder for vegetable and meat stocks.

I asked Jack if adding soluble nitrogen to his mokum carrots would make them grow faster. "No way. You'd just end up burning the shit out of everything," he said. "Adding Synthetic N is like adding a bomb—I mean, bombs *are* N, the same ingredient, so think what happens if you were to drop a bomb in the middle of a community of soil organisms."

"So let's say I'm a mycorrhizal fungus . . ."

"Kiss your ass goodbye," Jack said, chopping the air. "Gone, goodbye. N is ammonia, as in *ammonia*. It's burning, like the stuff you wash your floors with, only it's double, triple that in strength. If you're fungi, you're hightailing it out of there."

The Mexican carrots were from a large organic farm, an example of what Michael Pollan calls "industrial organic" and Eliot Coleman once described as "shallow organic." Such farms eschew chemical fertilizers and pesticides and technically abide by organic regulations, but they use every opportunity to operate in the breach. They grow in monocultures, they look to treat symptoms instead of causes, and, to cut to the real offense, they don't feed the soil.

"Carrots like these are all grown in sandy soils," Jack told me. "It's sand, water, and fertilizers." Organic fertilizers are the tools of the shallow organic farmer's trade. Like chemical fertilizers, they are applied in a soluble form, feeding the plant but not the soil.

Jack squeezed the juice and read the refractometer. "Whoa," he said.

"What did you get?" His expression had me imagining the Mexican carrot registering 20.9.

He shook the refractometer, squeezed more juice, and stared at the monitor. "Holy cow . . . zero."

"Zero?"

"Zero point zero," he said, flashing me the screen. "There's no detectable sugar."

"I didn't know a zero-sugar carrot was possible," I said.

Jack was silent a moment, holding the carrot up to the light as if it were a lab experiment. "Neither did I."

He said the Brix discrepancy could be attributed to several factors. Mokums were bred for outstanding flavor, for one thing, giving them a hereditary leg up on the Mexican organic variety, which was likely conceived for high yields or better shelf life. So comparing the mokum with the Mexican vegetable wasn't actually comparing carrots with carrots. And then there's the mokum's stress response, in this case to the cold snap we were experiencing. Freezing temperatures kick-start a carrot into converting starches to sugars. This neat physiological trick raises the internal temperature and prevents ice crystallization, helping the carrot survive another day. The Mexican carrots, in contrast, hadn't lived a day under a balmy sixty degrees.

But none of these excuses could disguise the essential difference between the two carrots. Jack's carrots were satiated with nutrients; the others were starved. By afternoon's end on this chilly fall day with Jack, I'd come to another paradigm-shifting realization about soil. Until then, I had held on to a remarkably simple misconception about conventional agriculture: that chemical farming kills soil by poisoning it (which it can) and that ingesting chemicals is unappetizing and harmful (which it probably is). But both miss the larger point if you're after a 16.9 carrot. Chemical farming—and bad organic farming—actually kills soil by starving its complex and riotous community of anything good to eat.

—

"To be well fed is to be healthy," Albrecht said. And he meant more than eating fruits and vegetables. He wanted to know what the fruits and vegetables were eating as well. (As Michael Pollan observed, you aren't just what you eat. "You are what you eat eats, too.")

Albrecht would not have been surprised by a 0.0 carrot, since he warned that soil microbes "dine at the first table" and should therefore have their plates filled with minerals. If not, plants couldn't be truly healthy. Nor could we.

In 1942, Albrecht proved his point. Just before World War II, when most Americans ate food grown close to where they lived, he came across Missouri's military draft records and discovered a correlation between recruits considered unfit for military service and soils lacking in minerals. Drawing a line across a map of Missouri, Albrecht overlaid the clusters of rejected men.

His hunch, that the washed-out soils of the southeast, close to the Mississippi River, would produce men of inferior physical capabilities, while the relatively drier (and therefore mineral-rich) soils of northwest Missouri would produce men of better health, proved to be exactly correct. Approximately four hundred men out of one thousand from the southeast were rejected for the draft, while only two hundred of a thousand from the north were. And, as Albrecht predicted, the area in between, where the soil was in fair condition, rejected three hundred recruits.

Alarmed by declining soil fertility at the end of World War II, Albrecht warned that our nation's future was at risk. He called for a major national initiative to restore the health and fertility of America's soils. Instead we went in the opposite direction, by industrializing agriculture. Not surprisingly, vegetables and fruits (and grains, milk, and even animal products) suffered. In the past fifty to seventy years, many vegetables have shown nutrient declines of anywhere from 5 percent to 40 percent.* Researchers now refer to large-scale "biomass dilution"—plants that have such low concentrations of certain nutrients that they do not adequately nourish the people who eat them.

⁓

I was once a member of a panel on sustainable food, where I made the point that these declines in nutrient density, especially in the mineral content of soils, were connected to a host of diet-related diseases. A nutritionist in the back of the room took me to task: Foods may have lost some of their trace minerals, he conceded, but because our bodies need so little of these to survive, we excrete those minerals anyway. Trace minerals—zinc, selenium, and copper—are named as such because we don't need a lot of them.

---

* This, according to Donald Davis, the author of one analysis, is at least partly genetic—the result of breeders selecting for high yield. "When breeders select for high yield," he explains, "they are, in effect, selecting mostly for high carbohydrate with no assurance that dozens of other nutrients and thousands of phytochemicals will all increase in proportion to yield."

"Instead of talking about the real issue here—how our modern diets include the wrong foods in bad proportions," he said, "you're bemoaning a food system that's succeeded in producing a plentiful, cheap supply of fruits and vegetables—fresh, frozen, canned, and even processed—by sacrificing a few minerals we pee out at the end of the day. I don't know what you're complaining about, and I suppose neither do you."

There was some truth to his critique. In a country where the leading dietary sources of energy are abundant carbohydrates and fats, within a world where 840 million people suffer from chronic hunger, it is difficult to get too worked up about food lacking in micronutrients. A carrot puffed up on nitrates and water is still a carrot with nutritive and caloric benefits. To compare it with a 16.9 mokum is to admit an embarrassment of riches.

But to say we need only a specific amount of certain micronutrients is exactly the kind of reductionist dietary advice that got us into trouble in the first place.

Several years after the nutritionist asked his question, I met with Joan Gussow, former chairwoman of nutrition education at Columbia University, who helped me with the answer. Joan is a longtime analyst and critic of the industrial food system and the woman who famously said, "I prefer butter to margarine because I trust cows more than chemists." She, too, was one of the Fertile Dozen at Laverstoke and served as part of the brain trust that guided the opening of the Stone Barns Center.

She affirmed that soil minerals are the building blocks of human nutrition, and at the core of healthy eating. "We're focused for some reason on single nutrients, on specific, magic bullets for our health," she told me. "But it's the mixture of foods—we call them *diets*, by the way—where the real nutrition comes from."

How would she have answered the nutritionist in the audience? She would have asked a question of her own: How did he know that we need only this much of X and this much of Y, and that we excrete the rest? After all, these days we're no longer concerned only with preventing deficiency diseases, like scurvy, which can be conquered with a magic bullet such as vitamin C.

Now, she said, "we're talking about degenerative disease, and degenerative disease takes a long, long time to develop. There are no magic bullets. There are only diets that appear to equate eating with a healthy life."

The Western diet does not appear to be one of them. Of the diet-related diseases that have spiked in the past century, the obesity epidemic would seem to have been impossible to predict. And yet, in the 1930s, Albrecht came close. He knew that cows grazing from well-mineralized soils ate balanced diets. But when kept in a barn and fed a predetermined grain ration, they never stopped eating, overindulging in a vain attempt to make up with sheer volume for what they weren't getting in their food. Albrecht believed our bodies would likewise stuff themselves for the same reason. Starved of micronutrients, he said, we will keep eating in the hope of attaining them.

Of course, obesity is influenced by many different things, but John Ikerd, a professor emeritus of agriculture and applied economics at Albrecht's alma mater, the University of Missouri, argues that Albrecht's conclusions about mineral depletion and obesity have never seriously been considered.

"If we humans have this same basic tendency as other animals, as Albrecht suggested, whenever our food choices are limited, we may well consume more of some nutrients than we need in an attempt to get enough of others to meet our basic nutritional requirements," Ikerd once said. "The lack of a few essential nutrients in our diets might leave us feeling hungry even though we have consumed far more calories than is consistent with good health."

Ikerd cites a damning statistic: from 1900 to 1950, Americans' physical activity decreased, as did their caloric consumption. In the second half of the century, they became even less active but ate more. "The sedentary lifestyles of many Americans obviously contribute to the growing epidemic of obesity," he conceded. "However, excessive eating and the resulting excessive weight also contribute to sedentary lifestyles. Many Americans may overeat because their food leaves them undernourished. . . . The human species obviously didn't evolve that much over 100 years, but the food system most certainly did."

The connection between depleted soils and obesity is rarely considered. And though scientists now see how well-mineralized soils beget healthy plants, there is still too little knowledge about how plants use those minerals.

"No one understands the mechanism for synthesizing minerals into molecules—no one," Joan told me. But this synthesis is, nonetheless, key to healthy plants and healthy people. "Foods are an evolving mix of metabolizing molecules. Diets represent a whole range. To separate the nutrients out of a diet is to render them—nutritionally, anyway—completely useless." Which is what most nutritionists do. They look at a vegetable or a fruit or a piece of bread and break it down into vitamin components: this gives vitamin A, this provides calcium, this contains the United States Recommended Daily Allowance of folic acid.

Albrecht did not. He worked backward from his observation of healthy people. "Rather than assuming what a healthy diet should be, he looked at healthy people and figured out what made them healthy," Klaas told me. "He could almost always trace it back to healthy soil."

When Joan said that diets represent the whole range of synthesizing molecules, she may well have been talking about how a plant creates flavor. Flavor, as Jack had pointed out, isn't about individual minerals. It isn't calcium or manganese or cobalt or copper. Flavor is the synthesis of all these things. The more minerals available for the plant to synthesize, the more opportunity there is for better flavor.

When we taste something truly delicious, something that is *persistent*, it most likely originated from well-mineralized, biologically rich soils. As it turns out, our taste buds may be far more sensitive than any chemist's tools.

Eliot Coleman would agree. "I'll never forget the night my wife and I sat down to a plate of carrots we had just harvested from the field," he told me once. "We dug in. And then I just stopped, fork in hand. There was a glow to the orange—it was incredible. I mean, it really glowed, like it was lit up. I just had to sit there and look at it. *Something* was going on. How do you prove a glow? A nutritionist would say, '*No, a carrot is a carrot is a carrot.*' A scientist would say, '*No difference.*' But taste the damn carrot."

A Gift from Nature

CHAPTER 7

STAND ON A ROOFTOP somewhere in rural America today (and, increasingly, rural anywhere in the world) and what you'll see is a large area of land farmed in monoculture. From that vantage, you might say we've won the war with nature, the one we've waged since the end of World War II, because it appears that nature has surrendered. Uniformity is everywhere. Look out from a rooftop in Blackfoot, Idaho, and you'll see hundreds of thousands of acres of Idaho russet potatoes. In Immokalee, Florida, you'll see tomatoes; in Castroville, California, artichokes; in Hereford, Texas, Angus beef cattle; in Sumner County, Kansas, wheat. If you're in Iowa—anywhere in Iowa—you'll most likely just see corn and soybeans.

These are, of course, extreme examples, paragons of agribusiness. By growing a heck of a lot of one thing, they make the modern American diet possible. On the other end of the production scale are small farms. They are the kinds of family operations local-food enthusiasts champion—the producers that line the aisles of farmers' markets or supply a community's CSA. Relative to five-thousand-acre monocultures and confinement feeding operations, they are, in fact, diverse, with different kinds of vegetables or multiple breeds of animals. But they're often guilty of specialization, too. They grow vegetables or fruits or small grains, or they raise animals for meat. Almost none incorporate all four. (Klaas's farm, with its elegant rotations of mixed crops, was an agricultural anomaly; and even it didn't include livestock.)

And yet, even before I visited Klaas's farm and stood in a ditch at Stone

Barns with Jack, appreciating his subterranean worldview, I found myself perched on a rooftop halfway around the world, in the Extremadura region of Spain, looking over a very different pastoral scene. It was a snapshot of how farming could sculpt rather than dominate a landscape. Stretched out before me like a chenille bedspread was a network of fields partitioned by thick, low-lying stone walls.

I was looking out at the *dehesa*, a system of agriculture that has existed in this part of Spain for more than two thousand years. Spanning over thirteen thousand square miles of grasslands and oak trees, it's a landscape that, despite millennia of persistence, would be relatively unheard of today were it not the birthplace of one the world's most beloved foods: *jamón ibérico*. The dark red, almost mahogany-colored cured ham is the result of a unique breed, the black Iberian pig, paired with a traditional method of free ranging in the famed oak forests that now stretched below me.

I already knew a thing or two about how the *dehesa*'s savanna-like landscape provides the perfect habitat for the pigs—an ideal marriage of wild terrain and controlled design—and on this clear day I could actually *see* it. I could make out how early peasants had selectively cut down the original forest, leaving enormous five-hundred-year-old oak trees spaced intermittently across the landscape. Rooting happily through the dense grasses, the Iberian pigs stumble upon pile after pile of acorns, yielding hams with a sweet, distinctively nutty flavor and incomparable fat.

But what I didn't know until my bird's-eye view from the rooftop was that the *dehesa*'s porcine paradise is home to more than just happy pigs. Before me I saw sheep and cows grazing in open pastures. Bisecting the pastures were thick stretches of forest, and off in the distance I could make out areas cultivated for grain—barley, oats, and rye. And I saw small Spanish homes dotting the landscape, too, with vegetable gardens and hanging laundry. It was perhaps the oddest sight of all: real community right there in the middle of the daily commerce of agriculture.

While the *dehesa* is rightly revered for its old oaks and traditional *jamón*, if not its ineffable place in Spanish identity, I realized I was looking at something even greater. We are told—in abstract, hopeful terms—that small, diverse farming communities are the best alternative to our industrialized food system. But rarely do we actually see them, a rooftop view that is so clear it looks like a blueprint for the future of sustainable agriculture. And perhaps the future of cuisine, as well.

My rooftop realization came at the end of several visits to Extremadura, a region of western Spain whose countryside is largely defined by the *dehesa*. I had not gone there in search of the perfect ham, nor had I gone to investigate a two-thousand-year-old landscape. I went to eat natural foie gras.

If that sounds like an unlikely occupation for a chef, it was. First, because flying to Spain to eat foie gras is as contradictory an idea as traveling to Canada for good barbecue. And second, and perhaps more to the point, because the idea of "natural" foie gras is, in and of itself, a contradiction.

Coveted by gastronomes, demonized by critics, *foie gras* is French for "fat liver," which is an apt description: prior to slaughter, enormous amounts of grain are funneled down a bird's esophagus, more grain in the span of a few weeks than most ducks and geese see in a lifetime. The glut of food causes the liver to swell to up to ten times its normal size. For a 175-pound person, it's equivalent to eating about forty-four pounds of pasta per day. It may sound delicious. But natural? Not so much.

At least that was what I believed, until one day when my brother, David, co-owner of Blue Hill, left on my desk a three-paragraph clipping from *Newsweek* magazine about a farmer in Extremadura named Eduardo Sousa. According to the article, Eduardo produced foie gras without force-feeding. His system, I read, was based on free-range feeding that took advantage of geese's seasonal instincts. Like any migratory animal, geese have a built-in response to cold: the temperature drops, and they gorge to store up fat in preparation for migration. The result, in this case, is *naturally* fatty livers—a feat I hadn't thought possible. I pinned the article to the board in front of my desk, intending to investigate further.

At the time, foie gras was quickly gaining attention from the press—even more so than usual. Chicago had already outlawed the sale of foie gras, and legislation was pending in California and New York.* The issue, vehemently argued by animal rights activists (and just as vehemently refuted by foie gras supporters), is that force-feeding, or "gavage"—the requisite part of the fattening, wherein a metal tube is inserted down a duck's throat to deliver food—is cruel and painful. Watch a Farm Sanctuary video of the weakened, bloated ducks and you can see why the practice is so controversial; it's not a convincing picture of animal welfare.

And yet, denouncing the product on moral grounds is easy; refusing to cook with it is not. The problem for us chefs, and it's not a small one, is that foie gras is so delicious. It's truly, indisputably luscious—fatty and unctuous, capable of transforming even the most humble dish. In other words, foie gras makes chefs look like better chefs.

~

I first tasted foie gras out of a can.

It was the mid-1970s. My father, a businessman, was approached by two Frenchmen who had an idea for a new children's board game. Their meeting took place in our living room, and as a house gift they presented a small black can of foie gras. My father brought out the melba toast. He insisted my brother and I stop watching television and acquaint ourselves with the delicacy. One of the Frenchmen removed the gray, wet slab of liver from the can with great fanfare, while the other spoke in near religious terms about the

---

* Though Chicago's ban was later repealed, California's foie gras prohibition went into effect in 2012.

rich culinary tradition of foie gras, and the superiority of French life. I remember taking a bite. It was awful—the smell, the texture, the whole idea of it—but while the displeasure was short-lived, the perplexing reverence for such a disagreeable food stayed with me.

Twelve years later, I came to see foie gras in a different light. Less than a week into my first apprenticeship at a high-end restaurant in Los Angeles, I discovered that a guest chef, Jean-Louis Palladin, would be preparing a special menu for the evening.

"Who's Palladin?" I asked Matt, the saucier.

"*Chef* Palladin," he said, with an expression of disbelief. "Chef Palladin is the greatest chef in America."

He wasn't overstating it. Palladin was just twenty-eight years old when he became the youngest French chef to win two Michelin stars for his restaurant La Table des Cordeliers in Gascony, an area of southwest France renowned for its Pyrenees lamb, its high-quality sheep's-milk cheeses, and, most especially, its foie gras. At an age when most chefs are still struggling to define themselves, Palladin had within his grasp the coveted third star, which at the time meant becoming one of the greatest chefs in the world. Instead, he abruptly left France and opened Jean-Louis at the Watergate, in Washington, D.C.

He arrived in 1979, at a remarkably bleak moment for American cooking. The great advances of modern life that large food corporations successfully sold to the American public—frozen, processed, and fast foods, out-of-season produce—had gone from novelty items in the 1930s to established fixtures on our plates by the '70s. Supermarkets, by this time well established, began competing heavily on price. The pressure to increase slimming profit margins affected not only how food was sold but how it was grown, too—pressure that would mitigate diversity, compromise quality, and forsake flavor for volume. In this new, consolidated era of agricultural efficiency, small farmers were left with few outlets—organized farmers' markets were just beginning, for example—and without direct access to consumers, many sold their farms to developers and cashed out.

Palladin didn't arrive at the bleak American culinary scene and call it a wasteland, which was the conventional attitude of most French chefs. "You'd hear it every day back then," New York restaurateur Drew Nieporent once told me, about his time spent working through the ranks of classic French restaurants. "'The butter is better in France, the beans are better in France'— it was the tyranny of the superior French product. And it was myth."

Palladin was one of the first to debunk the myth. He celebrated iconic American products such as Virginia ham and sweet corn, and distilled them through the rigors of French technique, elevating a mere baked potato or crab cake to culinary art. And he showcased more lowly (and often ignored) ingredients, such as barnacles, blood sausages, and pig's ears.

But the ingredient he most prized, foie gras, was at the time unavailable in America, and illegal to import (unless it came in a can, which it didn't take an epicurean to reject). Palladin was not deterred. He flew to France and shoved goose livers into the gullets of monkfish, predicting—correctly— that customs agents would avoid inspecting fish for contraband on his return. At the height of the illegal importation, Palladin was smuggling approximately twenty livers per week and serving them as off-the-menu specials to diners eager to taste the real thing.

His fearlessness made him a chef's chef, as well as a culinary star. Chefs and food lovers from around the country began making pilgrimages to taste the new American cuisine of Jean-Louis Palladin.

———

I didn't know any of this at the time. But when Palladin strode into the Los Angeles restaurant that day, in the manner of someone well accustomed to his own importance, it was hard not to be impressed. Tall and thin, with an impossible mop of curly hair crowning his large head, he barreled through the small kitchen with manic masculinity, thundering directives in a voice so low it sounded as though it came from his kneecaps. Behind oversize glasses, there were fierce, appraising eyes. He never stopped moving, especially in the throes of preparing a chicken sauce—a rich reduction of chicken necks, feet, and red wine—searing, stirring, whisking, smelling, and, every few seconds, it seemed, tasting.

By 6 p.m., with everything in place and the guests just starting to arrive, Jean-Louis paced up and down the long line of stoves, banging his hand and begging for action. "The orders!" he demanded. The orders came quickly, and suddenly I was introduced to pure high-octane kitchen action.

In the whir of it all, two memories of what he prepared stay with me. One of them I tasted, and one I did not. The one I did not taste was a chicken dish with only its lowliest parts—gizzards, cockscombs, and a cut from the thigh called the "oyster"—all bound together with that sauce. The one I did taste was foie gras and chestnut soup. I thought it was magic.

A FRENCH TRADITION

The French tradition of foie gras began, as one might expect, with peasants. In the rural south of France, the family goose was kept in a box under the domain of the *grand-mère*. Three times a day, she poured warm mash down the bird's gullet, gently massaging its belly with deep circular strokes. The goose was slaughtered in time for Christmas, reverence by way of ritual.

It was a chef, Jean-Joseph Clause, who in 1778, as the head cook to the governor of Alsace, conceived of "pâté de Contades," scented with truffles and cooked in a pastry crust. Struck by the flavor, and probably seeking attention from the throne, the governor sent the pâté to King Louis XVI, who claimed it to be "the dish of kings." The governor remained a mere governor, but Chef Clause was awarded twenty pistols (today's four stars?), and foie gras became the prize of the culinary world.

Foie gras might have remained one of the country's most coveted foods—to this day it's a fixture of the French Christmas table—but, like most traditions, it endured primarily because it changed so dramatically. Two transformations—one technical, one psychological—enabled its survival.

Improved production techniques during the Industrial Revolution—food sterilization, for example, and also corn used for feed instead of barley and millet—ensured standardization and faster weight gain. But the Industrial Revolution also helped to ensure a change in mind-set, subtle as it may have been, whereby animals came to be seen, like almost everything else, as commodities.

"The goose is nothing," wrote Charles Gérard in his 1862 study of Alsatian cuisine, *L'Ancienne Alsace à Table*, "but man has made of it an instrument for the output of a marvelous product, a kind of living hothouse in which there grows the supreme fruit of gastronomy." The goose is nothing; the process is everything. The increasing manipulation of nature and the ability to mechanize food production coevolved.

By the 1960s, the process had become increasingly industrialized, centralized, and—along with the rest of agriculture—specialized. Ducks, not geese, became the favored "instrument," as they conformed better to the new large-scale operations. Geese are high-strung, sensitive birds, prone to stress; you can certainly fatten a goose liver, but not without considerable care and effort. Ducks are more pliable.

The introduction of a new hybrid duck breed in the 1970s solidified duck livers as the foie gras of choice. Breeders developed a sterile cross between a male Muscovy duck and female Pekin duck, called a Moulard, or "mule" duck. Crosses of different breeds—in any animal, not just ducks—often produce a new generation with better characteristics. The phenomenon, referred to as "hybrid vigor," yields healthier animals that grow faster and, if you're lucky, taste much better than either of their parents. For anyone looking to raise animals, hybrid vigor often translates into higher profit—as proved true for the Moulard. The Moulards surpassed both Muscovy and Pekin in their ability to withstand factory conditions. They were more disease resistant and more docile. They were also quicker to gain weight, resulting in larger livers. And with the use of artificial insemination, these ducks could be bred on demand.

For chefs, there was another, more pressing advantage to the new hybrid.

Unlike the finely textured goose and Muscovy duck livers, which rendered much of their fat in a hot pan, the Moulard liver maintained its integrity under heat, allowing for that delicious, crowd-pleasing sear. (Until then, a thick slab of foie gras, roasted as one might a steak, would have been nearly unimaginable.) Farmers didn't have to watch a simple flu strain wipe out a third of their feathered profit, and chefs didn't have to stand at the stove while a chunk of their $80 duck liver lost more than half its size in the pan.

By 2007, there were thirty-five million Moulard ducks bred for foie gras in France, and only eight hundred thousand geese. Today, it's the way foie gras is produced everywhere in the world—France, the United States, and Hungary (Israel, too, before foie gras production was banned there in 2005). In one tiny corner of Spain, Eduardo Sousa was doing something radically different.

What began in 1812 as a quiet family tradition became headline news in 2006, when Eduardo's foie gras won the Coup de Coeur award for innovation at the Paris International Food Show (SIAL), beating out thousands of other entries. He was the first non-French foie gras producer in the history of the competition. Asked about it many months later, Eduardo said, "A Spaniard winning for foie gras? That really pissed the French off."

The French condemned Eduardo's liver—first accusing him of cheating, and then refusing even to call it foie gras. "This cannot be called foie gras," wrote Marie-Pierre Pée, secretary-general of the French Professional Committee of Foie Gras Producers, "because it is strictly defined as a product from an animal which has been fattened."

In other words, if there's no force-feeding, there is no foie gras.

A RUMINANT'S-EYE VIEW

For months, the article about Eduardo stayed posted to the corkboard above my desk, mostly forgotten. Foie gras wasn't always on Blue Hill's menu, but when it was, I never considered it controversial. Hudson Valley Foie Gras, an artisanal producer in upstate New York, provided impeccable and consistent livers to the best chefs in the country. For me, as for most of them, the foie gras debate consisted mostly of how best to prepare it.

It wasn't the mounting political controversy, or a heartrending PETA video, that sparked my conscience. Instead, it was an early July morning I spent with the lambs that finally had me reconsidering foie gras's place on our menu.

That morning, I walked out to the pasture and watched Padraic, the livestock assistant at Stone Barns, move the one hundred or so sheep to a new paddock of grass. I had thoughts, if not visions, of the Marlboro Man. Padraic is six feet four, with chiseled features and piercing eyes, and as he tipped his cowboy hat up to the sun, I waited for him to open a tin of Skoal or crack a leather whip to keep the sheep and their lambs moving. Instead he called out in a gentle coo, opened the fiberglass fence, and gently waved the first of the lambs onto the new grass. She excitedly trotted to her next meal.

"That a girl," he said, tapping her on the rump. The rest of the sheep crowded close together and herded themselves into the new paddock, bringing to mind a small bison charge.

Until that moment I thought I knew good lamb. I had sourced plenty from local farmers over the years, and I had roasted enough lamb chops and braised enough shanks to recognize a well-raised lamb when I ate it. What I didn't know, and—since Klaas had yet to introduce me to William Albrecht—what I'd never stopped to consider, was: *What does a lamb want to eat?*

It's a funny sort of question, but out in the field, watching the lambs excitedly drive to new grass, without being pushed or cajoled, it wasn't hard to recognize that they actually cared quite a lot about what they ate. You could even call them picky. Just like the cows I once observed with my grandmother at Blue Hill Farm, they moved quickly over certain grasses to get to others—to nosh on clover and mustard grass, avoiding horse nettle and fescue along the way. They resembled hungry, slightly aggressive diners at a

Las Vegas buffet, which is the point, really. Lambs on a grass diet don't so much get fed as work to feed themselves, and the distinction is not small.

I remember, when I was a young line cook, hearing the famed chef and owner of Le Bernardin, Gilbert Le Coze, explain the inspiration behind his seafood-only restaurant. "Nothing is more stupid than a cow," he said, launching into a well-polished diatribe against cud-chewing ruminants. "To just stand there, grazing all day long—there is no spirit to that. But a fish is such a wild creature—that gives him another dimension." Standing next to Padraic on that summer morning—the baby sheep dancing in circles around their mothers, their eyes bright, their fleeces shiny—it was tough to see Le Coze's point.

Padraic pointed to a sheep inches away from us that was running the bottom of its muzzle over the blades of grass—a rapid reconnaissance of what's available for breakfast. The hairs right below the jaw act as a kind of radar for what a ruminant is looking for, and what it's looking for depends on many factors, including the weather, the time of year, and even the time of day. Like us, they balance their diet, getting enough protein and energy by choosing which plants to ingest (only, as Albrecht's field experiments proved with cows seventy years earlier, they do a better job of it).

Padraic's job, under the direction of livestock manager Craig Haney, is to rig the game. It's to ensure that when the sheep get to a patch of grass— when they finally commit to a bite, and then another—they're rewarded with rich, nutritious diversity. This is the "take half, leave half" rule of grazing: flood the pasture with the sheep when the grass is at its perfect point of development—just before its adolescent growth spurt, when it's tender and full of sugars—but move them out quickly so the grass can recover before the next bout of grazing.

The sheep, of course, have no idea that the grasses have been carefully tended to, that their delicious spread has been painstakingly prepared by seeding certain varieties of grasses, rotating in other animals, and adding natural supplements to the soil. In fact, as I stood there with Padraic, it seemed clear—for the first time I was seeing, not just tasting, the difference—that much of the pleasure the lambs enjoyed was because of the hunt itself. They needed (and wanted) to work for their meal. Which is perhaps why they looked so purposeful. They weren't as wild as a fish evading a hook, but their drive gave them another dimension that Le Coze didn't recognize.

To be fair, Le Coze could have been referring to what we've managed *to do* to ruminants over the past several decades. Instead of allowing them to forage, we do the work of foraging for them. We feed them corn and other grains and generally restrict their urges by narrowing their diets. And their activity: in America, most ruminants start out on grass but finish their lives confined in animal feedlots. We dull them. And so, yes, we do sort of make them stupid.

Take Colorado lamb, famous for giving us those uniform and fatty chops. Since fat carries flavor and retains moisture, it's pretty easy to have a moist and juicy bite of feedlot-finished meat. But as Garrison Keillor said of the modern chicken, you can "taste the misery" in every bite. Great chefs will tell you the misery you're tasting is greasy fat. Greasy fat coats your mouth. It's sweet, soft, and nutty, tasting nothing like the animal you're eating. And it surrounds a kind of watered-down version of what lamb could be, which is ironic considering the Cadillac-size chops.

The added insult: most lamb recipes instruct you to "remove fat cap and discard." We do this without thought, as if we're unpacking groceries. When I was training to butcher meat at a restaurant in New York, part of my job included cleaning forty racks of lamb for dinner service. With each rack, the restaurant's old French butcher had me pull off the solid inch of fat covering the loin. A small incision near the bone, a quick yank, and the fat layer tore off, like the peel from a grapefruit. On the way to the dumpster to throw out the discarded fat, I thought about the irony. The restaurant paid the highest price for this part of the animal, only to toss 10 percent of it in the garbage? (Since that fat was essentially a mountain of corn feed, wasn't I really just throwing Iowa in the trash?)

When I asked the butcher why the chef wanted the fat removed, he simply said, "It's disgusting, so much fat." He was right. Growing up in France, the butcher had undoubtedly never seen a one-inch cap of fat on a rib of lamb. Feeding an herbivore grain (intensively anyway) is a recent invention, and despite the fact that the practice has become so ubiquitous—and in the case of Colorado lamb, so coveted—it's not actually delicious.

Farmers like Padraic and Craig, standing in the field shepherding their herd, might look like portraits of America's lost agrarian past, but, by giving the lambs what they want, they are in fact creating a modern recipe for delicious meat—complex and richly textured, without flab or a greasy aftertaste, and with a flavor that changes throughout the year.

❧

"The challenge of cooking in America," Palladin once said, "is to discover the newest and best products from the different states—baby eels and lamprey from Maine, fresh snails from Oregon, blowfish from the Carolinas and California oysters—and then to learn how to integrate them into your cuisine."

But Palladin did more than simply integrate regional specialties into his cuisine; he created markets for them, too. Thomas Keller, the renowned American chef who as a young line cook made several pilgrimages to the Watergate, has said that Palladin was doing a kind of farm-to-table cooking before there was a name for it, and that his style—highly technical, cutting-edge, and artful—influenced the industry.

"When you get the professional side involved," Keller explained, "it trickles down to everybody." Increasingly chefs in America saw an opportunity to focus their cuisine on cultivating relationships with farmers to supply their menu. Before Palladin, Keller said, "chefs weren't developing relationships with farmers, gardeners, and fishermen."

Joan Nathan, a cookbook author and former *Washington Post* writer, told me that Palladin's greatest gift was never squandering the chance to connect with a new farmer. "When he heard about a farmer growing something new or something great—it didn't matter what it was, a great ham, fresh zucchini blossoms—he'd hop on his motorcycle and think nothing of driving a few hundred miles to find it. By dinner service, it was on the menu." And when he couldn't find an ingredient, he persuaded other farmers to grow it.

Palladin can't take credit for persuading John and Sukey Jamison to raise grass-fed lamb, but he does deserve acknowledgement for discovering them. Long before farmers like Craig and Padraic got in the game, John and Sukey were perfecting the art of intensive grazing on their Pennsylvania farm. They quickly learned to celebrate the "inconsistency" of their product.

"Oh, yeah, you can taste the difference," John once told me of his lambs. "By age, by diet. You'll get stronger-flavored lamb in May and June, based on the young wild garlic and onions, and then a leaner taste in late summer from the wildflowers. In fall you start to see the cold-season grasses, giving you the most mature and delicious fat of the year."

When I first met John, I asked him how he and his wife got started. "We were a couple of hippies who didn't want Woodstock to end," he said. It was in the aftermath of the 1970s oil crisis. Prices of gasoline had quadrupled in just a few months, and the combination of falling supplies and international unrest caused grain prices to double or triple.

Along came the development of portable electric fencing that could be set up and moved by one person (a product of New Zealand, where grass-fed livestock was the only game going). "That was really the beginning of the *intensive* part of rotation grazing," John told me, differentiating the practice from free ranging, where the animals pasture in enormous fields and are rarely moved. Portable fencing gave small farmers the chance to mimic what bison herds had been doing on a continental scale for thousands of years.

The Jamisons set up a successful grass operation on two hundred acres in western Pennsylvania, but their philosophy never went mainstream. The oil crisis ended. The era of cheap fuel and grain returned, and so too did status quo methods of confined animal farming. We've been making our ruminants stupid ever since.

For many years, it was difficult for the Jamisons to compete with the grain-fed operations—and lamb in this country was already a tough sell—but in 1987 their luck changed. That's when Jean-Louis Palladin called to ask them to deliver a few lambs for a congressional dinner at the Watergate Hotel.

"We entered the kitchen carrying the lambs on our backs," John told me. "Palladin introduced himself—he was wearing Jordache jeans and high-top sneakers." The chef placed the lambs on a table in the corner. John could see his arms waving wildly as a swarm of white coats surrounded him. At last Palladin called John over while examining the organs. He guessed the age of the lambs based on the layer of fat surrounding the kidneys. (He was off by three days.) Then he ran his hand along the carcass and poked his nose deep into its cavity. "Even with that enormous mane of wild hair, and thick, oversized glasses, he stuck his entire head into the carcass, breathing in, as if he was about to taste a vintage Bordeaux," John said.

From that day on, Palladin began ordering the Jamisons' lamb for his restaurant, naming it on the menu, and soon John and Sukey were receiving orders from chefs around the country. Farmers began asking to visit, to learn how to incorporate the Jamisons' methods on their own farms.

"It's a funny thing," John said to me recently. "Here we were adhering to our ideals from the '60s—living simply, improving the land, making the world a better place—and trying to farm in the great French peasant tradition. Along comes a chef, feeding some of the wealthiest and most influential Americans, who helps make what we do suddenly famous in America."

Jean-Louis Palladin's contributions to gastronomy were enormous and well documented. But one of his most lasting and least heralded legacies might be helping to ensure the success of the Jamisons, who have gone on to inspire a small network of livestock farmers to wean their animals off grain. (*Small* is the operative word; true grass-fed lamb—which means the animal doesn't get a lick of grain—accounts for less than 2 percent of all lamb raised in this country.)

John credits Palladin with helping create a consciousness for a generation of chefs. "That first delivery day, Sukey and I stood there with Palladin in front of the lambs after the cooks went back to work. His eyes had welled with tears." Tearing off a piece of butcher paper, the chef quickly drew an outline of France, dividing the country into a kind of regional taste map, describing the different flavors of the lamb based on the grasses they foraged. It was the first time John met a chef who understood how feeding grain flattens flavor. Palladin celebrated the inconsistency.

"He got really animated again, pointing to the areas with the best grasses and wild herbs that produce the very best lambs in France. He studied the map, trying to situate what he had just tasted. These place-based flavors had been etched into his memory, and now he reveled in the thrill of adding new ones."

Palladin didn't buy in to the grain-fattening mania of animal farming, not because it was inhumane or because it was destructive to the environment, but because it never produced anything really good to eat.

A few months after my walk with Padraic and the sheep, I stood in the kitchen watching a cook devein an especially large foie gras liver. Suddenly I was thrown right back into the early July morning watching the Stone Barns lambs forage for grass. That pastoral scene—lambs hunting for their breakfast, the farmer masterfully orchestrating the right kind of meal at the perfect moment—was the antithesis to the fattened liver, which now reminded me more of that one-inch fat cap on a Colorado lamb rack.

Of course, the two aren't entirely analogous—geese and ducks are omnivores, for one thing, which means they're better able to digest grain than ruminants—and I wasn't about to haul my beloved foie gras off to the dumpster. But standing there in a moment of quiet contemplation, I wondered about the difference. How could I talk a free-choice game (as I did about all the animals we served at Blue Hill) and at the same time, on the same menu, support a system of not just corn feeding, but forced corn feeding? And a lot of it.

Luckily for me, it wasn't long after that moment that my friend Lisa Abend, a *Time* magazine journalist stationed in Spain, called to ask if I'd ever heard of a man named Eduardo Sousa. I looked to the right, and there on my corkboard was the *Newsweek* clipping my brother had handed me. Lisa had been assigned to write about Eduardo, to evaluate with a chef whether his natural foie gras was for real, and if it was any good.

Palladin, the greatest champion of foie gras this country has ever known, might have simply hung up the phone at the preposterous (and, for a native of Gascony, offensive) idea. Or, when Lisa asked if he was interested in seeing his farm and tasting the livers, he might, as I did, have simply said, "I'll come."

CHAPTER 8

W<small>E ARRIVED</small> at Eduardo's farm late in the morning, after an overnight flight from New York. Lisa picked me up at the Madrid airport, and we drove southwest toward Badajoz, traversing Extremadura, an arid region that looked the way I imagined El Paso might look if El Paso had a winter.

Extremadura has two provinces—Badajoz to the south and Cáceres to the north—both sparsely populated. Lisa, a former European history professor, explained that when the Christians were reconquering their land from the Muslims in the Middle Ages, they referred to this area as the Extrema Dorii, Latin for "far side of the Duero River." It's a literal translation; Lisa said it was used in the same way Americans called everything outside the thirteen colonies "the West."

Less likely, though it would be technically accurate, Extremadura could also have referred to the "extra-hard" environment: hot, bitterly dry summers, cold winters, and high plains intersected by steep mountain ranges. Despite its difficult terrain, and undoubtedly because of it, the region was home to the original conquistadores, the famous soldier-adventurers who set off for the Americas. As I imagined the rugged television cowboys of my youth, Lisa's comparison didn't seem too far off.

The scene outside my window was every bit a portrait of the Spanish wild west: vast expanses of open land were intercut with towns that revealed their Moorish influence—homes with plastered white walls and thick archways. Much of the land we drove through was barren, but by the time we pulled past Fuente de Cantos and began nearing Eduardo's land, in Pallares, it had changed dramatically. Suddenly we seemed to be in the African savanna, but with greener pastures and healthier tree cover.

An unmarked dirt road led up to Eduardo's farm, or we guessed it did. No one was around. A furious barking dog tied to the side of a shed greeted us. The place looked deserted. We found Eduardo lying on his back in a small, open field, his cell phone raised above his head. Two dozen or so geese circled him in a raucous chorus of quacking and feather shaking.

"*Bonita!*" I heard him say as we approached a bright orange fence. "*Hola, bonita!*" Thinking he was on speakerphone, we slowed down, only to realize he was snapping pictures of his geese. A black eagle flew threateningly low. Eduardo didn't seem to notice.

"*Hola*—Eduardo?" Lisa said. Eduardo snapped more pictures. By now I was close enough to see that he was laughing.

"Eduardo?" she said, much more loudly. The geese shrieked and ran for the other side of the fence line, and Eduardo stood up quickly, his carefree air marred briefly by concern. After whispering something in the direction of the geese, he beamed even more brightly, then turned to acknowledge us with a gentle wave. Eduardo is large but not fat, with small eyes, puffy cheeks, and very thick black hair. His rounded belly, green sweater-vest, and brown loafers called to mind a building superintendent.

Lisa introduced us. "*Vale*," she said. "*Dan es* chef *de* Nueva York." Eduardo raised an eyebrow in my direction.

"It's an honor to meet you, sir," I said, awkwardly formal. In the instant of that raised eyebrow, I was overcome with a feeling that our trip was doomed. Foie gras without gavage? Who was I kidding? More to the point, who was *Eduardo* kidding? You didn't need to be Columbo to question this guy's story. He looked nothing like a farmer, and this didn't look anything like a farm. There were no tractors, no barns, and no silos. There was only a smiling, slightly chubby man in a green sweater-vest and a phone filled with morning portraits of his geese.

A long moment of silence followed. I fought the urge to speak bluntly.

Chefs suffer from this often. It's a regrettable but identifying trait acquired from years of working in a restaurant kitchen. The dialogue is curt. It skewers subtlety to get to the point. The point is *to get to the point* before the hot food turns cold. It's a survival tactic. And it works. But out of the kitchen, it is often difficult to regulate.

"How often are you moving the geese to new grass?" I asked abruptly. Lisa, startled, repeated the question in Spanish, drawing it out for the sake of politeness.

Eduardo shook his head. "I listen to the geese," he said. "I give them what they want." We began walking the perimeter of the fence.

"And what are you feeding them?" I asked.

"Feed? No, we don't feed," he said.

"He doesn't feed his geese?" I asked, looking at Lisa. Having traveled halfway around the world to learn about what was considered an impossibility— foie gras without force-feeding—I wasn't prepared for a farm that . . . didn't feed at all?

Eduardo smiled and held out his hands, palms down, moving them up and down briefly as if to say, *Slow down, the understanding takes time.* "The geese eat what they want. They feed off the land," he said. "Very simple."

We continued walking around the fence. The geese followed, slowly at first, their movements nearly imperceptible, but within a few minutes they were in a neat phalanx, marching across the paddock until they arrived, a few feet from where we stood, quacking and ruffling their feathers in delight.

Eduardo pointed at the power source for the orange fence, a solar box that converted the sun's energy into electricity. "The geese avoid getting too close to the fence. It feels foreign to them, I think. And anyway, it doesn't much matter, because only the outside of the fence is electrified."

"Only the outside?"

"The outside is electrified and not inside of the fence—there's no current running through the inside."

I looked at Lisa and laughed. "A fence for the animals that's not electrified. Does he mean they're free to leave?"

"Free!" Eduardo said, his arms flapping wildly to show me just how free.

His job, he explained, was to give the geese what they wanted, and if he succeeded, they wouldn't leave. Part of what they wanted, apparently, was to not *feel* fenced in, because if they felt fenced in they would feel manipulated. "They eat less when they feel manipulated," he said.

"But they're still fenced in, even if it's not electrified," I noted, a little too pointedly. Lisa struggled to find the right wording, to avoid offending Eduardo, but he anticipated the question and cut her off.

The fence, he said, is used only when the geese are too young to protect themselves from predators. And even then, "The geese don't feel fenced in. They feel protected." To be fenced in, in fact, didn't exist on Eduardo's farm. Until now, I hadn't thought of a fence as much more than an instrument of enclosure and control. But for Eduardo it is a means of protection—a physical one and a psychological one, too. In feeling unmanipulated, the geese felt free, and a goose that felt free was, according Eduardo, a hungrier goose.

My mind went to Padraic and the lambs at Stone Barns. Is it possible that the animals were so delicious not just because they grazed on grass at the perfect moment, but because they felt free to do so? Perhaps the secret to natural foie gras was similar to that for superior lamb. Allow the goose to feel free, give it the opportunity to eat what it wants to eat, and nature will take care of the rest.

Eduardo insisted we drive to another area of the farm. The fencing system was important, he told us, but the freedom to roam and forage was essential to the success of the foie gras. He wanted us to see the adult geese at work.

He drove us along a back road, so slowly that I wondered if he had a flat tire. The effect, if not the intent, of this meandering was a shared appreciation for our surroundings. Winding through these open fields, intercut with enormous oak trees, it struck me for the first time (I didn't admit to Lisa or Eduardo that I hadn't bothered to consult a map of Spain before coming on the trip) that we were driving through the famed Spanish *dehesa*. I had seen pictures and heard about the history of this land, but for most chefs (and Spaniards, too), coming to the *dehesa* is a pilgrimage to a sacred place, the source of the renowned *jamón ibérico*.

Food writers tell us that chefs are obsessed with superior ingredients, especially ingredients that make their cooking sing—Périgord truffles, artisanal olive oils from Italy, sea salts from Brittany—and it's true. Like anyone devoted to their craft, we are drawn to ingredients that help us elevate a dish. Which is another way of saying we're drawn to anything that makes our food taste better. But there are a select few products that inspire complete subservience—which arrive at and depart our kitchens unchanged and without garnish. These foods fall into the category not of ingredients on a chef's palette but of fully formed works of art. A perfectly ripe cheese, for instance, or a just-picked heirloom tomato, still warm from the sun. Or *jamón ibérico*. Even the most talented chefs (perhaps *especially* the most talented chefs) agree that they're better left alone.

But unlike the perfectly ripe cheese or the heirloom tomato—which can be produced with exceptional results most anywhere—no one has been able to replicate the taste of *jamón ibérico*. It is unquestionably the finest ham in the world. With a taste that is both rich and dry, as nutty as Spanish almonds or aged sherry—it's almost indescribably mouth-filling and deeply satisfying.

I first saw *jamón ibérico* a few years after leaving Los Angeles, while working as a line cook in Paris for the great French chef Michel Rostang. Chef Rostang was famous for his modern interpretation of classic French food, but in the industry he was also known for his massive eruptions during a stressful dinner service—outbursts that would often bring young line cooks to tears. I witnessed one such explosion that I still think about, nearly twenty years later. Guillaume, a lovable but absent-minded vegetable cook, had used the wrong potatoes for a fricassee, something he apparently had done many times before, and Chef Rostang, on seeing the potato dish leave the kitchen this one night, nearly split in two with anger. He screamed Guillaume's name, then hurled a barrage of curses and insults at Guillaume's attitude, his intelligence, his appearance—in a way that was so personal and so intense, I was sure Rostang's heart would just give up from the thunderous beats required to pump the rage through his body. (In fact, he had already suffered two heart attacks, both in the middle of dinner service.)

It went on so long that it stopped conversations in the dining room, which is when the maître d'hôtel, Bruno, appeared in the kitchen carrying a leg of *jamón ibérico* in a tong, the traditional metal clamp that holds the ham for proper slicing. Up to that moment, I had only seen pictures of *jamón ibérico*, and I never imagined that my first sighting would be in a famous Parisian restaurant. Nor did I ever imagine that the *jamón* would act as a pacifier. Bruno placed the ham (purposefully, I was told later) next to the unhinged chef, who stared down at it and immediately, almost reflexively rested his right hand on the front of the leg. He stopped yelling and looked down at the ham as though he were looking down at the crib of his sleeping newborn. It was as if Chef Rostang, suddenly in the presence of something so perfect, felt embarrassed by his behavior.

———

As we drove, I admired the oak trees outside my window—the source of the pigs' famous acorn diet. The oaks were green and gray, gnarled through the trunk. They looked ancient but powerful, as though they'd risen up from the thick grassland through sheer force of will.

I mentioned how thrilled I was to finally see the *dehesa* for myself. "It's more beautiful than the pictures," I said. "Amazing."

"And this is the ugliest time of year!" Eduardo said, his right index finger pointing straight up in exclamation. "You have to come back when it's green, and the light is fading. Right now, well, I'm sorry for the way it looks."

Lisa explained that locals, in their attachment to the *dehesa*, always bemoan your not seeing the landscape at its peak. "I've been coming to the *dehesa* for a long time, and I swear I get the same reaction whenever I marvel at the beauty. It's like you're visiting someone's home for the first time who's apologizing for the condition it's in."

We took a sharp right onto a dirt road and drove slowly through a maze of trees until we hit an opening. The lush grass and the scattering of oak trees came into quick focus. I asked whether Eduardo raised any of the famous Iberian pigs.

"Pigs? Sure, I have some pigs," he said with disinterest, as though shrugging off a litter of barn cats.

Suddenly Eduardo screamed—"LOOK!" He slammed on the brakes, threw his body forward, and pasted his hands on the windshield. (I thought to myself: *dinosaur?*) Eduardo could see his beloved geese in the distance, doing what I imagine he saw them do every day, which was waddle in the grass and hunt for food. We were at least eight hundred feet away, but he leapt from the car and began walking very slowly, crouching slightly and humming something I couldn't hear. I followed close behind. Suddenly, with what I would have mistaken for theatrics if I wasn't seeing it up close for what it was—love—he fell to the ground and began to crawl.

"*Hola, bonitas,*" he said, and Lisa translated for me as we followed him. "Lovelies," he was saying. "Oh, my lovelies. How are you, my lovelies?"

He stopped and showed us that they were scavenging olives from a collection of trees. He was smiling in the way a father might when he sees his children sitting down together to a well-rounded meal. Eduardo acknowledged that it was an expensive lunch. He said he probably makes more money selling his olives for first-press olive oil than he does from his livers.

"In the end they eat 50 percent and I sell the other 50 percent." It was the "take half, leave half" rule of rotating herbivores onto fresh grass, except here the geese dictated the terms. He paused in a vain attempt to calculate the math, adding simply, "They're always quite fair."

"If you make sure the geese are relaxed and happy, you'll be rewarded with the gift of fatty livers. That is God's way of thanking us for providing so much good food for the geese," he said, in a pronouncement that somehow sounded neither mystical nor evangelical, just likely.

Or was there false modesty at work here? I pressed Eduardo about intervening more than he let on. Didn't he, and his father and grandfather before him, face any challenges from the environment? Eduardo shook his head. His challenge had nothing to do with the landscape, he said. It had to do with the marketplace—with the chefs, distributors, and consumers who all demand yellow foie gras.

❯━❮

The quality of a liver is determined by several factors. Among the most important is its color. The yellower, the better. A pale liver commands a much lower price.

Chefs learn early in their careers to be vigilant about avoiding pale livers. My education came in cooking school, when we visited Ariane Daguin at her famous specialty-foods distributing company, D'Artagnan. We learned about how all the finest foods—caviar, truffles, and of course foie gras— were imported and arranged for distribution to the best restaurants in America. Toward the end of the warehouse, we passed a small refrigerated cold room. I looked inside. Three signs, marked A, B, and C, were spaced apart and hanging along the back wall. Livers were arranged on a long table below each rating. Off in the corner, underneath a piece of notebook paper with a hastily scribbled A++, a small table held perhaps a dozen livers. I walked over for a closer look. They were the smoothest, brightest yellow livers I had ever seen.

I asked Ariane if these golden livers were separated as A++ because they were going to the most famous chefs. She looked at me. "*Mais non*," she said, "these are for the chefs who know the difference."

The problem for Eduardo is that the coveted yellow color comes from corn. The higher the concentration of corn in the feed, the better chance you have for brighter livers. Since Eduardo only occasionally allowed for free-choice corn, his livers were naturally pale gray. For many years he embraced this idiosyncrasy in a vain attempt to celebrate his process. "I cannot tell my geese to make their livers more yellow," he would say. It didn't matter. People wanted yellow livers and were willing to pay more for them. Eduardo had a difficult time competing.

As luck would have it, several years ago Eduardo's geese spent their last few weeks in an area of his farm inundated with lupin plants. Lupins are a good source of protein, popular in livestock feed. They grow wild throughout the *dehesa*, often densely concentrated in certain areas. They also happen to be bright yellow. Eduardo's geese didn't especially care for the plant until it matured and went to seed. Then, he said, they nearly attacked it, gorging on the seeds and devouring the entire pasture.

"They went wild!" he said, fondly remembering the sight. He forgot about how much of it they ate until after the slaughter. That's when he discovered that the livers had turned yellow, as if his geese had consumed enormous amounts of corn. The next year, he maneuvered them into the same lupin-dense field, which again made the livers bright yellow. It's become a routine.

Thinking of *jamón ibérico*, I asked Eduardo if he wouldn't prefer that his geese eat a diet of the famous acorns alone. They may not provide the same bright color, but surely the flavor would be compensation enough. He shrugged his shoulders. "That's for them to decide."

"Acorns," he added, suddenly impatient. "Why is *jamón ibérico* always just about the acorns? Acorns—'the best feed in the world!' Acorns—'the best fat in the world!' Did anyone ever consider that there are acorns all over the world, but no one can replicate *jamón ibérico*?" He paused to indicate that the answer was obvious. "These geese eat tons of acorns, but if they don't move around, if they don't eat all this grass"—here he raised his arm to outline the lush pasture—"without grass, the acorns are nothing."

The grass, he explained, makes the acorns taste sweet, which means the more grass his geese have access to, the more acorns they eat. Their systems are primed, in essence, because of a chemical reaction that occurs between grass and acorns. Eduardo claimed that the reaction causes an increase in weight much faster than if the geese ate acorns alone.

Just then, twenty or so pigs walked into our view. With their lumbering torsos, Iberian pigs tend to resemble beer kegs with legs. They have large ears that stick out like the bill of a baseball cap, shading their eyes from the fierce Mediterranean sun. Their snouts are unusually long, making it easier to root around for acorns.

It was the first I'd seen them up close, and it was a thrill—and not just because I was standing just a few yards from the most famous pigs in the world. I was inspired because, until that moment, Iberian pigs and their famous hams had always been synonymous with thick, undulating fat—the product of their indulgent acorn diet. They were, in my mind, the hog equivalent of couch potatoes. Up close, looking at the muscular, long-legged animals, I realized my mistake. Having spent a lot of time with happy pigs, pigs that live outdoors and eat organic grain, foraging and farrowing in a kind of porcine bliss, I shouldn't have been so surprised. But I swear I saw something I'd never seen in a pig. I saw *proud* pigs.

Eduardo didn't look inspired. He looked irritated. "My geese eat more acorns than those pigs," he said, waving his hand in the direction of the famed Iberians. "And my geese are half the size!"

A REVOLUTIONARY TAX

Just as we turned to head back for lunch, Eduardo seemed to reconsider, suggesting that we see a few more geese first. He thought we'd find a group in the general vicinity, but as we walked around, stopping at several clearings, he admitted that he really had no idea.

It was another odd moment at his farm. How could he not know the whereabouts of his animals? If the geese were a hobby, a sideshow to the famed Iberian pigs, misremembering their location might make sense. But a foie gras company that didn't keep track of its livers? And seemed to take pride in not knowing?

We continued searching. Eduardo's hands were clasped behind his back as he walked. He looked, I noted, very nearly like a goose. His head rotated back and forth, and he kept his nose pointed skyward, as if following a scent. In forty minutes, we didn't see a single goose.

Another black hawk swooped low over a hilltop. I asked Eduardo about them.

"Lots of hawks," he said. "Lots to eat."

"Like what?"

"Goose eggs!" he exclaimed cheerfully. "We lose more than half our eggs to hawks."

"Half?" I repeated to Lisa. "That's unbelievable."

"Yes! The geese lay once a year, forty to forty-five eggs. In a good year, I'd say eighteen to twenty survive. So yes, more than half."

Chicks die all the time—from disease, from predation, from a drenching downpour—but to be wiped out of half your stock (and 50 percent of your potential profits) before they even hatch is a staggering handicap. I looked up to see two more hawks fly off in what I understood to be the direction of Eduardo's brooding stock.

"Would it be fair to say that the hawks are your biggest obstacle?" I asked.

"I don't think so," he said politely. "It's why nature has a goose lay so many eggs. There has to be enough to pay the revolutionary tax for living outside."

We came through the thick brush and into an open field. A finch with a bright yellow chest was in full song from the uppermost branch of a nearby oak. The sun, golden on the horizon, suffused the pasture with soft light. The oak trees cast long shadows that looked like rows of fallen soldiers.

I noticed that Eduardo was again looking up at the sky, a few hundred yards to the left. In place of the hawk, there was a small flock of wild geese, flying in our direction. As they got closer, Eduardo's geese began honking more loudly. By the time the wild geese got to within fifty yards, you could clearly hear them honking as well. They sounded, to my untrained ear, like they were having an argument. I couldn't tell which group was louder.

"The wild geese come to visit?" I said.

Eduardo shook his head. "Sometimes they come and they stay."

"Stay . . . ?"

"Sometimes they never leave," he said.

I tried to convey my disbelief, offering the analogy of a wild pig happening upon an American hog confinement farm and choosing to stay. Eduardo didn't seem to understand the point, and not because of the translation. It was the concept of ten thousand pigs in confinement that he found hard to believe. At first he thought it wasn't possible. Then he just seemed uninterested in learning more.

"But Eduardo," I said, "isn't the DNA of a goose to fly south in the winter and north in the—"

"No," he interrupted, shaking his head. "No, the DNA of a goose is to seek conditions that are conducive to life, to happiness. When they come here, that is what they find."

◆───◆

Twenty minutes later, I was sitting in the back of a restaurant in Monesterio, a quiet town just north of Seville. Monesterio is a town that, strictly speaking, lacks a town. There are a few small stores, but not much else.

The light-filled room was spare, with faux country-western furnishings and a television at the bar playing a Spanish soap opera. Amid the empty tables, Eduardo sat heavily, his hands in his pockets. He was alert, raising his chin often, with an anxious vitality that said he'd prefer to be outside in the cold. As we waited to be served a plate of his goose liver, he smiled nervously, squinting his eyes as if enjoying a brisk wind.

The waiter passed our table several times, empty-handed, and finally shrugged a bit as if to say, "I'm not the chef." Eduardo watched him pass with a nod of respect that, to go by the waiter's backward glance at him, suggested this was a familiar routine.

Finally the waiter arrived with the foie gras. "Voilà!" he said, and then turned to me and slowly, in his best English, said, "Freedom foie gras." On the white plate sat a pâté of the liver, with three sprigs of chives sticking out from its center (which Eduardo, either offended or embarrassed by the gratuitous garnish, removed with a quick sweep of his hand). Next to it the waiter placed small ramekins of sea salt and black pepper and a plate of thinly sliced baguette. He hung there expectantly, as if he were waiting for Eduardo to approve a bottle of expensive wine.

Eduardo lifted the plate of foie gras to his nose and inhaled. He put it back on the table and then suddenly lifted it again to his nose, bringing it so close this time that it nearly touched his nostrils; he jerked the plate in small clockwise circles, agitating the foie gras to release its aromas.

It was a funny ritual—it was liver, after all—but it struck me as especially strange because the pâté had been prepared by Eduardo and his small staff the previous winter. After his flock is slaughtered, the livers are preserved, either as a pâté like the one we had sitting in front of us or in jars as confit—individual slices of liver stored in their own fat. Eduardo was not simply evaluating his liver; he was evaluating how well he himself had prepared it.

He inhaled, this time very hard, and his shoulders shot up toward the ceiling. With a nod to the waiter, we were left alone.

"Last season's foie gras," Eduardo said, scrunching his nose apologetically as he returned the plate to the table. "It's all I have." If the restaurant's atmosphere and decor made an unlikely setting for a culinary revelation, the room-temperature slab of Eduardo's foie gras pâté kept with the trend. There was an air of low expectations.

Eduardo waited for me as I dug into a small section and brought it over to my plate. "Last year's liver," he repeated, smiling.

I took a bite. The smell was what got to me first, because as I chewed I was struck by the smell of meat. I most especially smelled liver. Foie gras, as a rule, is never described as delicious *liver*—you wouldn't describe white truffles as perfumed fungus, either—but here I was, unmistakably tasting liver. And not metallic, muddy-tasting liver, but sweet, deeply flavored, *livery* liver. It occurred to me, as I took another bite, that foie gras is essentially a small amount of liver flavored by a whole lot of fat. I had never thought about it that way because I had never known any other foie gras. Eduardo's foie gras was very different: it was a whole lot of liver enhanced by a small amount of fat.

When I mentioned this to Eduardo, I noticed he wasn't eating. He nodded in agreement. "To taste just fat is to taste nothing. The fat should be integrated, to carry the flavor." I took another bite, amazed now at the texture—it cut like room-temperature butter but again tasted like deeply flavored meat. It was incredibly delicious.

The waiter appeared with another plate. This time, a jar of Eduardo's confit of foie gras sat in the center. Using a spoon, I dug into the center to reveal what looked like a marbled prime rib, leveraged by yellow, glistening fat. I took a bite. And then another. I tasted cloves.

"Eduardo," I said, "the cloves are perfect here."

"Cloves?" he said. "No, no cloves."

"Really?" I said in disbelief, because I truly didn't believe him. "Star anise?"

"No, no star anise."

If you want to irritate a chef, start by questioning his palate. "No cloves, no star anise," I said, a little testily.

Eduardo shook his head, spreading the meat on a piece of baguette. "No seasoning." He qualified that he sometimes used salt and pepper—but even those were superfluous if the geese had the correct diet. He quickly listed certain plants that provide salinity, and others that impart peppery qualities. "If you have these in the right proportion," he told me, "your meat will, too."

"You season your livers in the field?" I asked.

"The geese eat what their heart tells them to eat," he said, letting his fingers do a quick dance on his chest to show what an easily understood, animal thing the heart is. "I just make sure what they want is available to them."

I took another few bites and watched Eduardo eat. Even when he stopped chewing, his lips moved silently. He appeared to be lost in thought, or prayer.

"Eduardo," I said, dipping back into the jar for another bite. "How many chefs are serving your foie gras?" He shrugged and shook his head.

"Which chefs?" I asked again as I went for more foie gras. "Are they only in Spain?"

He shook his head again, jutting out his lower lip for emphasis. "No chefs."

I put down my fork. Some of the most famous chefs in the world are in Spain. Chefs like these demand only the best ingredients. This was the best foie gras. How could he not be selling his livers to chefs? It seemed impossible.

"Chefs?" he said, gently wiping his mouth. "Chefs don't deserve my foie gras."

CHAPTER 9

A CHEF'S WORTH is largely determined by his interpretation of great ingredients.

During one of my first nights in the kitchen at Chez Panisse, in 1994, a dessert leaving the pastry station caught my eye. Actually, I more or less gasped in disbelief, and that's not because the dessert was so beautiful (it was) or because I hadn't seen a dessert like it before (I hadn't). I gasped because it was so *crazy*. It was a single peach on a dessert plate, no sprig of mint, no swish of raspberry sauce. It was Peach, unadorned.

I walked my New York City attitude over to the pastry chefs. The peaches were stacked on the counter, delicately wrapped in cellophane and lined up like soldiers, awaiting deployment. The chefs lovingly cradled each peach on its way to the plate; the waiters ferried them away, walking gingerly as though they were carrying soufflés. Everyone acted as though this was nothing remarkable at all, as only Californians can act around things like fresh fruit and the weather.

I began to laugh. "Wow, tough night," I said to the pastry chef. She looked at me but did not respond. So I picked up a dessert menu and was introduced to my first Californian farmer. "Mas Masumoto, Sun Crest Peach," it said, and nothing more.

It's hard to imagine there was a time, not so long ago, when chefs didn't name the farmers they purchased from. *Organic* and *local* weren't buzzwords on restaurant menus. The sign of a serious restaurant was to present impossibly large imported raspberries in the middle of January. Chefs in training back then didn't generally go to California to learn; they went to France. I went to France by way of California, as a young line cook in various kitchens. Alice Waters's famed Chez Panisse was a last stop. A few weeks of observation in the kitchen turned into several months of exploration on the farms that supplied the restaurant.

I stayed largely on account of that peach. When I took a bite of it later that night, did the lights dim and the warmth of a religious spirit come over me? No, but I've never tasted something quite so peachy. In that sense, it was not unlike my experience of tasting the Eight Row Flint polenta ten years later. As I bit into it, I remember thinking that the peach had a fullness of flavor to it—bold, like a stew of meat—that made you think you had in your mouth something much richer than fruit. I was struck as much by the acidity as by the sweetness. It was like a nicely balanced wine. The juice ran down my face and chin. One bite, and then another few bites, and pretty soon all that remained were bits of flesh sticking to my face.

It was the best peach of my life, but I have to qualify that, because, like most Americans born in the past fifty years, I didn't know what a peach should taste like. Breeders in the 1970s and '80s created new varieties for functionality, not for flavor—low-acid, high-sugar peaches that could be picked while still hard, capable of withstanding the rigors of cross-country travel.

Masumoto's peaches were incredibly delicious. But more than that—as if a peach needs to be more than that—they got people to consider good food as inseparable from good farming. You'd think that would be obvious, but chefs often make it difficult to see. When we cook ingredients—whether peaches or foie gras, or most anything—we transform them. Foie gras is seared and paired with mangoes and sherry wine vinegar; peaches are peeled, poached, and perfumed with lemongrass and vanilla. The cooking technique, or the flavor combinations, can be surprising and delicious. (The more aggressive the technique, or the more far-reaching the combination, the more it's likely to taste merely surprising.) Either way, when it's in a chef's hands, all the vectors point back to the chef. Process trumps product.

Alice was saying: *Taste what Mas Masumoto created; I can't do better.* She did not say what most chefs say: *Taste this dessert I made with Mas's peaches.*

A NOUVELLE CUISINE

The surprising thing about the prerogative of the chef is that until recently it didn't exist. The chef's authority (and celebrity) is such an accepted fact of fine dining today that it obscures the fact that for most of the past century, it was the diner—not the chef—who held power over the menu. Restaurants were the public's domain, places where patrons could count on a familiar repertoire of recipes with the luxury they lacked at home. They came for entertainment, yes, but also for convenience and comfort. Restaurants, after all, are named for a *restorative*, a large bowl of soup.

Like early musical compositions, classic dishes were often passed on without attribution. Chefs cooked in obscurity, handcuffed by the weight of tradition. That's a bit simplistic, of course—Fernand Point, Auguste Escoffier, and César Ritz are famous exceptions—but for the most part, to be a chef was to be a practitioner of recipes that had been worked out long ago. The Guide Michelin, the preeminent guide to the great restaurants of France, introduced the star rating in 1926 to recognize good cooking, but it almost never highlighted chefs.

Paul Bocuse, the legendary French chef, once said that in 1950, a chef's life amounted to being "locked up, cloistered in his smoke-filled basement . . . at command, and without real power of creation." With no sense of agency or reward of public recognition, it meant the life of a chef was one of toil and hardship.

The hardships loomed especially large in the kitchen, where the conditions were laborious, oppressive, and, for the most part, unkempt and dangerous. George Orwell famously described his descent into culinary hell in his memoir *Down and Out in Paris and London*. "The kitchen was like nothing I had ever seen or imagined," he wrote, "a stifling, low-ceilinged inferno of a cellar, red-lit from the fires, and deafening with oaths and the clanging of pots and pans."

The professional kitchen was foul, furious, and crude. To compare today's chefs—highlighted and exalted as we are in magazines, television, and fund-raising events—with the anonymous, hard-drinking, hard-living laborer chefs of the past is to recognize a change not only in lifestyle but in vocation.

How did it happen? It's difficult to point to any one person or event as a pivotal turning point, especially since the influence and prestige of chefs have continued to evolve. But it's worth giving credit to Paul Bocuse for redefining how the world perceived not just haute French food but what it meant to be a chef.

Long before the Food Network, before endorsement deals and signature frozen food products, Bocuse was an unapologetic promoter of his image. He refused to toil anonymously. He named his restaurant after himself, which was not a commonplace practice, as it is today. Bocuse was the impresario, taking the place of the maître d'hôtel. It was an idea so radical it might have failed had he not been such an effective self-marketer.

Within the context of the 1960s and '70s (and even by today's standards), Bocuse was a pioneering global chef—the first to export French products to Japan, for instance—earning himself the cover of *Newsweek* in 1975. He became the most famous chef in the world.

Bocuse's rise to prominence coincided with the birth of a new cuisine in France. Chefs like Michel Guérard, the Troisgros brothers, and Alain Chapel, with Bocuse as a kind of ringleader, conceived of *la nouvelle cuisine française* as a reaction against the confines and extravagances of classical gastronomy—the so-called *grande cuisine*.

It was food critics Henri Gault and Christian Millau who first identified and celebrated the movement, in 1973. "Down with the old-fashioned picture of the typical bon vivant," they wrote in their restaurant guide, *Le Nouveau Guide Gault-Millau*, "that puffy personage with his napkin tucked under his chin, his lips dripping veal stock, béchamel sauce, and vol-au-vent financiere. . . . No more of those terrible brown sauces and white sauces, those espagnoles, those Perigueux with truffles, those béchamels and Mornays that have assassinated as many livers as they have covered indifferent foods. They are forbidden!"

Nouvelle cuisine, by contrast, stressed lightness and simplicity. Inspiration for dishes came from around the world, but chefs connected the food to their own regions, adapting traditional recipes with new cooking technologies (like microwave ovens and vacuum-pack cooking) and revised techniques.

More than any other technique, sauce making was reconsidered. Sauce (which starts with a great stock—bones, meat, and vegetables simmered in water) is the most important French contribution to gastronomy. It is a reduced version of the thing you're eating, as opposed to a condiment or chutney, which serve to complement or counterbalance the thing you're eating. Sauces enrich, deepen, and layer flavors. Until nouvelle cuisine, sauces like béchamel and béarnaise were easily recognized and timeless. But the new guard of chefs insisted that these classic sauces masked the flavor of the protein on the plate. Their new variations were an attempt to instead enhance it with a lighter style. There was less butter and cream, and limited use of flour as a thickening agent. Such changes may sound insignificant, but most of France saw them as heretical, a challenge to culinary and cultural heritage no less dramatic than a group of American chefs questioning the culinary wisdom of serving a bun with a hamburger.

Even the ingredients themselves were reconsidered. Nouvelle cuisine was also, less famously, the first farm-to-chef movement. Chefs like Alain

Chapel developed direct connections with the morning markets and created menus around ingredients they found and sometimes specifically requested from farmers. The goal was to let the ingredients speak for themselves. Gone were the elaborate platters and ritualistic table-side carvings. Instead chefs began plating food individually, with an interest in a few unique flavor combinations. (Nouvelle cuisine can take credit for the *menu dégustation*, a tasting of multiple distinctive, smaller-portioned courses.)

What emerged was a modern, innovative, and highly personal style of cooking. From sourcing to preparation to presentation, nouvelle cuisine forced diners to concentrate on the aesthetics and aromas of what they were about to eat. It made them experience flavor in a new context, thereby redefining the role of chefs—as artists, masters of their own creations—as much as it revolutionized cuisine.

And the effects were global. These kitchens were the incubators for the great chefs of the 1980s and '90s, like Wolfgang Puck, Jean-Georges Vongerichten, Daniel Boulud, David Bouley, and Jean-Louis Palladin. They used the Bocuse model—talent and increasing prestige—to demonstrate that a chef could continue to step out of the kitchen. They showed how chefs could engage with the public to nurture a wider appreciation of food and culture. And they did it while opening multiple restaurants throughout the world.

Jean-Louis Palladin, whose food included fusion, fruit in savory preparations, and sometimes ornate presentations, never liked the term *nouvelle* to describe his cuisine. He nonetheless was the first to modernize American food in the same vein as the nouvelle cuisine chefs in France, by foraging for only the best ingredients and highlighting those ingredients with imaginative cooking.

There's no better example than rewinding to the plate of chicken Palladin prepared that night in Los Angeles, with all the ignoble parts. The real brilliance of the dish wasn't the use of underappreciated cuts, nor was it that Palladin had the birds raised by a farmer following his precise directions. The brilliance was the sauce, which Palladin himself labored over all afternoon.

Those secondary cuts could sing only because they were bound together by superior technique.

—

Culinary historian Paul Freedman has written that nouvelle cuisine "in a sense extends on either side of the borders of simplicity and artifice, for if on the one hand its advocates praised freshness and integrity, they also wanted to expand the imagination of the chef, embracing such French 1968 slogans as 'everything is permitted' and 'it is prohibited to prohibit.'" Which is what truly great cooking is all about.

When Eduardo said that chefs don't deserve his foie gras, he was really saying that unless you serve his foie gras like Alice Waters served that peach, and call it what it is—Mas Masumoto's Sun Crest peach—you're arguing that the translation of the work is better than the original. Chefs don't deserve his foie gras because the livers are already brilliant.

But if we want to respect nature's gifts, we needn't simply serve a lone liver (or a peach, or anything else on a plate). By transforming superior ingredients into something transcendent, chefs like Bocuse and Palladin showed how it was possible to make a gift from nature even more brilliant.

CHAPTER 10

NOT LONG AFTER RETURNING from Extremadura, I was standing at Blue Hill's kitchen delivery door when Craig Haney pulled up to unload Stone Barns' weekly slaughter of 150 chickens. Craig looked disheveled, exhausted, and—having just completed the nine-hour round-trip to the nearest slaughterhouse—rightly disturbed. "Running late?" I asked.

"Running on empty," he said, sounding like a country song.

The subject of chickens can turn Craig into a temporary killjoy. It's not that he doesn't enjoy raising them. Or that he lacks a market: of the 150 birds, Blue Hill will buy 100, sometimes more if Craig is willing to sell them. The rest get sold at the Stone Barns farmers' market, and since the chickens are delicious, demand is overwhelming.

The problem is profit. The 150 chickens were being unloaded at a retail price of $3 per pound, at an average bird weight somewhere in the neighborhood of three and a half pounds. That puts the gross revenue per bird at about $10.

But if gross receipts are misleading, in the business of farming they are often grossly misleading. Taking into account the cost of purchasing the chick ($1), the feed during its seven weeks of life (organic, in this case), electricity, gas (though not mileage and wear and tear on the vehicle, or on Craig's mental state), the $2.25 to get the bird slaughtered, and on and on (and on and on), Craig will net, exclusive of labor, about $3 per bird—which is to say $450 per week for raising chickens.

"Add in labor," he said to me as I jotted down the numbers on a piece of paper, "and I think this is pretty close to a not-for-profit business."

"What about raising more chickens?" I asked, in a tone that suggested, *Have you thought of that?*

"We could, but something else would have to give."

I thought of the egg-laying chickens. Since the 1,200 laying hens essentially add the same benefits to the farm as the broiler chickens we raise for meat—namely, that they follow the sheep, spread the manure, eat the grubs and insects, and provide nitrogen for the grass—wouldn't reducing their number, or maybe even eliminating them altogether, be one place to start?

"Yeah, I've thought of that," Craig said. "But then we'd have less eggs . . ."

I nodded my head but quickly calculated adding another thousand broiler chickens to the bottom line. All of a sudden the net per bird increased by nearly fifty cents. Not bad.

I offered some other ideas, like filling the moving pens with just ten more birds each. I had been in one of the pens the day before with Craig. It wasn't crowded. By New York City standards, you might call the living space generous. Could the chickens suffer a little less elbow room for the sake of the bottom line? I decided they could. As Craig put away the last of the delivery, I added ten chickens for every movable chicken pen, which when fed through the spreadsheet disproportionately bolstered the bottom line.

My brother, David, once helped me understand the restaurant business by explaining that it was much like the airline industry: the flight's going to take off, and for the most part the costs are fixed, so once you fill the requisite number of seats to break even, every passenger after that is essentially pure profit. Was there a farming equivalent to airline vacation packages and a restaurant's early-bird special? According to the spreadsheet, there was: Raise more chickens.

"What about the sheep?" I asked as Craig got ready to leave. For the moment I was kidding, and then somehow, just as quickly I wasn't.

The amount of real estate required to raise sheep on an all-grass diet is significant, and though I knew the intention was to grow the herd to the point at which Craig could process several per week, potentially turning a profit, maybe the capital investment could be better spent elsewhere. We'd gotten an average of one lamb a month since Stone Barns opened—one lamb that, on a busy night, would last about an hour. It was excellent lamb, but so was the lamb from other farmers. Considering the numbers, I had to ask myself, why wouldn't we do this? Without lambs, I was proposing a tenfold increase in chicken production, maybe much more.

As Craig packed up to go home, I took the time to ask him more about the costs of raising the birds. Airlines try to get as many people to fly with them as they can, but they cut costs, too, and since, what the hell, I was already into these numbers, I could push a bit further. Where was the fat?

It looked as though the largest costs, aside from labor, were definitely the organic feed and the processing costs. A 40 percent bump to serve the chickens organic grain seemed, at first glance, awfully steep. But since Craig can label the chickens organic, it allows him to raise the price per pound. I figured it was a wash.

But $1 just to buy the chick? And $2.25 to process? Staring at the numbers, these choices seemed a little like serving hot food on a short flight—overkill. I made a note to investigate on-site breeding and slaughtering. ("Vertical integration," my brother would later explain. "But vertically integrating a non-profit farm raising a few hundred chickens sounds weird. The way to do it would be to contract with other farmers," he said, scribbling some numbers on a notepad in an apparent gene-determined quest to push profits, "and use Craig's protocol for raising chickens, which is unique. Less hassle, and much less risk.")

The experience of studying the numbers so closely was putting me in a kind of trance. To earn more without adding any real costs was exhilarating to think about. The simple categories I sketched out—costs on one side, profits on the other—were no doubt the same Excel spreadsheet computations made by an industrial animal feedlot. Inputs in, profits out. Leaving all of nature's frustrating complexities off the spreadsheet (how would the grass degrade with a spike in the nitrogen-rich chicken manure, for instance?) wasn't my intent, but for the moment it sure made the calculations easier to contemplate. I found myself really enjoying the exercise.

Which is my point. Why? Why, in this ten- or fifteen-minute, off-the-cuff exchange, did I fall into the trap of looking at the world of chickens through a spreadsheet? I didn't own the farm. I wouldn't have profited at all from this calculation. We weren't talking about raising a better-tasting chicken. In fact, we were almost assuredly talking about raising a less tasty chicken, which, coming from a chef, made it all the more curious. What I wanted—ridiculously, now that I think about it—was to get to an impressive profit number for raising broiler chickens. Not for any shareholders or investors, and in the end not even for Craig. The spreadsheet was merely for me.

In *The Unsettling of America*, Wendell Berry mentions a reporter who writes about "the bigness of modern agriculture—which he approves of, so far as one can tell, because it amazes him." I had, if only for a few minutes, pushed the idea of turning Stone Barns into a chicken factory because the numbers amazed me.

I handed the spreadsheet to Craig. He studied the numbers before smiling. "So it's either get big or get out," he said, quoting Earl Butz, President Nixon's secretary of agriculture in the 1970s. "We just retraced the last sixty years of American agriculture."

AN INSULT TO HISTORY

Later that week, I read over my notes from the visit to Eduardo's farm. There were exclamation points, double underlines, and sketches of the farm. Near the end I came to a page where I had written in the margins. I had asked Eduardo what he thought of conventional foie gras. Since 99.99 percent of the fattened goose liver in the world is produced with gavage, I asked, what did he think of normal foie gras?

I underlined his response: "It's an insult to history."

Of all the things that can be said about the way foie gras is produced, "an insult to history" struck me as oddly abstract. Thinking that some research might help decode his statement, I looked up the history of foie gras. It turns out, what I knew about foie gras didn't go back far enough. In characteristic American fashion, my history only looked back a few hundred years. In fact, the practice dates back five thousand years, to when ancient Egyptians observed how wild geese along the Nile gorged on figs before migrating. The meat from these birds was understandably more delicious, and quickly became popular among the ruling class.

Unfortunately, nature couldn't keep up with human demand. Force-feeding was the answer, allowing the same results to be achieved year-round. Ancient frescoes discovered in the tombs of Egyptian officials show servants forcefully gorging geese with pellets of grain. Whether the force-feeding was intended to enhance the liver specifically or for the benefit of the meat, it's difficult to know. (Some believe that the liver itself might not have been eaten until Roman times.) But the practice quickly spread throughout Europe, thanks in large part to the Jews, who used the rendered goose and duck fat as schmaltz.

Either way, Eduardo had a point. The discovery of fattened goose liver came from an observation about nature, a scene along the Nile so pure you can picture it. For Eduardo, to take what was natural—liver from geese gorging for the long migration—and turn it into a year-round industry is an insult to its history.

TRANSFORMING THE CHICKEN

The story of the chicken in this country might just be the closest thing we've got to a living, breathing example of Eduardo's insult to history.

Not that long ago, nearly every farm had a small flock of chickens, of numerous varieties. They were laying hens, as opposed to broilers, as eggs were a dependable source of income, and broiler chickens had not yet been bred for meat production. At one time, at least sixty known breeds thrived across the United States. Raising only one breed, and raising only chickens at the expense of other animals, is an entirely new concept, and an industrial one at that. A farm raising only chickens would have been as unique and unlikely a hundred years ago as a multispecies animal farm like Stone Barns is today.

It all changed quite by accident, in the Delmarva Peninsula of Delaware, where in 1923 Mrs. Cecile Steele ordered fifty chicks from a hatchery, which misread the order and instead sent five hundred chicks. What do you do with five hundred laying chickens when you're not set up to handle a mountain of eggs? You either send back the order or, in the case of Mrs. Steele, you build a small shed and raise the birds for meat. Eighteen weeks later, she sold the birds for sixty-two cents per pound—equal to more than $5 per pound today. The following year, her husband left his job and stayed home to raise the thousand chicks his wife ordered—this time intentionally. Within three years they were raising ten thousand birds per year, and their neighbors soon followed suit. Ten years later, this tiny, two-hundred-mile peninsula of Delaware was producing seven million broiler chickens a year.

Success meant that more farmers started raising only chickens. Looking at a map of what, after World War II, became the official "Broiler Belt"—extending from the Delmarva Peninsula down through North Carolina and across the South into Mississippi and Arkansas—you get the sense that the domino expansion of the chicken industry unfolded the old-fashioned way: farmers saw the success of their neighbors and jumped on board. It was the farming equivalent of keeping up with the Joneses.

Specialization certainly helped farm profits (my spreadsheet was clear on that), but it didn't last. The stunning increase in production meant an eventual price reduction as supply outstripped demand. When profits fell, farmers looked for ways to cut costs and become more efficient.

The Perdue chicken company, of Salisbury, Maryland, came to epitomize that efficiency. In 1920, founder Arthur Perdue went into the poultry business. At first, selling chickens was for Perdue a mere by-product of egg production. It wasn't until his son, Frank, joined the business, in 1939, that the Perdue company radically transformed itself with a few key decisions. The first was to give up the egg business. When a rare disease, leukosis, wiped out their entire flock of egg layers in the 1940s, Arthur and Frank decided to switch to meat chickens, gambling that the price of meat would rise. It did—dramatically.

In part, the burgeoning success of the broiler industry was due to better broilers. By the 1930s, breeders were developing chickens that simultaneously grew faster and required less grain feed. The same advances in genetics that reshaped the foie gras industry achieved a kind of über-success with broilers and laying hens. Craig still benefits today. His broiler chickens grow from chick to market weight in an astonishing seven weeks, less than half the time Mrs. Steele's chickens took to fatten and with less than half the feed.

As Steve Striffler notes in his book *Chicken: The Dangerous Transformation of America's Favorite Food*, this wasn't just progress. It was a fundamental change in how we raise animals: "The barnyard chicken was made over into a highly efficient machine for converting feed grains into cheap animal-flesh protein."

Determined to maximize the profitability of this new machine, Frank experimented with mixing his own feed for the chickens—an exercise that quickly proved lucrative. His mixes were cheaper and put weight on his birds faster than anything he could purchase. Within ten years, Perdue owned its own feed mill, one of the largest grain storage facilities on the East Coast. It owned the processors, too—the companies that slaughtered, cleaned, and packed the chickens—after Frank concluded that Perdue paid other companies too much for the service. (For what it's worth, that was my conclusion, too, when Craig and I looked at the $2.25 cost to slaughter each bird.)

But Frank Perdue's most important innovation was yet to come. In 1968, Perdue became the first poultry company to differentiate its product with a label. It was one of several transformative moments in modern agriculture—on a par with what the nouvelle chefs did to gastronomy by breaking with tradition and cultivating their own distinctive cooking styles. Frank bet that consumers would favor a particular kind of chicken and pay more for it, a radical idea at the time. Paul Bocuse might have helped shape the business of haute cuisine, but Frank Perdue ultimately revolutionized the business of food.

How did Frank transform his chickens from a commodity into a real brand? He saw that Maine processors got an extra three cents per pound for their yellow chickens. The marketplace deemed yellow chicken to be of higher quality—just as chefs have always deemed yellow foie gras to be superior.* Frank added corn gluten and marigold flowers to the feed, which indeed gave the chickens' flesh a yellow hue but did little, if anything, to change their flavor.

The tactic worked. Sales began to rise, and then in the 1970s Frank became Perdue's official spokesperson. Appearing in hundreds of ads for more than twenty years and coining the famous line "It takes a tough man to raise a tender chicken," Frank Perdue broke yet another mold. Corporate leaders had never dared hawk their products, for fear of cheapening the brand. But Frank's honest demeanor (and perhaps the fact that he resembled a chicken) made people take notice. An *Advertising Age* marketing analyst told *People*

---

* In part, the distinction was merely symbolic. In her classic *Much Depends on Dinner*, food historian and anthropologist Margaret Visser describes how culinary preparations have always capitalized on the innate appeal of "golden" food: "From the Middle Ages on they doused food in saffron and marigold sauces, daubed egg-yolks over meats, yellowed their pastries, and even gilded joints of meat and large pieces of confectionary. Gold is still high in our mythology; its appeal has been brilliantly harnessed by the marketing of crumbled chicken chunks under the word nuggets." There's scientific evidence to suggest that our preference may have pharmacological benefits as well. The yellow color in poultry meat and eggs comes from xanthophylls—the same carotenoids found in certain fruits and vegetables and the prized yellow-colored milk of Jersey cows. They're a proven antioxidant.

magazine, "He had a weird authenticity that made you want to believe there was actually something special about his broilers."

Perdue profited as well from the trend toward healthier diets. Increasingly blamed for nutritional ills, meat declined in popularity, especially in the 1980s and '90s. As beef and pork consumption fell, sales of chicken rose by nearly 50 percent. Perdue, now the number-three producer of poultry, helped meet and create demand by breaking the whole chicken into prepackaged parts. It wasn't a new idea—Don Tyson, of Tyson Foods, Perdue's longtime competitor, convinced the U.S. Army in the 1960s that precooked, portion-controlled chicken was a suitable and cost-effective way to feed the troops. That was another transformative moment in the poultry business: To break the chicken into parts was to create infinite opportunity. By the 1980s, unshackled from the convention of selling the whole bird, poultry companies were creating thousands of value-added products—precooked, frozen, marinated, pulverized (into the most infamous poultry incarnation, the McNugget)—where the parts netted a substantially greater profit then the whole.

In a funny way, Perdue wasn't just giving up on the whole chicken; it gave up farming chicken in the first place. What was apparent to Frank early on in his career—I say that with a little confidence only because I was aware of it by the time I had completed my own rough spreadsheet—was that the basic business of chicken farming is beset by a problem: it's not a good business. Today, Perdue's annual sales are more than $4.5 billion. According to its website, the company "partners with more than 2,200 independent farm families." That means they essentially outsource risk, contracting farmers to raise birds for them. In doing so, they avoid overhead and tying up capital in land ownership. They avoid the burdens of disease and bad weather. In other words, they avoid farming.

For the farmers that raise the birds, Craig had it right: get big or get out. Poultry companies contract with a variety of different-size farms. But the numbers speak for themselves, and they say it all: a small grower will net about $18,500 per year; the larger growers are double the size but earn nearly quadruple the income, netting almost $71,000 per year.

And my brother had it right, too. Vertically integrate the entire business—in Perdue's case, that meant control the breeding, the processing, and the marketing. Own everything, but don't own the farm.

CHAPTER 11

Not so long ago I visited a highly regarded avant-garde restaurant. The menu was cutting-edge, the dishes small and exacting. After a thirty-course meal, the chef brought me back to the kitchen for a tour.

Standing at the pass, he signaled to a cook, who carried a freshly plucked chicken carcass swaddled in cheesecloth. The chicken, the chef explained, had been sent to him from a farming cooperative in France, which had raised the rare bird with the hope of preserving its superior genetics. He admitted that it was probably the best chicken he had ever tasted. But then he turned to me, almost apologetically, and said, "What the hell am I going to do with an entire chicken?"

That the chef could even ask this question is largely due to men like Frank Perdue, who found profit in breaking up the bird and taught us to cherry-pick only the most desirable parts. But it's due also to the legacy of Fritz Haber; without the endless supply of cheap grain feed allowed by his synthetic fertilizers, our modern meat-eating ways could never have materialized.

Americans have now arrived at a point that was once unthinkable: there is no upper limit to the amount of meat we can consume. That's because the choices we make about which parts of the animal to eat—the breast over the legs (or the chops over the hocks)—determines how many animals are produced. When we can afford to eat "high on the hog"—the costliest cuts of meat on any animal—the relative worth of the other parts plummets. There's no ceiling on the number of animals raised because suppliers—producers, processors, retailers, and, yes, we chefs—then throw the bulk of the carcass away.

This happens all the time. Supermarkets in the United States offer mountains of cutlets and steaks and loins—restaurant chefs feature them in seven-ounce portions—but unless you venture to a specialty market (or dine at an ethnic restaurant), you'll find getting your hands on liver, heart, or tripe difficult to do.

Once a senior position in kitchen hierarchy, the butcher's job has been diminished by the convenience of ordering meat in primal parts. (During my first apprenticeship with that French butcher, I was only cleaning racks of lamb, not breaking down the whole animal.) A restaurant butcher is like the guy behind the meat counter at your local supermarket. He may look the part of someone who just hacked up a steer to fill your order of steaks; more than likely he's simply unwrapped pre-packaged meat delivered directly from the processor.

It's a classic example of what Paul Roberts, in his book *The End of Food*, calls the "protein paradox." Meat production has outstripped people production. With all the stupefying advances in breeding and grain feeding, the cost of one pound of meat is cheaper now than at any time in history. That hasn't alleviated world hunger or led to any kind of meat-eating democracy. In fact, if anything it's enabled—and at this point it encourages—not just the American appetite for eating high on the hog, but a kind of pork chop dictatorship (call it white meat supremacy). We eat too much meat, all right, but we also eat way too much of the wrong parts.

Agriculture's great efficiencies are partly to blame for this. They've become too good at producing a lot of animals. Or they've become too good at producing a lot of animals too cheaply. But American consumers are also to blame. As more women entered the workforce, the responsibilities of the kitchen weren't so much reassigned as they were abandoned. Today the average American spends about thirty-three minutes a day preparing food; that's half the time spent when Frank was first hawking his tender birds. Eighty percent of chicken sold in the 1970s and '80s was in its "unprocessed" form (read: natural, with bones and skin), and 16 percent was sold in processed varieties. By the end of the 1990s those numbers had completely reversed.

Since pre-prepared and heavily processed products flooded the market, kitchens have become more like assembly rooms. In other words, Ranch Flavored Chicken Fries became both possible and inevitable. We didn't become a nation of precooked eaters so much as we became a nation of eaters who don't cook.

The ones who *do* cook, chefs, are to blame here, too. We've helped shape American cuisine, particularly when it comes to eating meat, by putting center cuts at center stage. The seven-ounce slab of protein on your dinner plate is as much an American invention as it has become an American expectation. Compare it with ethnic cuisines, where, when meat appears, it does so modestly.

Our excess has been made possible by the largesse of industrial agriculture, but nothing about it has been inevitable, and in the end there's very little that's truly delicious. The best parts, as most chefs will attest, come from the supposedly lesser cuts. These nonprimal parts suffer from a disadvantage, however, which hobbles their stardom: they require you to chew. And in some cases, to chew and chew.

Chicken breast, lamb loin, filet mignon, rack of pork—in other words, the cuts we covet most—come from muscles that rarely work when the animal is alive. It's what makes them desirably tender, but also bland. With little activity, these cuts develop very little intramuscular fat; without the fat there's nothing to carry flavor. By contrast, all the hard labor imposed on lowly muscles—the exercise required of the legs, or the cleansing work of the liver and kidneys, or the requisite pumping of the heart—lead us, the eaters, to have to work as well. If prepared correctly, the payoff is meat that's pungent, rich, and dense. And, yes, tender too. But that payoff requires a lot more *cooking*, and a lot more craft.

"It's easy to cook a filet mignon and call yourself a chef," chef Thomas Keller once wrote. "But that's not real cooking. That's heating. Preparing tripe, however, is a transcendental act."

We can, and perhaps sometimes we should, expect transcendence from chefs, in the same way that we expect it from artists. To lift us out of our usual understanding of things, of what we know. Which is why Palladin's chicken dish that night in the Los Angeles restaurant, with its lowly, unrecognizable parts, is so often replayed in my mind. If the story of America's beloved bird is like that of Humpty Dumpty—fallen and broken apart—Palladin showed us one way to put it back together, deliciously.

Of course, that's easier to say than to sell. I can tell you that a chicken gizzard (the thick, muscular organ that pulverizes the bird's food before it gets to its stomach) has more flavor per half bite than what you might get from an entire breast, but I can hardly get anyone to order it.* It's no wonder so few chefs buy whole animals.

The poultry industry would argue, What's wrong with that? What's wrong with Americans eating the parts of the animal they want to eat? Isn't that what being in the food-service business means? If a man wants boneless, skinless chicken breasts for dinner, the culture of American business is to figure out how to let him do it. And that may be true, up to a point. We've reached that point.

In just the past thirty years, the poultry industry has tripled its production of chickens, from 11.3 billion pounds to 37 billion pounds. But we don't *eat* 37 billion pounds of chicken, and the excess chicken has to end up somewhere. Processed food—Chicken McNuggets, for example—heroically soaked up the imbalance for a time. So did pet food.

More recently, excess chicken is fed to cattle (an eyebrow raiser, since

---

* Culinary historian Betty Fussell once told me that Americans have a singular preference for blandness. In her research, she discovered that in taste tests for beef, anything gamy in flavor was described in terms of disgust. We have become so dissociated from food in its natural form, she argues, that being reminded of it is unpleasant, if not unpalatable.

cattle aren't carnivores) and now increasingly even to fish. Why? Because while the price of wild fish—the traditional feed for aquaculture—has risen dramatically in the past decade, the relative cost of chicken has gone down. It's cheaper to feed chicken. What emerges is a nightmare version of the loaves and fishes—an agriculture system out of control, in which the loaf (grain, in this case fed to chickens) is produced in such excess that it gets fed to fish. I once asked an aquaculture biologist what was sustainable about feeding chicken to fish. There was a long pause. "Well, Chef," he said, "there's just too much chicken in this country."

Another solution has been to look abroad—to dump unwanted parts onto other countries. The United States is the largest exporter of chicken products to China because, luckily for Perdue and other poultry companies, the Chinese prefer dark chicken meat. By the 1990s we were producing 90 percent of all the chicken China imported, which led Chinese authorities to argue that U.S. poultry companies were dumping their excess production at falsely low prices. Russians did as well, calling the chicken parts *nozhki Busha* ("Bush's legs") after then U.S. president George H. W. Bush's aggressive efforts to export them to Russia.

In recent years, Mexico has become another popular outlet for America's overabundance. In 2008, the country eliminated tariff protection on imports, opening the floodgates to hundreds of thousands of tons of chicken legs. The decision immediately wreaked havoc on small states like Jalisco, one of Mexico's largest areas for poultry farming. Some Mexican poultry producers were forced to consolidate to lower their costs, in effect repeating the path Perdue pursued in the 1960s. In the meantime, Jalisco's poultry workers, displaced from their communities, began entering the United States illegally in larger numbers. And they've found employment where they have skills to match: at poultry-processing facilities like Perdue.

Wouldn't it be easier to cook every part of the bird?

Instead, the end result is a food system that plays out like an ongoing national tribute to Rube Goldberg. The overproduction of grain helps enable the overproduction of chicken, which lowers the price of chicken, which means even more chickens are raised to make up for declining revenue. That leads to even more unneeded chicken. So it's fed to other animals it probably shouldn't be fed to, like fish (which are increasingly farm-raised, in part due to the offshore pollution caused by producing too much grain). And then the overproduced chicken gets dumped to places like Mexico. To compete, Mexico turns to the same kind of system, the get-big-or-get-out system that feeds on itself: produce more chicken at lower prices. Laid-off poultry workers seek work in America, often illegally, which drives down wages and helps poultry companies produce . . . more chicken.

CHAPTER 12

I N E A R L Y O C T O B E R , Lisa called to tell me Eduardo had been in touch to invite me to see the slaughter of his geese. He asked that I be in Spain on the tenth of November to witness "the sacrifice, with gas."

I remembered Eduardo emphasizing the importance of the slaughter to the success of the livers. Done wrong, the livers would be ruined. He explained that his killings happened all at once and then added, cryptically, that they were performed "in a state completely absent of stress, like when a person slits their wrist in a bathtub—the sweetest form of death." I wasn't sure I agreed with his comparison, but I was definitely intrigued.

But a few days before I was to leave, Lisa called to report that the slaughter apparently wasn't going to happen as planned. She said Eduardo had given no reason. Too late to cancel, I went anyway. I figured the adult geese, prepped, if not fully loaded with the requisite fat for slaughter, were still worth seeing.

When I arrived at the farm, Eduardo greeted me with surprising affection, like someone reconnecting with a childhood friend. I suspected he felt badly about changing the date of the slaughter.

"Eduardo, it's no problem about the geese," I said, wanting to put him at ease. "I'll see the processing another year."

He raised his eyebrows when Lisa finished translating. "*Sí, sí,*" he said, without apology or explanation. He suddenly launched into a description of the slaughter, perhaps thinking I wanted a reenactment of events I couldn't witness firsthand. "They go to sleep," he said. He closed his eyes and tilted his head slightly, resting it on his right hand to mimic a soft pillow.

"Together?" I asked, more skeptical than inquisitive.

"No, no, it's true," he said. "They go to sleep together and they feel nothing. Nothing."

Eduardo described a mazelike system of fiberglass fencing that he erects in front of a large, insulated room. He showed me how he sprinkles corn to lure the geese and how he calls out to them, clapping his hands. Then he ran through the invisible maze, carefully imitating a goose's waddle.

"When one has gone in, the others see that it's okay, that they're eating, which is when they follow," he said. "They must enter the room with free will." The geese don't struggle, they never lose self-control. Free will, he explained to me patiently, is why the livers taste so sweet.

How did he know for sure that the animal hadn't suffered?

"The taste!" he said. "That liver you had—did it taste like a goose that struggled?" Of course it hadn't. It tasted like a goose that had been coddled and massaged to death, but how was I supposed to know the difference? Until I met Eduardo, I didn't even know it was possible to produce foie gras without at least some discomfort. I pushed him for proof.

He nodded. "The year my son was born, I was feeling a little unsure myself. I wanted a clean conscience. So after I gassed them, I opened up the door and let the air circulate. Twenty minutes later they woke up. They were dazed . . ." He tilted his head back and shook it very slowly from side to side, looking convincingly befuddled. Then he righted his head, his eyes closed for a very long blink. When he opened them, he looked at me excitedly. "Then they went right back to gorging. They acted as if absolutely nothing had happened!"

Eduardo's methods may sound extravagant, but many studies have shown the connection between suffering and degraded meat. If an animal is stressed—during its life, and especially in those last moments before death—it manifests, just as Eduardo claimed, in the flavor and texture of the final product.

It's something I saw firsthand when we opened Blue Hill at Stone Barns. Back then, Craig was taking one Berkshire pig a week to the slaughterhouse, but what came back was often far from delicious. José, our butcher, pointed out red streaking in some of the muscles, and I found the pork mostly dry, tough, and much less tasty than expected. Craig realized the lone pigs were unduly stressed by the travel, so he decided on a few changes. He took two pigs along for the ride (a macabre version of the buddy system) and made sure to provide them with extra feed. And he hung enlarged photos of the farm's woodland inside the trailer. Craig still had only one pig slaughtered; the other traveled back to the farm. The following week, the same pig, now acclimated to the trip, accompanied yet another pig back to the slaughterhouse for his last ride. The streaking disappeared, and the Berkshires' incredible flavor returned.

I asked Eduardo if his efforts were meant to ensure the highest-quality livers or to guarantee the welfare of his geese. He shook his head slightly and smiled, a sign that he didn't understand the question. I tried again: "What motivates you? If you had to choose, is it sweet livers you want, or a painless end to life?"

Eduardo raised his eyebrows. "What's the difference?"

Eduardo had arranged for us to stop for lunch that afternoon, but first (naturally) he suggested that we stop to visit the almost-ready-for-slaughter geese.

We drove in and out of different pastures for about twenty minutes, until we came to a group of Iberian pigs. On a hunt for acorns, heads low to the ground, they moved in unison across the landscape. Eduardo had me turn off the car. The geese, he said, should be close by. We stood under an oak tree and waited. Carpeting the ground were cracked acorn shells, like spent bullet casings.

"The pigs were just here," Eduardo confirmed, bending over to inspect the leftovers. I found an intact acorn that the pigs had missed and held it in my hands. It was enormous.

"The geese eat this whole thing?" I asked.

He smiled and pointed off in the distance. Twenty or so geese, in single file, emerged from a patch of tall grass, quacking loudly.

"They're huge!" Lisa gasped.

It was true: the geese looked almost prehistoric, like small dinosaurs. They must have tripled in size since I'd seen them in spring. The birds honked loudly, suddenly gathering together and flapping their wings as they came to an area of untouched acorns.

"Do they ever fight over acorns with the pigs?" I asked.

"When the pig gets nasty, the goose will slap him across the face with its wing," Eduardo said, jutting an elbow out from his body and flapping it back and forth. "The pigs are scared by this." He bent down and called out: "*Hola, hola*, my ladies."

The geese dropped their heads to forage. "See? See how they're carrying a big knapsack of fat?" Eduardo pointed. Rings of fat were visible around their necks. "And look down," he said, grabbing my arm and pulling us down to our knees for a goose-eye's view. "The bellies are dragging on the ground."

Eduardo said that another trick to determining their readiness is to observe the geese in the rain. "Geese sweat fat," he explained. He pointed to a goose's chest. "Right there," he said. "They use their beaks to spread the fat on their feathers, like putting on a raincoat. So look at how well the water repels off their bodies and you'll get an idea of how much fat the liver will have."

Raincoats or no, I wanted to say they looked obese and ready for slaughter—the slaughter I had flown halfway around the world to see. But Eduardo only sighed. "It's been a bad year for acorns," he said. "But not the worst. There were some times when it wasn't worth slaughtering."

Eduardo explained that years ago he'd sometimes reluctantly fed them grain to supplement their diet. ("Free-choice grain—no *yah-yah-yah*," he said, pumping his fist down an imaginary goose's throat.) I couldn't tell whether he gave them grain to ensure he had a product to sell or if he was trying to please his distributors, who still preferred the addition of grain. Eduardo said it reminded them of the French livers they were used to selling, and what they knew to be the best quality.

"I say to them: You know where I had the worst foie gras of my life? Paris! Paris was where I had the worst foie gras. It was garbage."

Eduardo blamed the bad livers on the corn itself, not the gavage. He said it made them predictable. And not in a good way.

"The livers should have a similar structure, but each liver in the end is different. They should taste different," he said, sounding much like John Jamison did when he praised the inconsistency of his grass-fed lamb. I told him that most chefs look for the opposite, for uniformity. He kneeled on the ground again and, raising his loosely curled fists to his eyes like binoculars, leveled his sights for a last look at his departing geese. "Chefs are wrong," he said.

A DEBT TO *JAMÓN*

Later that afternoon, we were back in Monesterio, in the same restaurant where eight months earlier I'd tasted Eduardo's foie gras. As I was gathering up my coat and bag after lunch, I heard Eduardo speaking to Lisa and turned around to find his right arm extended up in the air. He was holding a thin slice of *jamón* between his thumb and forefinger. The sun, golden and diffused in the waning hours of the day, streamed through the restaurant's window, backlighting the ham like an X-ray.

Only then did Eduardo finally acknowledge his debt to the pigs. "My goal in life is to have my livers remind people of this," he said, the ham's

incredible weblike striations of fat clear to see. He used his left index finger to carefully trace the lines, back and forth and looping around, following the glistening white veins as purposefully as if he were driving on
the winding roads of the *dehesa*. It was an extraordinary gesture, in part because until that moment Eduardo had essentially dismissed the pigs as
incidental.

"You know," he added, looking at the translucent slice of ham dangling in
the air, "*jamón ibérico* is the best ham because it's the perfect expression of the
land."

Lisa later told me that Eduardo's use of the word *land* was probably quite
intentional. "Land" in Spanish is *tierra*, which means more than what is under your feet. *Tierra* is defined holistically, meaning the soil, the roots, the
water, the air, and the sun.

Jamón ibérico's significance, Lisa explained, is as much cultural as it is gas-
tronomical, with deep ties to Spanish identity. Throughout most of Spain's
history, Catholics differentiated themselves from the ruling Muslims, and
from the thriving Jewish community, by eating pork. Eating it "proved" you
weren't Jewish or Muslim (read: infidel).

I was reminded of once hearing a young Spanish chef describe what
ibérico ham meant to him: "Ham?" he said with a broad smile. "Ham is God
speaking."

"You know," Lisa said to me, "the whole time with Eduardo and his
geese, I kept thinking how his livers are piggybacking on a two-thousand-
year-old tradition of *jamón*. I was really glad to hear him acknowledge it."

The Spanish obsession with *jamón* is about much more than food. It's
about an old, almost forgotten way of relating to what it means to be Spanish.
Perhaps Eduardo's reluctance to acknowledge the pig was his reluctance to
articulate what Lisa said—in some ways his geese were getting a free ride.
To understand Eduardo's foie gras, I realized I had to better understand
jamón ibérico. And in order to understand the *jamón*, I'd have to learn more
about the *dehesa*.

The next morning, Lisa got in touch with Miguel Ullibari, the former director of Real Ibérico, an organization devoted to promoting *jamón ibérico*. A soft-spoken man in his forties, Miguel was intrigued by Eduardo Sousa's work and, like Lisa, believed that Eduardo's system was deeply indebted to the *ibérico* model. Miguel agreed to be our guide for the day, arranging for Lisa, Eduardo, and me to visit Placido and Rodrigo Cárdeno, the two brothers who run Cárdeno, one of the best and oldest *jamón* producers in the region.

As Miguel explained it, there is a whole taxonomy of *jamón*. *Jamón ibérico* can technically refer to any cured ham that comes from Iberian pigs, but there are several classes within that category. True *jamón ibérico*, as I had always understood it, requires the designation *jamón ibérico de bellota* (acorn) or *jamón ibérico de montanera*. It is significantly more expensive than other versions, which come from pigs that are fed grain (a less expensive diet) and are often cured for less time.

Placido and Rodrigo produce only the best *jamón ibérico de bellota*. Their business is neatly divided, with Placido managing the curing process and Rodrigo in charge of raising the pigs.

When we arrived at Cárdeno, it was Placido who greeted us first. He wore tinted glasses and had the faint outline of a mustache. He ignored the suggestion that we tour the seven-hundred-acre farm and instead brought us to a picturesque field littered with oak trees. The savanna-like expanse looked like a postcard picture of the *dehesa*. (In fact, later that afternoon I was handed a Cárdeno postcard showing the exact place where we had been standing.)

"Beautiful," I said.

"Oh, no, no," Placido said, shaking his head and looking down at the ground. "This is the ugliest you'll ever see the *dehesa*. You have to come back when it's green and lush."

Placido explained that Cárdeno had produced *jamón* as early as 1910, on land that was originally owned by his maternal grandmother. "My mother, she came from aristocracy because she had the land," he told me. Raised in a small town, she met Placido and Rodrigo's father in grade school. They fell in love, which, according to Placido, was when things got complicated. Their father didn't own land and, though his family operated a *jamón*-curing business, without land he was deemed unworthy of their mother. "They became a little bit of a Romeo and Juliet story," he told me. "But they were in love. So they won."

Their father raised the pigs and cured them himself, until Placido and Rodrigo joined the business in the 1960s and split the chores. The brothers' deep devotion, and perhaps their sibling rivalry, brought the quality of the hams to new levels.

We walked to an embankment and saw fifty or so Iberian pigs dunking themselves in a pond and running out to dry on the surrounding grass. The older pigs lay in the sun, dozing. It looked like a porcine version of a Sunday afternoon in the Tuileries. No one appeared more content than Placido.

"Funny, isn't it?" he said, as if seeing the scene for the first time. We all pointed and smiled, except Eduardo, who was off examining an oak nearby, his large frame diminished by the width of the old trunk.

As we walked among the trees, Miguel and Lisa discussed the history of the *dehesa*, a place that until recently, I discovered, had been more famous for producing wool than hams.

THE *DEHESA*

The *dehesa* system originated during the Middle Ages. By 1300, the Christian Reconquista of Spain had reclaimed the area of Extremadura from the

Muslims. For the burgeoning wool industry, the uncultivated landscape represented thousands of miles of potential pastureland. Suddenly the victorious Christians had a vast new resource for their prized Merino sheep. The subsequent influx of livestock prompted the destruction of the area's thick forests, resulting in the sparse oak population we see today. Peasant farmers built stone walls around the sheep's grazing areas—the word *dehesa* comes from the Latin *defensa* (defense), which refers to lands protected against wild animals and predators.

To be a Spanish sheep breeder at that time was to belong to a powerful union, with outsize influence in creating policy. Between the fifteenth and eighteenth centuries, the number of sheep in the *dehesa* grew from 2.5 million to 5 million. The importance of wool to the economy allowed for the organization of sheep breeders into an elite guild called the Mesta. Laws were passed to protect the grazing areas. Removing grass from any part of the *dehesa*, for example, was outlawed. These laws—a reflection of the economic and political power of the Mesta—did more than establish a legal framework for protecting the sheep breeders' assets. They helped inculcate a reverence for the land.

A law passed in 1548 codified the rights of the oaks. Harvesting even a branch was illegal:

> Any person caught chopping down, conveying or loading a holm oak tree or coppice in our dehesa will be fined five hundred maravedis in benefit of the council; and for a branch as large a man's body, the person will pay a fine of three hundred maravedis; and for a branch as large as a man's thigh, two hundred maravedis, and for a branch the size of a man's calf, one hundred maravedis; and for a branch the size of a man's wrist, twenty-five maravedis, and for smaller branches ten maravedis.

The eye-for-an-eye philosophy makes it clear that the Spanish people had come to see the region as a living organism, a part of themselves. Its success was their success. To diminish the *tierra*—to remove any part of it for personal use—was to diminish the rising fortune of the Spanish people.

It was the prosperity of the *dehesa*'s wool industry that eventually subsidized Spain's explorations of the New World. An intense period of colonization followed—in the Americas but also in Australia, New Zealand, and South Africa, all of which quickly grew into major centers of wool production. The Spanish wool industry eventually collapsed as the new colonies came to produce it more cheaply. But the investment in maintaining the *dehesa*'s ecosystem didn't stop paying dividends. It adapted to the changing times with other breeds of animals—especially the black Iberian pigs, which were exquisitely suited to the land.

"If the pigs weren't famished by the fall, none of this would work," Miguel told me, describing the region's long, dry summers.

Evolved to thrive on very little, the Iberian pigs feed on grass, seeds, and grain naturally found in the *dehesa*—just enough calories to grow larger but far fewer than in conventional operations. By late October they're ravenous, which, conveniently, marks the beginning of the *montanera* phase, roughly between November and March, when the acorns have fallen from the trees. In the space of four months, the pigs gorge enough to put on 40 percent of their slaughter weight, acquiring a band of fat as thick as a down comforter.

Eduardo raised his eyebrows as if to say, *Sound familiar?*

"It's very simple, really," Miguel said, and then looked out over the oak-filled landscape and added, "but it's also very complex." In his effort to demystify the process, he was careful not to debunk the aura of the ham.

The black, bristled, stocky animal inhaling the undergrowth in voracious pursuit of acorns is not just a hungry pig. It's a pig that's found the perfect conditions for gorging. And the perfect conditions for gorging are specific to the *dehesa*.

It starts with the quality of the acorns, Miguel explained. There are two common types of oak trees in the *dehesa*: the holm oak and the cork oak. Holm oaks produce the sweeter acorn—the pig's preferred meal—but cork oaks produce later in the season, extending the supply. Both acorns are unique for their large size—another credit to the *dehesa*'s ecology. The original intent for spacing the oak trees may have been to provide shade cover, but it also allowed for the development of deep root systems. Without competition for soil nutrients and water—and this is key, since water is a scarce resource—the oaks grow to be enormous and robust, producing larger and sweeter acorns. To further encourage acorn production, the trees are pruned periodically.

A ravenous pig working a carpet of sweet acorns is the endgame of a long series of preparations. But the pigs have to work for the feast. Because the trees are so well spaced, a lot of grass foraging happens along the way, which makes the acorns taste even sweeter.

"There's a physiological component to this," Miguel said. "Grass plus acorns makes acorns inordinately sweeter for pigs. So they eat more acorns."

Eduardo shook his head. "Not just for the pigs, my friend. For the geese, too."

The requisite hunt, a trot from tree to tree with mouthfuls of grass along the way, is intense exercise for the pigs. By the standards of conventional American pork production, that's wasted energy. More exercise means more expended calories, which means more feed. Fortunately for confinement producers, Americans have industrious spirits, heroically overcoming dry, stringy hams with sweet glazes and candied pineapple. But in the *dehesa*, it's the exercise that creates oxygenated muscles, and thus the deeper flavor.

It's a lesson I learned from Craig Haney long before I ever laid eyes on the *dehesa*. The year Stone Barns opened, I asked Craig to sit for a tasting of pork from a favorite local producer. The couple who raised these pigs were exceptional farmers. I hoped Craig would come away determined to raise pigs at Stone Barns that surpassed the flavor.

"I'm not convinced," Craig reluctantly said, after thoughtful bites of several different cuts. "It's nice and soft, but my pigs have more flavor." I gathered several cooks later that night and did something I had never thought to do before: a side-by-side comparison of Stone Barns pork versus the local producers' pork. Craig was right; it wasn't even close. The soft, almost flabby meat from the local producers could be cut without a knife, but if you ignored the butter-like texture—and that's key, because a soft texture is so often mistakenly confused with quality—there was little to savor. Craig's pork, on the other hand, had a deeply rich pork flavor.

Many years later, I learned that the local farmers intentionally curtailed the pigs' exercise, creating muscles that were less oxygenated, and therefore less chewy. Which is why Craig's pigs, raised in the forest surrounding Stone Barns, were so superior.

Exercise not only deepens flavor; it also creates space within the muscles for fat deposits. It's that space, and the muscle mass itself from all the exercise, that enables the meat to integrate the oleic acid from the acorns.* When Eduardo held the *jamón* up to the light that afternoon and declared it the inspiration for his foie gras, he was celebrating the process by which the fat becomes integrated into the meat. That rich tapestry of fat and muscle is essential for creating the ham's extraordinary flavor.

———

We were interrupted on our walk back by Rodrigo's son, Ringo—Placido's nephew—who approached us in full stride from across the field. He was tall and thin, with long, wavy brown hair and a duffel bag slung over his shoulder. With the forest as a backdrop, he looked like a medieval hunter.

---

* Studies have shown that more than half the ham's fat content is oleic acid (the same type found in olive oil), which, as fats go, is one of the better ones to consume. It's been proven to lower total cholesterol, particularly LDLs (also known as "bad" cholesterol), and actually slow the development of heart disease.

"My name is Ringo," he announced as he approached. "I have come to welcome you to Cárdeno."

I introduced myself and thanked him for having us. Eduardo did the same. Upon hearing Eduardo's name, Ringo straightened his shoulders. He stared. "Eduardo Sousa?" he asked. "From Fuentes de León?"

Lisa laughed as she translated the formality of Eduardo's response: "Yes, I am he."

As we arrived at Cárdeno's official tasting room, Miguel explained that Ringo was training with his uncle Placido to learn the art of curing *jamón*, and that Placido's son, currently in veterinarian school, might one day join Rodrigo to learn the art of raising pigs. I asked Ringo if he thought a nephew training under his uncle was more sustainable than a son training under his father.

"What I know," he said, artfully dodging the question, "is that I have the very best teacher in the world."

Miguel nodded and turned to me. "This is the classic *dehesa* tradition, and absolutely necessary for the long-term sustainability of the whole system."

Rodrigo appeared. He was larger than his brother, with a weathered, leathery complexion that hinted at long days out in the sun. Whereas Placido was mild-mannered, with a laconic demeanor, Rodrigo seemed brusque and uninhibited. He shook hands with everyone in the room, including several of his coworkers, helped himself to a beer, and sat at the end of one of the couches to smoke a cigarette. We gathered around, small white plates of *jamón* placed on a coffee table in front of us.

Rodrigo spoke first, assuming we wanted to know how such great ham was made. "You need to start with great legs!" he blurted out to no one in particular, and then took a long, slow drag of his cigarette.

Embarrassed by his younger brother's abruptness, Placido said softly, "It's true. I could never cure a brilliant *jamón* without a great pig. It's impossible."

Rodrigo ignored him. "All you need to add is a little paprika and some salt; otherwise you're masking the flavor." Eduardo raised his glass to the air in a silent toast of solidarity as Rodrigo pushed on. "Look, when a great leg is born, and we get to raising it, Placido's eight-year-old niece could make a delicious ham!"

Placido smiled uncomfortably. He leaned over to me as Rodrigo continued speaking about the importance of raising pigs the right way. "Not exactly. Give her forty-five years to learn, and all the right weather conditions, and yeah, sure . . . it's easy."

The color of the *jamón* was very near to purple, and even more marbled than the piece Eduardo had held up at the bar in Monesterio. Miguel handed me a piece with extra fat and told me to hold it in my hand. Unlike the fat from a Berkshire pig, the kind we cure at Blue Hill at Stone Barns, it began to melt right away. "Seventy percent of that fat is unsaturated," he said. "These pigs are like an olive tree on four legs."

I let the fat melt on my tongue for a moment before swallowing. Its aroma was distinct, incredibly nutty and aromatic. I felt as though I was tasting the *jamón* equivalent of Bordeaux's finest, the Lafite Rothschild of hams. And yet my surroundings spoke nothing of luxury. The tasting room was dark and musty. Cigarette smoke wafted through the air. We drank warm beer in plastic cups. The *jamón* was on simple white plates, piled on and hastily arranged. While the ham was delicious, it was also a little dry—a minor complaint, maybe, but true.

I turned around and saw the *jamón* in the tong. The man slicing was cutting near the bone, struggling to remove whole pieces intact. We were sampling the very last of the leg. It had been cut into dozens of times, probably over many weeks, if not months. Another leg rested nearby, untouched and at the ready in case anyone wanted more.

The expression on my face must have given me away. Miguel tapped me on the leg. "This is exactly the point," he said quietly, waving his arm around the room as everyone laughed and drank beer. "The life he leads—he looks and he acts like a peasant farmer. It's not the picture of riches, but I'm telling you he's very rich. And I'm also telling you that Juan Carlos himself could come to visit here, and if there was still meat on an old *jamón*, that's what they would serve. Because *jamón* is not about luxury. It's a poor product; it comes from poor land. I believe this is why it has survived."

RETURN TO A ROOFTOP

We finished the tour on the rooftop, which was where I finally got that bird's-eye view of the *dehesa*'s pastoral scene.

Placido's distinctively Mediterranean home rose out of the ground like the ancient oaks surrounding us, towering over the family's property. It was such a clear day you could see for miles. Eduardo quickly walked to the corners of the roof and again cupped his hands in front of his eyes like binoculars. He was looking for geese. I was looking for pigs, but instead I saw a cluster of veal calves grazing in the field. I asked if the pigs followed the cattle on the grass.

"This is Spain," Miguel said. "Pigs don't follow anything. They lead."

That was an understatement. To ensure an abundant supply of acorns for their binges, each pig requires about four acres of *dehesa*—sometimes more, sometimes less, depending on the acorns available in a particular season. During a below-average acorn year, the number of pigs allowed to fatten is restricted. It keeps the price of the best *jamón ibérico* high, but it also acts as a kind of insurance against overwhelming the balance of the *dehesa*.

Without the pigs moving, Placido explained, the grasses between the trees would suffer. Pigs eat just some of the grass, a collateral feast on their way to the next targeted acorn tree. It's the cattle—who follow just behind the pigs—that do the real grazing. After all, they're herbivores, and they happily trail the pigs and eat what the pigs have overlooked. The same is true of the sheep. While they no longer make up the dominant industry (and are not included on Placido's farm), sheep are still scattered around the low-lying areas of the *dehesa*. They eat leftovers, the grasses the pigs and the cattle don't want or can't get to on account of the size of their mouths.

Like at Stone Barns, all that grazing actually improves the grass. Spread out, the manure from the animals fertilizes the field. The trampling of the hooves—whether it's sheep or cattle or pigs—also helps break down materials such as fallen leaves. Returned to the ground, that organic matter helps to sustain the billions of soil organisms and conspires to make the famously delicious assortment of grasses. It's the variety that's key to the health of the animals, and to the larger ecosystem. All that grass diversity increases the numbers of butterflies, beetles, ants, and bumblebees, which in turn supports the animals that prey on those insects, like lizards and snakes.

The thick forest stretches I saw in the distance provide habitats for wild birds. They flourish here, too—red kites, booted eagles, and short-toed eagles all provide a built-in population control for insects and rodents. They're also integral to seed dispersal in the *dehesa*. In their hunt for grubs and insects, they'll peck at the manure left behind by an herbivore, spreading it across the field and helping it to fertilize more efficiently, which means healthier pastures the next time the pigs come around to graze the grass.

Herein lies an essential fact about the famous *jamón ibérico*: it's not just about *jamón*, in the same way that *tierra* isn't just about the thing you stand on.

"You know, it's a funny thing," Miguel said to me. "What you realize looking out from up here is very obvious, and it's something most of the world has yet to realize: *jamón ibérico* is just one product of the *dehesa*."

He named two rich sheep's-milk cheeses from the area: the Torta del Casar, with its gamy, acidic, and somewhat smoky flavor, widely distributed throughout Spain, and La Serena, considered one of the finest sheep's-milk cheeses in the world. Both cheeses are made from the milk of Merino sheep. (Merinos are not much of a milking breed—the yield is particularly low—but the sheep were so integral to the culture of southern Spain and the health of the *dehesa* system that when the price of wool fell, milk replaced the wool.) But, to Miguel's point, I had never known that they were products of the *dehesa*.

Miguel pointed to the cows, a relatively unknown Morucha breed, which he said might be more flavorful than any beef in America. The Moruchas were once used as fighting bulls (Ferdinand, of the beloved children's book, is the most famous example), and since they originated from Black Iberian cattle, they've evolved to forage in the *dehesa*. The cattle are constantly in motion, which, just as with the Iberian pigs, oxygenates the muscles and yields bold, flavorful meat, darker than most steaks.

At the mention of Morucha beef, Eduardo made a tight fist and brought it to his lips. *"Fantastico,"* he said, wincing at the thought that I had never tasted it.

"There's a history of poor communication and a lack of modern commercial networks in this area of Spain in particular," Miguel continued, explaining that these products aren't well-known because they've historically been produced for personal consumption. "That's all changing now. You need only to look at yourself as an example. Here you are discovering Eduardo's foie gras, but Eduardo's foie gras is a very old thing."

I asked what else provided an economy for the *dehesa*, and they pointed to the oak trees. "Not the acorns, but the cork. That's really the economic engine of this whole system," Miguel said. Stripping the bark without harming the trees is a highly skilled task, done with extraordinary precision. Peeled like bananas, the dark orange trunks are a familiar sight after harvesting. Nearly a quarter of all wine corks in the world originate in the *dehesa*.

Placido pointed to the more open areas in the distance where barley, oats, and rye were grown—for animal feed but also for the table. "I wouldn't say this is a highly profitable undertaking. The land is suited more for grazing. But we do produce grain, and it cuts down on what we have to import for feed."

Dissecting the open pastures were thick forest stretches, thinned appropriately for charcoal production—another "industry" I didn't know existed here. And the most surprising feature of the landscape—the homes scattered throughout, the hanging laundry, and the sounds of children playing in the distance—seemed just as central to the scene as the cork and the pigs.

Wendell Berry once described the land as an "immeasurable gift," and he wasn't just referring to food. In the rush to industrialize farming, we've lost the understanding, implicit since the beginning of agriculture, that food is a process, a web of relationships, not an individual ingredient or commodity. What Berry refers to as the *culture* in agriculture is as integral to the process as the soil or the sun. Here in the *dehesa*, culture and agriculture seemed not only intertwined but interchangeable.

——

I'm sometimes asked what is meant by *sustainable agriculture*. I've never arrived at an easy answer.

When I was on that panel with Wes Jackson in California several years ago (before he schooled me in the root systems of perennial and annual wheat), Wes was asked for examples of sustainable farming. He didn't offer any.

"Small societies have pulled it off here and there, but it has been beyond the cultural stretch of most of humanity to do agriculture right century after century," he said, his words injecting the room with an air of defeat. "Thinking of agriculture as a mistake is a good place to start." Wes's work to make agriculture perennial—mimicking natural ecosystems like the prairie—is intended to remove annual planting decisions, and thereby the wild card of human shortsightedness.

At dinner that evening, I told Wes about Klaas and Mary-Howell. I described their conversion to organic, the complex rotations for soil fertility, the propagation of ancient varieties of wheat—everything pointing to farming's potential to do a lot of good. "What's not sustainable about that kind of agriculture?" I asked him.

"Because it won't last," he said. "Klaas sounds great. His son might prove to be even better. But sooner or later someone is going to show up and do something stupid to degrade the land. That's been the history of agriculture."

An organic farmer like Klaas might be important and inspiring, and he might produce delicious food while improving the fertility of his soil. But don't call him sustainable, said Wes. Biologically speaking, he is more like a historical blip: here today, gone tomorrow. ("What can I say?" he said. "We live in a fallen world.")

In the last few minutes on the roof at Cárdeno, I stood near the edge and looked out at the *dehesa*. The afternoon sun remained hidden behind a thick curtain of clouds. Then, just as we were leaving, a spectacular flood of light splashed across the fields. Once again, I saw the entirety of the astonishingly diverse landscape. But what struck me wasn't the diversity so much as the permanence of the scene. I was enjoying the same view that Placido's grandfather enjoyed. It was, when I stopped to think about it, the same view that Placido's grandfather's grandfather enjoyed, which was enough to call Wes's conviction into question. A two-thousand-year-old agricultural landscape isn't a blip.

~

The *dehesa* has survived (thrived, even) despite its poverty. In fact, as Miguel argued, the *dehesa*'s survival may be *because* of its poverty.

"The land of Ibérico was a poor part of Spain until just the last few decades," he told me. "Understanding and respecting nature was not a choice, but the rule to survive."

Unlike American settlers, who were spoiled by our country's natural abundance, Spaniards couldn't simply drop their plow and move on to better land. Agribusiness never capitalized on the *dehesa*'s wealth because the land isn't quite good enough.*

* This hasn't stopped them from trying, especially recently, as *jamón ibérico* has become internationally renowned. "Meat companies see an opportunity to raise a cheaper ham. And they're doing it," Miguel told me. "Of course, instead of under trees, the pigs are fed indoors, crammed into a small space, and along with the acorns, grains have become part of the diet as well." Thank-

But the impoverished land is only part of the story. For a place to produce such remarkable products, including *jamón ibérico*, and for a semiarid region to hold on to its remarkable biodiversity—so high it has been compared to that of a tropical rainforest—my rooftop view couldn't fully explain the embarrassment of riches, nor why the riches have endured.

I thought back to John Muir, and his statement that "when we try to pick out anything by itself" in nature, "we find it hitched to everything else in the Universe." That much was clear even from the rooftop. Remove, say, the sheep from grazing and the pigs would begin to feed on denuded grass. Eventually the *jamón* would change, too.*

But *how* these things are connected—the hitching itself—is just as important. One way to measure the strength of these connections—which is to say, the sustainability of a place—is to look at how deeply they penetrate the culture.

Aldo Leopold, reflecting on America in the mid-1900s, believed that our culture had it mostly wrong. His anthology *A Sand County Almanac*, published shortly after his death in 1949 and considered by many scholars to be the bible of American environmentalism, includes his now famous essay "The Land Ethic," in which he argued that our idea of a community, prefaced on the interaction between human beings, was simply too limited. A broader definition of a sustainable community was needed to include soil, water, plants, and animals—"or collectively: the land."

In other words, he defined community the way the Spanish use the word

fully, the *jamón* equivalent of a Frank Perdue hasn't yet emerged, because those deeply embedded striations of fat running through the ham cannot be replicated on a larger scale. "It doesn't work," Miguel told me when I asked about intensively raising *jamón ibérico*. "Less exercise means less intramuscular fat, which means less distinctive taste. Large meat companies try, but they will fail, because it reduces the complexity of the *dehesa* to the amount of acorns they need to feed a pig to get it fat. If you want to raise pigs in confinement and bring them acorns from Portugal, fine, but don't call it *jamón ibérico*. Because it isn't."

* In a sign of how deeply connected the world's become, even for a place as remote as the *dehesa*, the trend of switching from cork bottle stoppers to screw tops has diminished the supplemental income of the Extremaduran farmers, threatening their livelihoods.

tierra, viewing each component of the system the way Klaas viewed his wild plants—as so critical to the other parts, they fit together like a living pyramid:

> The bottom layer is the soil, an insect layer on the plants, a bird and rodent layer on the insects, and so on up through various animal groups to the apex layer, which consists of the larger carnivores. . . . Each successive layer depends on those below it for food and often for other services, and each in turn furnishes food and services to those above.

But Leopold was not entirely satisfied with this metaphor. Any complete understanding of the land, he argued, required an ethical component as well.

"That land is a community is the basic concept of ecology, but that land is to be loved and respected is an extension of ethics," he wrote. Leopold saw it as our responsibility, as members of the community, to protect nature's greatest gift: its capacity for self-renewal. "A land ethic, then, reflects the existence of an ecological conscience, and this in turn reflects a conviction of individual responsibility for the health of the land."

Without an ethic, the connections invariably weaken. To Leopold's American readers, this was a radical concept, but a sense of ethics is implicit in the culture of the *dehesa*. Farmers are raised to respect the land as nearly sacred ground. As Lisa told me, Spaniards find the *dehesa* nothing short of profoundly beautiful, in the same way that an American might find Yosemite or the Rockies beautiful—not only because it is, but because so much in their history and education tells them it is.

"It's very much a question of *values*, not just value," Miguel explained to me. "That's what explains how the traditional farmers and producers have behaved for generations, and why still today they put tradition, nature, or instinct before technology, choosing to produce better, not just more."

Today, farmers routinely plant new oaks to replenish natural loss. It's not done for personal gain—in their own lifetimes those trees will never produce an acorn. The practice is largely based on what their parents and grandparents did before them, a tradition in line with the Mennonite belief that you start raising a child one hundred years before he is born.

To truly see how deeply these values penetrate the culture, one need only look at what people are eating. Extremaduran food is unadorned and simple, reflecting its peasant origins and the poverty of the land.

Start with ham. (The Spanish always do.) As Miguel explained, *jamón* is, in essence, a poor product. The meat is sliced paper thin. It is served sparingly. And it's merely one preparation of one part of the pig. There are other regional *embutidos* (cured meats), like the *morcillas* (blood sausages, which come in numerous forms), *lomo* (cured tenderloin), and the famous chorizo. The ribs are often served in the regional variation of *migas* (a traditional Spanish dish of fried day-old bread crumbs). And then there is the *secreto ibérico*—a prized cut of meat near the shoulder blade that is prepared simply, cooked over a high flame.

Since great *jamón* cannot exist without the sheep, there are delicious uses for sheep's milk (such as the renowned Torta del Casar and La Serena cheeses) and also the meat from older sheep once they no longer graze—*caldereta de cordero* is a slow braise of mutton, stewed with garlic and potatoes. The Extremaduran *chanfaina* is braised second cuts—brain, heart, kidneys, and liver—mixed with boiled eggs and bread crumbs.

The abundant wildlife of Extremadura—partridge, rabbit, deer, and wild boar—are also incorporated into the cuisine, and they are often served with locally foraged mushrooms and greens. Bee populations thrive, feasting on the plant diversity to produce exceptionally sweet honey. And of course there is the local olive oil, which makes its way into almost every dish.

The land never suffered from settlers imposing their dietary preferences on the ecology, as ours did during American settlers' disastrous westward march across a fertile and undeveloped continent. It was just the opposite. People's diets (as well as pigs') evolved from, and with, the ecology.

Eduardo's foie gras, while not a traditional product of the region, is informed by the same set of values. It's difficult to argue that it is a poor product. One of Eduardo's livers costs about $700—an exorbitant price, when you compare it with the $80 Moulard livers I can buy from suppliers. And yet, as Eduardo pointed out to me, he almost never sells his livers whole. Instead he makes them into pâté (a technique owed to chef Jean-Joseph Clause, the French culinary genius awarded twenty pistols by King Louis XVI), or confits single slices, cooking them in their own fat. In this way, he can preserve them for weeks or months, whereas a fresh liver has a lifespan of a few days. And, more important, he stretches them over many servings. Eduardo's geese require quite a lot of time and care, and copious amounts of natural feed. Stretching one liver over multiple meals—over multiple weeks—means the liver can be savored.

Compare that with the average restaurant preparation. With foie gras (especially Moulard duck liver) faster and easier to produce, the result has been a radical change in the way we're able to enjoy foie gras. Most American chefs today lob off a large slab of the liver and sear it in a pan, just as they would a seven-ounce steak (though the fat is so meltingly tender, you can eat it with a spoon). In the words of Thomas Keller, that's not cooking, that's heating. The "transcendental act" of good cooking turns out to be more than just culinary. It's an ecological act, too.

It took that rooftop view to get me to see something about Eduardo I hadn't fully considered. His foie gras is so brilliant because of what it owes to the *dehesa*, a system of agriculture that is almost antithetical to farming as we know it.

Eduardo's conviction that "all you need is to give the geese what they want and they will reward you" may sound sentimental, but he is referring to a universal truth about nature. When we allow nature to work, which means when we farm in a way that promotes all of its frustrating inefficiencies—when we *grow nature*—we end up producing more than we could with whatever system we might replace it with. In a finely tuned system like the *dehesa*, it gives you *jamón* and beef and cheese and figs and olives, and enough will be left over to fatten your geese, too, if you let it.

When Eduardo said that conventional foie gras was an insult to the history of foie gras, he was not speaking about force-feeding. He was commenting on the affront to the natural world, on the destruction of what nature can provide. It is free for the taking, so long as you play by the rules. Playing by the rules means you act quietly on the land—some profits from olives and figs are forfeited for the geese, some geese are sacrificed for the hawks, cattle and sheep share the land with pigs, pigs share the acorns with geese, and so on. Conventional American agriculture mostly does not play by the rules. In its relentless drive for higher yields—more corn, more chemicals, more monocultures—it fixes the game. It de-natures nature. That's the insult. It makes a mockery of what was a gift.

Wes is right: we live in a fallen world, and no matter how hard we try—no matter how brilliantly we feed the soil, or how ably we adopt Klaas's masterful rotations—agriculture itself will always come into some sort of conflict with nature if we seek to control it. But what if we could coexist with natural systems rather than dominate them? What if the model for the *dehesa*—disrupt nature (flood the system with pigs, strip the cork off the oak tree) but also act with restraint (limit the pigs to whatever the acorn harvest dictates, harvest the cork sparingly and replant trees for future generations)—could become the template for the future of agriculture?

Looking out over the land, I realized that in front of me was an answer to a question at the core of Aldo Leopold's land ethic: how do we keep in check our incentive to abuse the land for maximum economic return, and transition from a "conqueror of the land-community" to a "plain member and citizen of it"? We should start, he thought, by erasing any divisions between eating and farming. From my perch, that much seemed obvious. And yet the dictates of the American diet have done exactly the opposite, demanding that the land produce what we most want to eat.

Our current template for changing the system is to opt out of it: eat seasonal, buy local, choose organic whenever possible. For all the virtues of farm-to-table eating, a rooftop view of the *dehesa* makes the shortcomings of that ethos easy to see. Our job isn't just to support the farmer; it's really to support the land that supports the farmer. That's a larger distinction than it sounds like. Even the most sustainably minded farmers grow crops and raise meats in proportion to what we demand. And what we demand generally throws off the balance of what the land can reasonably provide.

Jamón ibérico might have done the same if the culture didn't dictate slicing the meat paper thin. Or if *jamón* was the only product of the region. Instead, a larger cuisine grew out of negotiations with the environment, and it has helped maintain the delicate ecological balance ever since.

A rooftop view of the *dehesa* almost inevitably raises the question, What if our ways of eating—not merely a plate of food, but a whole pattern of cooking—were in perfect balance with the land around us?

Eduardo hinted at the answer when he held that piece of *jamón* up to the light as the model for his foie gras. Until that point, I had understood *jamón* only insofar as I understood that the white lines of fat running through the ham meant the pig had been blessed with a diet rich in acorns. Now I was realizing that Eduardo really held up the entirety of an ecological system, a road map for a Third Plate. Chefs can narrate that message in the same way that those striations of fat, and the red meat that surrounded them, narrate a story as intricate, complex, connected, and—to borrow from Jack's description of how soil works—*mysterious* as the landscape that made it.

CHAPTER 13

JUST BEFORE I left Cárdeno and returned to the airport, I asked Eduardo a question that had been lingering in my mind all day. Despite their best efforts, no one has been able to replicate *jamón ibérico* outside of the *dehesa*. Did he think it would be possible to replicate his natural foie gras elsewhere in the world?

"Yes, yes, of course," he said, though he cautioned that the livers wouldn't taste the same.

I asked him how they would taste. "It depends on what the geese chose to eat. They decide. In England, I know a producer that raises geese close to the sea. I hate this liver! It tastes like fish!"

"What about the lack of acorns?" I asked. "Wouldn't that be a problem?"

"No, really, you don't have to have acorns. In Denmark, there are no acorns; the geese eat tubers similar to a wild potato and the liver is beautiful, with the aroma of root vegetables. Raise them on a coffee plantation in the right way and the liver will taste like coffee."

I wondered what a Stone Barns version would taste like—and then, almost as soon as the question occurred to me, I was already plotting how I would persuade Craig to sign on to the experiment.

❧

"I have to tell you about the most amazing Spanish farmer!" I said when I bumped into him a few days later in the courtyard of Stone Barns. Craig opened his mouth and his eyes widened, in what I thought was genuine interest. By the time I mentioned foie gras, he looked like an Edvard Munch painting.

"Can we raise geese?" I asked. The question put him at ease.

"Actually, we have fifty or so, arriving any day now," he said. Apparently Craig had decided earlier in the year that geese would work well in his animal rotation at Stone Barns, and that their meat—he hadn't considered fattening the livers—would be popular at the farmers' market for Christmas.

I described Eduardo's farm and the natural foie gras. "Interesting," he said, looking at the ground and modulating his heavily stressed voice with frequent throat clearings. "I suppose it wouldn't hurt to try." *It wouldn't hurt to try* meant that it would actually hurt a lot. But he would try.

He began in August. Craig and I settled on dividing the daily chores. Craig's team would bring water; the cooks and the managers were assigned to the feed, twice a day. Though I hoped to mimic Eduardo's system, allowing the geese to forage for themselves, Craig reminded me that our pasture wouldn't provide the same assortment of acorns, figs, olives, and lupins that Eduardo's geese enjoyed. Without additional feed, he said, the geese not only stood zero chance of producing fatty livers; they would eventually starve. Grass alone (even a diversity of grasses, treated to Craig's methodical rotations) doesn't fatten a goose any more than grazing a salad bar would prepare us to play right tackle for the Broncos. Craig was determined to feed corn, but free choice corn, without forcing anything.

The first day, he escorted me out to their sequestered paddock, deep in the corner of the main pasture. The geese moved about in a loose phalanx, heads held high, eyes darting around at their new surroundings. I took notes on his instructions for the feeding. There wasn't much to it, Craig said, showing me the box that controlled the electric current for the fencing, and the bins where the grain should be dumped.

"Field gras," Craig's assistant Padraic said as he passed by the geese. "I can't say I've ever heard of it, but it sounds delicious."

Over the next several weeks, the feeding rotation went smoothly. After filling two large buckets with corn, the cooks would walk to the goose paddock, pour the grain into the troughs, and return the pails. There was never anything unusual to report, until one day our butcher, José, came to my office. He very seriously asked if he could talk about a "disturbing trend" and shut the door behind him.

"Chef, it's the geese. They are not happy about the grain," he said, his head bowed.

José has permanently disheveled black hair and sloped shoulders. Diminutive and painfully shy, he's the antithesis of every burly, testosterone-fueled stereotype of a young butcher. He works near the entrance of the kitchen delivery door, an oxford shirt under his chef's jacket to keep warm from the draft, hacking, sawing, and, most often, maneuvering his small boning knife around and in between the muscles of large carcasses. He's patient, deliberate, and very talented.

"It's not like it used to be," he explained. "They used to run to you as you entered the fence. It was like they smelled hot food coming or something. And then they'd run all over the grain bin, fighting for a good spot. Now when I come they sort of, I don't know, they ignore me or something."

When I saw Padraic later in the day, he told me that he, too, had noticed a change in the geese. "Yeah, it's the darnedest thing," he said, removing his cap and scratching his head. "I'm no goose-ologist, but you'd think they'd just inhale the sweet stuff, like the pigs. Maybe they will when the cold kicks in—that's what your goose whisperer guy said they'd do, right?" He didn't wait for me to answer. "For now, they remind me of eating at one of those fancy Japanese restaurants, where they serve you a bowl of plain white rice at the end of the meal to make sure you're satiated and all. That's how they're treating the grain—like filler."

I went to investigate myself the next morning. On my way to the field, I passed a truck unloading the monthly order of fifteen thousand pounds of grain. There should have been nothing strange about standing in the shadow of the hulking load as I stopped to speak to Craig. I had seen dozens of deliveries before. The pigs and the chickens had been eating their fair share of those deliveries for many years. But on that morning, I peered over Craig's shoulder and stared up into the back of the flatbed. The mountain of yellow corn looked imposing, formidable. It was not unlike the first time I took notice of the mound of white flour in the middle of Blue Hill's kitchen. Like the flour, the corn was its own kind of landscape, and not a kernel of it had been grown on the farm. With two buckets, one in each hand, I carried the grain over to the goose paddock. It trickled off the tops of the buckets and left a trail behind me.

Aldo Leopold believed that land should be defined as "a fountain of energy flowing through a circuit of soils, plants, and animals." But my role in transporting another farmer's already exported energy was more like a circuit breaker, not a conduit. If we were feeding these geese—and, for that matter, the chickens and the pigs—grain grown however many thousands of miles away, on land that had nothing to do with the land I was standing on now . . . where was the flow?

I dumped the buckets of grain into the troughs. The geese approached the feed politely, pecking at it like day-old Chinese food, a take-it-or-leave-it kind of meal. When given the option of grain—normally a candy bar for the goose palate—they preferred the free forage. By late afternoon, when I returned with two more buckets, the geese had eaten almost all of the morning's delivery, but they couldn't have been less excited to see me. Several of them flapped their wings and turned away. They were eating the corn, but they were no more choosing it than an addict chooses his drug. We might as well have stuffed it down their gullets.

This clearly wasn't "field gras." This was grain gras, just not forced. It didn't matter how noble our philosophy had been, or how big my idea had started out; we'd ended up raising those geese in exactly the conditions I'd been trying to avoid. Cover the paddock with a barn, supersize the bird, and you basically had a recipe for a Perdue chicken.

———

"Why do you listen to this man you call Eduardo? Why?" I heard the familiar voice of Izzy Yanay as I bent over a case of turnips at the Union Square farmers' market, a few weeks after my afternoon with the geese.

Of all the proponents of foie gras, Izzy is among the most active—and certainly the most convincing. Raised in Israel, he moved to America in 1980, determined to do something with his agriculture degree. He partnered with Michael Ginor, a chef by training, and opened Hudson Valley Foie Gras, becoming the country's first producer of fresh duck liver. Before Izzy, chefs used imported foie gras from Canada or France, usually preserved in tins (unless they were smuggling fresh livers via monkfish, as Palladin was known to do).

Izzy, who is in his early sixties, wears black T-shirts that cling to his frame. With broad shoulders, bulging biceps, and a thin waist, he resembles a younger Jack LaLanne, were Jack LaLanne a member of the Mossad. At Hudson Valley Foie Gras, he fields all the questions pertaining to the humaneness of their product. If you're a foie gras devotee and look forward to enjoying it in the future, you're glad to know Izzy is defending your rights. He's intense, assured, and dogged about refuting claims that foie gras is torture.

"If what I was doing was unethical," Izzy said once, "if someone could come to our farm and show me one thing we do that's inhumane, I would be the first one out the door. I don't torture my ducks, because tortured ducks don't make good foie gras."

Izzy told me once how a group of skeptics visiting the farm (Izzy allows visits from anyone with a real interest in seeing foie gras produced) came to an area where the ducklings were gathered. Many of the guests reached down to pet the babies, cooing over their cuteness. He told them that petting does not make the ducks happy. He said that dogs and cats like to be petted; ducks do not. His point? Fowls and mammals are different species. "Do not equate how you would feel getting a tube inserted into your throat with how the duck feels."

In trumpeting the miracle of Eduardo's natural foie gras (not to mention our own foie gras experiment at Stone Barns), I had unintentionally aligned myself with Izzy's adversaries, which is why on this early Saturday morning he had come to find me.

"Why not go to France if you're so interested in foie gras?" he demanded as I stood to greet him. "You know? Why a Spaniard? You want to learn about great cars, you don't go to Turkey, do you? You go to Germany. No? Is this not true?"

"Hello, Izzy," I said.

"No, come on now, tell me why if you want to know about foie gras you don't go to France? I tell you I want to cry when I go to France. I go to France and I visit a man like Marcel Guachie's foie gras plant, anywhere like this, and I want to cry. I want to cry, but I have to catch my tears because God forbid my tears touch the floor. It is a shrine, these places. These people are gods. They know foie gras. They look at a baby duck and they'll tell you what kind of foie gras you're going to get. Everything Eduardo knows, they've forgotten—that's how much they know. Me? Next to these guys? I'm nothing. I'm a speck. Dirt. They are king. I'm a baby with my thumb in my mouth. I'm nothing!" Izzy put his thumb in his mouth and sucked loudly. "Nothing!" People passed by us and stared.

"Tell me, Dan Barber, what's wrong with raising a goose inside? Because of this, now I can make the foie gras of my dreams. That's what we do. We can make the foie gras of everybody's dreams by putting them in one location under one roof."

"Why is that better?" I asked.

"CONTROL!" he screamed. It sounded as if his throat were coated in broken glass. "I can wake up and walk over and see how they're doing. Eduardo makes a big lie, a big, big lie, I'm telling you. Why? I'll tell you why. Geese in the wild—that's why."

"But I saw it," I said.

Izzy ignored me. "Geese in the wild!" He threw his head back and laughed severely. "Do you want to know what? Geese die. They die here. They die there. They die of laughter. They die if they're sad. They die if you touch them." With his index finger he reached out and touched me on the shoulder. "Die," he said.

"So, okay, capisce, it works like this, this is what Mr. Eduardo says: he says they eat acorns, yes, right, lots of acorns, and they eat some yellow grass, and they go free—free like this, *la, la, la*, in the forest." Both of Izzy's arms were in the air, and he danced on his toes like a ballerina. "Like this, you know, in the wild, and the geese, they are HAPPY." Izzy smiled a broad, fake smile, hands still in the air. "And they laugh, they listen to Schubert, and then one day . . . foie gras!"

He called out to a farmer standing next to us. "Hey, guess what? Let me tell you what. Do you know you can turn a goose's liver yellow if you feed it yellow? Yes! You can! Feed it yellow something and poof—the liver is yellow." The farmer looked confused. "Yes, true." Izzy went on, "And did I tell you? One time I feed my chicken popcorn. And do you know what? When I go to cook this chicken in the pan, it flips, on its own. *Flip, flip.* I go to cook, and like this." He turned his hand back and forth in the air to imitate a chicken breast flipping over. "It flips. Like this. *Flip!*"

———

Between Izzy's accusations and the geese's persisting ennui, I was not feeling optimistic about the fate of our "field gras." In fact, I was suffering from my own ennui when Lisa called one afternoon to inquire about how the experiment was going. Why not fly Eduardo to Stone Barns, she suggested when I shared my bleak predictions. He could inspect the geese and also pay a visit to Izzy at Hudson Valley Foie Gras. (She was curious, as was I, to see Eduardo's reaction to a more conventional approach.)

I wrote Izzy and asked if he might be open to the idea. "Of course, come and see our small operation," he wrote. "Having Señor Eduardo here will be interesting." He followed up with a phone call, adding that he had one condition: Would the *señor* be kind enough to bring a sample liver for all of us to see? I asked him if he wouldn't want to taste them as well? "I want to see them. Sure, taste too, why not?"

Three weeks later, Eduardo left his farm at 3 a.m. and drove through the darkness to the Seville airport, where he boarded a flight to New York City, with a stopover in Lisbon to change planes. As he passed through customs, security guards demanded that he open his refrigerated bag. They took one look at the contents and, according to Eduardo, went into high alert, yelling at the bewildered farmer for not carrying proper documentation and threatening to arrest him. "They treated me like a terrorist," Eduardo said. "And it was just a couple of livers." They eventually let him on the flight, but not before making him throw the livers in the garbage.

Eduardo and Lisa arrived at Stone Barns early the next morning. I first spotted them in the courtyard, standing off to the side. Eduardo smiled politely as I explained the history of the barns. After a few minutes, Lisa turned to me. "I really think he just wants to see the geese."

We met up with Craig and Padraic and walked to one of the pastures. "This is it. You've got the land. This would be perfect," Eduardo said, his head tilted back, admiring the tall trees of the woodland surrounding the pastures. "Not only geese. You could have monkeys living in these trees!"

Behind me, I heard Craig say, "Don't give Dan any ideas."

Kneeling at the sight of the geese, Eduardo beckoned just as he had that first morning in Extremadura. *"Hola, bonita! Hola!"*

Despite his affection for the birds, his diagnosis for our project was far from optimistic. The fence surrounding the flock had to go. The geese were spoiled, he said simply. Craig and Padraic looked on, befuddled by the complaint.

"If you want to raise Rambo," he said, "he cannot be coddled and fussed over." He went on. We suffered, he said, from our anthropomorphic tendencies: humans (perhaps especially Americans) will gorge on unlimited calories, but geese won't. At least not if they're pampered. We fed the geese grain twice a day, expecting that when the weather turned cold they'd gorge on the corn in preparation for the long winter.

"But what's winter to a goose that's had food delivered to him for six months?" he asked. "Geese are too smart for this. Why gorge when they know the next meal is coming? They cannot be tamed. They have to feel wild to kick-start that instinct for gorging."

Eduardo pointed to the woods, insisting that the answer to fattening the geese lay in foraging the forest, not only the pasture. "You can have a great liver. Once the weather changes, they'll eat compulsively. Boom, boom," he demonstrated, pecking the air.

Craig explained that there were some acorns in the forest, but not like in the *dehesa*. Eduardo waved his hand. "The problem is not what you feed it. The problem is convincing it that it's wild. If you create the right environment, the fattiness will take care of itself."

———

We set off to see Izzy. Craig and Padraic and several Blue Hill chefs came along as well. I watched Eduardo in the backseat as he wiggled his legs in excitement. (I later learned he had never been to America, and rarely traveled anywhere outside of Spain, which explained some of his enthusiasm for the perfunctory drive.)

Passing a group of tall trees, Eduardo craned his neck, turning around in his seat to get another look at the thick woodlands. "Wow," he said to Lisa, reiterating what he saw as our ecological advantage. "If I had a farm here, I'd make really good foie gras."

Arriving in the small town of Ferndale, New York, we pulled into Hudson Valley Foie Gras and parked our cars across from a row of long, white barracks. Izzy came out to greet us, introducing himself to Eduardo.

"Welcome, Señor Sousa," he said warmly. "I am happy to have you." He put his hand on Eduardo's shoulder. "But please, not a word about geese around here. I don't want to hear even the word!" I watched as Eduardo listened intently. "Every time one comes around, something bad happens. Every time. I'm telling you—every time. I once bought twelve hundred geese. Then the barn collapsed. Twelve hundred geese, and I only got six livers." Eduardo smiled and laughed, agreeing by way of a thumbs-up sign.

We began the tour with Marcus Henley, the farm's operations manager, who led us into a narrow building swarming with young ducks. He pointed to a wire rise on one side of the room. "We keep their water on top of that, and their feed on the other side, so they get exercise," he said. "Just like on your farm, Eduardo." Lisa and I looked at Eduardo, who raised his eyebrows up and down like Groucho Marx.

By the time we arrived at the gavage room, we'd met plenty of happy workers (so far as I could tell), seen clean conditions (Marcus told us proudly that, should any indication of pathogens become present, "we can blow the whistle and immediately lock the whole place down"), and observed ducks that proved Izzy's conviction that he didn't torture his animals.

Even in the gavage room, where the controversial twenty-one days of force-feeding takes place, I found the ducks comfortable, in a sleepy, after-brunch, Sunday-afternoon kind of way. Eduardo looked on impassively as a woman inserted a tube deep into the throat of a bird and poured in the grain. She worked with military precision. The exercise lasted five seconds or so, and then she moved on to the next duck in line.

I had imagined this moment for weeks. Here was Eduardo, the representation of two centuries' worth of free-ranging fowl, face to face with the insult to history—the process many animal activists call torturous and inhumane. How would he react? With outrage? Or tears? I envisioned him tackling Marcus, throwing open the doors, and shepherding the ducks back into nature. Instead he only shrugged his shoulders.

"The feeding is not a problem." He shook his head at the idea that the ducks were experiencing pain. "The problem is that the ducks don't even know they are ducks." And that was that. No drawn-out philosophical observations or shrill diatribes. Just the facts. The ducks lacked self-awareness.

Marcus walked over. He must have misread the expression on our faces as disapproval. "We need to be very careful here about anthropomorphizing," he said, looking at me. "And we need to be very, very careful not to extend our own preferences."

I didn't bother correcting him. It occurred to me that in focusing on the cruelty of gavage, we make ducks and geese human, and their treatment becomes intolerable. But this more than misses the point. What's intolerable is the system of agriculture that it reflects.

———

On the drive home, I watched Eduardo press his wide smile against the passenger-side window. He whistled and pointed at the packed parking lot of a shopping mall, the wonder of the drive-through bank, the thick forest interrupted by the valley's iconic pastures. He kept turning around to make sure I hadn't missed it. *"Mira, mira, mira,"* he said—look, look, look—in exactly the same way he had pointed out a goose about to home in on an acorn.

I realized then that Eduardo's work has a lot to do with creating a consciousness—not only in his geese, but in us, too. To taste his foie gras is to kick-start a chain of understanding about the geese (their natural instincts), the ecology that supports them (the *dehesa*), and the centuries-old culture that supports the whole system (the Extremaduran way of life and its varied cuisine).

Our modern way of eating supports the opposite. It dumbs down nature. It makes a duck liver—or a loin of lamb, a chicken breast, or a cheeseburger—taste the same whether you're in Scarsdale or Scottsdale, in June or January. Which, in a way, dumbs us down, too. I saw that soon after we left Hudson Valley Foie Gras, as we drove past a fast-food drive-through. It was lunchtime; cars idled in the line to order, inching forward together like widgets moving down an assembly line. The people in the cars waited in silence, their heads facing forward. No one looked any more, or less, excited about their impending meal than the ducks who had just been lined up in front of me, or, for that matter, than the geese at Stone Barns I had brought grain to a few weeks earlier.

⸺

I had to leave for Blue Hill in the city late that afternoon. Eduardo told me he was desperate to see New York and our sister restaurant, and though he would love to spend more time with Craig, he asked if he and Lisa could accompany me on the ride to the city. He shook Craig's hand and wished him luck. "Remember," he said gently, "we're not raising geese. We're not their caretakers. We *have* geese. They raise themselves."

It was just before 4 p.m. when we arrived at Washington Place. We planned on a 5:30 p.m. dinner at the bar—allowing enough time for Lisa to show Eduardo around the West Village.

"Eduardo had his camera out, snapping pictures every few seconds of the 'real New York policemen' and the 'real New York basketball players' on the West Fourth Street courts," she recounted later. When they finally returned to Blue Hill, Eduardo was still buzzing with excitement. They sat for a drink at the bar, and Lisa asked if he was ready to start the meal.

"Lisa," he said to her gently, "right now all I want is a *real* American meal. I want a hamburger."

They left a note for me—"Off for a walk," it said—and crossed the street to the Waverly Diner. Eduardo ordered a hamburger, fries, and a Coke.

"He loved it," Lisa told me after they'd both returned to Spain. "He loved everything about it. He loved the bad lighting and the loud music. He loved ordering at the counter and waiting for our number to be called. He loved having to fill his own paper cup from the soda dispenser. He loved the roasted peanuts, still in the shell. But mostly he loved his burger. He licked his fingers as he ate it. 'The food Dan makes isn't bad,' he said. 'But this is really *delicious*.'"

It was the happiest she had ever seen him.

FAILED GRAS

Craig slaughtered the geese the following December. José soaked the livers quickly in milk and salt to remove any trace of blood and brought them to the meat station for the evening's service. The cooks gathered around, staring at the liver lineup as if they were artifacts from an archaeological dig. They were the size of Ping-Pong balls.

"Failed gras," one of the cooks said, breaking the silence. José shook his head in disappointment as a line cook patted him on the back, a gesture that said, *We'll get 'em next time, son.*

Craig has been raising geese for several years since then, and, not sharing my Ahab-like obsession with natural foie gras, he's been untroubled by the consistently small livers. He likes geese because they work well in his animal rotation on the pastures, and because they're profitable. Having lost hundreds of chickens and turkeys to coyotes, he never adopted Eduardo's advice to remove the electric fencing and activate their built-in ability to gorge in the fall. "It wouldn't make them eat more," he told me. "Around here, it will just get them killed."

Meanwhile, back in Extremadura, Eduardo's geese have faced tougher times over the years. Predation from wild animals has increased significantly, wiping out larger percentages of the flock. And more recently, global climate change has meant milder winters. Without a jolt of cold to jump-start their natural instincts, the geese are less inclined to gorge during those critical weeks.

"It's very strange, this laid-back attitude of theirs," Eduardo told me. "It's lazy. They sort of just sit around like—what's the expression? American couch potato? Except they don't really eat."

Nonetheless, the last I heard from Eduardo, he was looking to sell his foie gras in South America, possibly even the United States. He's also methodically downloading what he knows of natural livers to his two children.

"Every day," he told me. "But no more than fifteen minutes. I tell them stories, mostly, about the geese, about the *dehesa*, about their grandparents. More than that, they will resent me for burdening them. The key, I think, is to tell them something new, and to do it every day."

———

I have to admit that after five years of failed gras, I started to lose hope of ever carrying on Eduardo's tradition. But then one afternoon I heard that Chris O'Blenness, Craig's new assistant, was headed out to a farm in Kansas to investigate a heritage breed of goose called Toulouse. Less domesticated than the modern breed we had been raising, the geese were purportedly large and fiercely protective, their natural instincts more firmly intact. The first time I got a glimpse of them, in the back pastures at Stone Barns three months later, I thought there was a striking resemblance to Eduardo's geese.

Still, it wasn't until later that same year that my excitement returned in earnest. Chris approached me in September, just as the days turned cooler. What did I think of letting the geese out of their fences to roam the back pasture? Of course I thought it was a great idea. But Chris really wanted to know if I'd be willing to pay if the coyotes got to the flock.

"Pay in advance?" I asked him.

"Yes," he said. "Pay to play." I said I would.

The next week, with no fence for protection, the geese roamed the back pasture. I don't think I was reading into things when I observed a certain swagger I hadn't noticed before. If freedom wasn't making them hungry, it at least had kick-started a confidence gene.

Chris furtively spread some corn feed on the grass in random places around the field, to encourage their appetites. "They come on it suddenly," he told me after a few days of trialing the new method, "and all hell breaks loose. They really believe that they've discovered spoils from a hunt."

In early January, Chris picked three of the largest geese for the first slaughter of the year. They were enormous birds, to José's delight. He went to work. Foie gras? No, not really. Instead of Ping-Pong balls, we got small lightbulbs. They were strikingly red and looked distinctive, almost regal compared with previous years, with streaks of yellow fat clinging to the liver. It wasn't integrated, but the lobes appeared *fattier,* more in line with foie gras.

Izzy didn't agree. After I described the livers to him on the phone, he said, "Let me tell you: it is not foie gras. Call it a nice liver. Call it a delicious, lovely liver. Call it anything you want, just don't call it foie gras, because it's not."

Instead of our usual preparation—a quick sauté, which meant enough for only a few lucky diners—I decided we would stretch the livers by preparing them as pâtés. They were preserved in small glass jars and sealed, as hundreds of years of culinary tradition dictates, by melting some of the excess goose fat and pouring a thick layer of it over the pâté. A few days later, I dug my spoon deep into one of the jars and, without giving it any thought, spread the meat and the glistening fat onto a warm slice of bread.

Some of the most memorable moments as a chef—at once revelatory and revealing—are provoked by tasting something delicious. This was one of them. Though I want to say I was brought back to the tiny restaurant in Monesterio where I first tasted Eduardo's foie gras, I wasn't really. The liver didn't taste as sweet, nor did it benefit from Eduardo's natural seasoning in the field (I admit to adding some salt and pepper).

But the flavor was nonetheless deep and pronounced, and the fat itself was unctuous and not the slightest bit greasy. At the risk of anthropomorphizing, it tasted more assertive, more sure of itself, than previous seasons' livers. And more of its place, too.

Izzy was right after all. My blind pursuit of foie gras had ignored something fundamental in all of this: why had I insisted on calling it foie gras? I had, unwittingly but assuredly, set the geese up to disappoint us. It was me who had failed the geese, not the other way around. Demanding that the livers live up to the archetype, or to a standard based on the riches of the *dehesa*, was not only impossible, it was a fool's errand. The ham we make from Craig's pigs at Stone Barns is delicious. Is it *jamón ibérico*? No, it's not, but we don't call it *failed ham* because of it.

Looking at the pâté, I was reminded again of the brilliance of that well-marbled slice of ham Eduardo held up to the light. In a way, both represent what's possible when gifts from nature become filtered through culinary tradition. *Jamón*, as with good pâté—as with almost anything produced by good cooking—relies on simple craft, attained and applied. And yet sometimes, with any luck, it transcends craft. It becomes more than the sum of its parts.

In the same way that Eduardo stimulated the geese's consciousness, a recipe or a meal or even a single plate of food can stimulate our own consciousness—about the animals we eat, the system that supports their diet, and the kind of cuisine a chef needs to create to support it.

The Heart Is Not a Pump

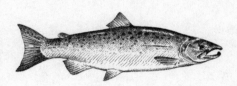

THERE IS A PHENOMENON in nature called the edge effect. The edge is where two distinct elements in nature meet and thrive. A continent's coastlines are a good example. The most productive and diverse habitats for marine life are where the vast sea finally meets the shore. These edge zones are hotbeds of energy and material exchange, thriving with life in a way that makes the deep sea or a large stretch of land seem dull by comparison. "Fragile ecosystems," as these areas are often called, is something of a misnomer. They are fragile because they're so full of life.

Think of the area at the perimeter of an open field, just before the thick forest begins. There's a flurry of extra activity and growth in that little bit of land. It's not beautiful growth; bushes, brambles, and big, rough ferns are fixtures on the perimeters, and they stand in somewhat unflattering contrast to a picturesque field of tall grass. But these corridors, unsightly as they may be, are nonetheless highly diverse and productive.

I first noticed the edge effect as a young boy at Blue Hill Farm, observing the fields from my perch on the Massey Ferguson tractor. Beginning to mow a new field meant a tortuously long first loop around the perimeter, which also meant more time to study the edges, the demarcation between what was to be mowed for hay—sweet, supple grass, standing tall and neat at attention—and what was essentially wild: a stretch of not quite forest, a no-man's-land of messy vines and wild berry plants. It was forest and field bumping into each other.

I've since learned from Klaas that this bicycle lane of semiwilderness, where both ecosystems interact and flourish, is recognized in ecology as an ecotone. But back then, as we circled the field and I stared down from my perch on the tractor, I imagined grass and forest engaged in a heroic turf battle for continuity and control.

There is a restaurant equivalent of an ecotone. Known as the expediter's table, or the pass, it's a narrow landing pad between the dining room and the kitchen, the meeting place of two very different ecosystems. The pass is the demarcation line between the calm of the dining experience and the relative chaos of the kitchen.

The orders for your food arrive here first, and these orders are organized and then communicated to the kitchen. As such, the pass is often filled with tension, and, in a sense, it's the site for that same kind of heroic battle for continuity and control I imagined as a kid—this time between the back of the house and the front of the house.

A GOURMET DISASTER

Several years ago, a group of writers and editors from *Gourmet* magazine booked a table for dinner at Blue Hill at Stone Barns. It is rare and exhilarating, and frightening, too, to cook for such a concentration of powerful food writers. I remember standing at the expediter's table, waiting for their ticket and thinking about how badly I wanted to impress *Gourmet*. What chef didn't? The editor, Ruth Reichl, was considered the high priestess of modern American cuisine, a woman whose critiques and observations helped mold a generation of chefs and food writers.

Blue Hill at Stone Barns, at that point, had not made much of an impression on Ruth. Though she had dined at Stone Barns soon after we opened— a little too soon, as we were unsure of ourselves, and the dishes lacked any kind of daring—we were never mentioned or reviewed in the magazine. I heard through a mutual acquaintance that Ruth had wished the meal had felt more farmlike—less in the mold of high-end city restaurants and more celebratory of what was harvested around us. Having met her in the kitchen after the meal, I didn't need to hear it secondhand: we had underwhelmed her. Here, two years later, through her surrogates, was a second chance.

Philippe Gouze, the general manager of Stone Barns since our opening, chose table 45, a freestanding table with an unobstructed view of the main vegetable field. And he assigned Bob, a wild card of a waiter—personal and engaging, but also erratic and, from time to time, inexplicably odd. One night, he alternately charmed and offended two tables right next to each other. One was a small group of older Westchester women. For that evening, Bob became a young southern gentleman with a drawl (he was from Teaneck, New Jersey) who had the women so enamored, they left a large tip and asked Philippe if he could be hired out for parties. At the next table, he enraged a couple of young Brooklyn hipsters—the table complained to Philippe that he wouldn't stop interrupting their meal with stories of his work as a conceptual artist in Williamsburg and asked to be left alone to finish their meal in peace. Bob obliged, and then, when he delivered the check, challenged one of them to a fistfight in the courtyard.

But we were so short-staffed the evening of the *Gourmet* editors' visit that Bob was the best choice. He knew the food and seemed to be gaining confidence, becoming less erratic, more even-keeled. Philippe promised me he would be at Bob's side the entire evening to make sure everything went smoothly. When Bob followed the ticket into the kitchen to meet me at the expediter's table, Philippe was behind him, an arm's length away, like a secret service detail.

"Okay, they're already thrilled," Bob said, bouncing up and down at the other side of the table. "Love the view, love that they don't have to choose from a menu. Blown away. They want it all."

Eating at Blue Hill at Stone Barns means giving up the freedom to choose your food. We don't offer menus, at least not the traditional kind. We abandoned them out of frustration, soon after that night when Craig Haney's grass-fed lamb chops sold out just as the dinner service had begun.

In place of an à la carte system, we started serving only multicourse menus, in the style of a Japanese *omakase* sushi bar: it's the chef's choice, built around the day's harvest, and you don't know what you'll be getting until it arrives at the table. If you think that sounds a little precious and high-handed, you are not alone. Many guests felt that way, and still do. It's one thing for a master sushi chef to pick the day's best catch. It's another to have the day's best vegetables and cuts of meat picked for you.

Or is it? The sushi chef is not just picking the freshest catch. He's in a conversation with you, often literally, analyzing what you'd most enjoy (and sizing up what he deems you're worthy of enjoying). If you've ever visited a sushi bar and sat near a group of Japanese aficionados, you've likely seen the yawning divide between your offerings and theirs.

At Blue Hill at Stone Barns, captains like Bob perform this kind of detective work with each table. They ask questions about restrictions and allergies, and then about preferences and aversions. A conversation begins, and with any luck the captain will get a sense of the table—whether they're adventurous and, if so, how adventurous. Are they Level I adventurous, which for the kitchen means they would enjoy something like braised lamb belly? Or are they Level III adventurous, which means they'll not only try anything—like lamb brains—but they expect it to be a part of their meal? Would they prefer a menu heavy on farm offerings, or are they looking for what are considered luxury items (foie gras, lobster, and caviar, which get written on their ticket in the kitchen as "FLC")? Given the choice, is it the kind of table that would prefer a just-dug carrot from the farm or a filet mignon?

There have been questions about the fairness of such a quixotic approach. A blogger once called it "gastronomic profiling"—making assumptions about what people want to eat based on a captain's drive-by impression—and there's merit to the accusation. But overall, we're more successful than we were under the old system. Freed of responsibility, diners are spared the anxiety of making a potentially bad menu selection. (Why did I have the salmon when I knew I should have ordered the lamb?) If a course falls short of their expectations, there's no guilt, no self-flagellation at having picked the wrong thing. I swear it makes our diners less critical.

More relaxing for the diner; mayhem for the kitchen. A party of four might have one Level III diner and one conservative diner. The other two might be allergic to shellfish, and one of the two might want no meat at all (but only one fish course, salmon preferred). The ticket arrives at the expediter's table—that edge between calm and chaos—and sometimes, in the course of trying to predict what the diner most wants, you write a menu with new dishes and spontaneous combinations of flavors that seem to flourish in the inspiration and tension of the moment. David Bouley, a chef who often prepares impromptu menus for special guests, calls this "kicking the ball around." He claims his most lasting dishes come out of the moments when his back is up against a wall.

Other times, in the pressure-filled seconds you have to decide a table's menu, you choose a dish that feels right in the moment but ends up proving disastrously wrong.

We hadn't yet abandoned the printed menu the night of *Gourmet*'s visit, but we were leaning that way—more gastronomic profiling, where the waiter would ask if the table wouldn't prefer putting aside their personal selections and "let the chef cook for you."

Which is why the *Gourmet* table's ticket arrived at the pass with six blank lines for me to fill in their six savory courses. Below this, Bob typed the notes. "Adventurous, Level III. Love vegetables. Love anything from the farm. No allergies. Want to be wowed."

It was the "Want to be wowed" flourish that reminded me to worry about

Bob. I looked up at him. "Bob," I said, "no bullshit, right? We're playing this straight."

"Hell, yes," he said, standing tall and squinting his eyes. "Straight like an arrow." He chopped his right hand through the air and disappeared into the dining room. I swore I'd heard him say "like an arrow" in a southern accent.

I began writing their succession of courses on the ticket. It was a muggy early July evening, so I started with a green gazpacho made from all the green farm vegetables. I listed the vegetables for Bob, who dutifully recorded them on his notepad. "Jade cucumbers, Zephyr zucchini, Malabar spinach . . ." I wanted to nail the theme from the first course, which was, back then, really the theme of Blue Hill at Stone Barns—resurrecting lost varieties and flavors and incorporating them into a new kind of modern cuisine.

"We're all about the farm," I said to Bob, my voice straining like a television evangelist. "But more than just a farm-to-table restaurant, we're committed to celebrating ignored varieties of vegetables, the ones that have been saved for generations, the kinds that were bred for flavor; the ones that have been all but lost to our big food chain." Bob scribbled furiously as I ladled the gazpacho into the bowls. Philippe, just behind him, raised an eyebrow but said nothing as Bob excitedly counted out loud the number of farm vegetables in the soup.

One of the longtime senior writers at the table, Caroline Bates, would report to Ruth on Monday morning—I was sure of it. The farm, and our message, would be infused into every course. And I would not be accused of underplaying my hand.

━━━

"Soup equals home run," Bob announced, proudly showing me the empty bowls. "The woman at the head of the table said she would just love it if the whole meal were made with farm stuff."

I prepared a salad next, with Regina dei Ghiacci, one of the oldest varieties of iceberg lettuce that Jack had been able to secure the seeds for. "This is what iceberg should taste like," I said, handing a leaf of the dark green lettuce to Bob. "It's been dumbed down to white, tasteless nothing— but this is what it used to be."

Bob smelled the lettuce as if it were a fine wine, and then chewed it thoughtfully. "A little bitter," he said.

"That's right—tell them that. Tell them this is the mother of iceberg, full of flavor, bitter and sweet, and the kind of variety we're really excited about reintroducing."

I sent Bob out to the table with the salads and looked down at the order, tapping my pencil against my forehead. I saw I needed four more courses to complete the meal. I had planned on serving an egg from the farm, followed by Craig's chicken and a taste of his Berkshire pork. I was still missing a fish course.

Our fish supplier had phoned the day before, excited by a particular piece of bluefin tuna belly he was holding—if I was interested. A highly migratory fish, bluefin run in local waters near Long Island just once a year. On a whim, I convinced myself it was worth the expense to show off this magnificent fish to our diners. It had arrived that morning.

Bob appeared with the clean salad plates and a boastful smile. "You're killing them, chef," he said. "Killing." We were busy now, backed up with orders from the dining room. Captains lined the expediter's pass to discuss their tables. I needed to decide on a fish course and move on, so I committed to the tuna belly.

The belly cut was from the midsection (the *toro*), one grade removed from the area closer to the head, the *o toro*, or great *toro*, which is the most expensive piece of fish in the world. You wouldn't be blamed for comparing it to the finest *jamón ibérico*: densely rich, with an equally stupefying penetration of sweet fat. Laying the belly across the cutting board, I sliced off a small piece and popped it in my mouth.

I've often described *toro* as buttery rich, a description I considered apt until I read Jeffrey Steingarten's much more evocative account of eating blue-fin belly: "At first it was like having a second tongue in my mouth, a cooler one, and then the taste asserted itself, rich and delicately meaty, not fishy at all. The texture is easier to describe—so meltingly tender as to be nearly insubstantial, moist and cool, not buttery or velvety as people sometimes say. Have you ever tasted a piece of velvet?"

I cut into a small, well-marbled section of the belly that had a deep, almost purple-red color. As I portioned it into long strips, the fat melted on my fingers. I seared the tuna quickly in a pan, plating it on a simple stew of spring onions and peas. The fish was the star, but I wanted the farm on the plate, too. I brought it to the expediting table. "Bob, this is local bluefin toro," I said.

"Hell, yes, it is!" he yelled, his face pink with excitement.

I looked Bob in the eye. "Local bluefin," I repeated, "with early summer vegetables from the farm."

Bob departed with the dishes. He returned several minutes later, sullen and confused, but feigning an air of nonchalance. "They're deep in conversation," he offered, and then abruptly left the kitchen. By the time he reappeared, with six tuna plates—the vegetables were picked at around the fish and mostly eaten, accentuating the untouched tuna—I knew what was wrong.

There are times when a plate of food is allowed to leave the kitchen that should never have even been plated. A terribly overcooked steak, a congealed sauce, a mess of wilted greens—these slips happen in the rush of service, and although (or maybe because) the offending plates are rarely returned, chefs sleep a little less soundly. Call this slip an error of judgment rather than an error in execution. Or just call it moronic. Even though I tried to blame Bob—had he said anything strange or offensive to make them lose their appetite? "Like what?" he asked me innocently—I knew I had been the lone architect of this little disaster. You don't wear the high ideals of sustainability on your sleeve—you don't gloat about saving old, forgotten seeds of lettuce—and then serve a plate of bluefin.

Because, like beluga sturgeon and Chinook salmon, bluefin tuna are going extinct.

———— ·> ————

In the mid-1990s, ocean conservationist Carl Safina wrote an epic ode to the
bluefin, following it throughout the world's oceans and documenting its demise. It was a landmark book, hailed as a call to action on par with Rachel
Carson's *Silent Spring*, the transformational exposé about the effects of pesticides on the environment.

I read *Song for the Blue Ocean* as a line cook at David Bouley's eponymous
restaurant in New York City. I didn't read it because I was a burgeoning environmentalist; I read it because I was working under Bouley. Along with a
loose tribe of other chefs like Jean-Louis Palladin, Gilbert Le Coze, and
Jean-Georges Vongerichten, Bouley is credited not just with reimagining
fish cookery, which had remained classic and tired even among nouvelle cuisine chefs, but, more important, actually reinventing how fresh fish was handled, from the moment it was caught to the time it was delivered to the
restaurant. Inspired by his obsession, I began reading cookbooks devoted to
seafood, and then books about fishing and the oceans. *Song for the Blue Ocean*
was the most memorable because it presented convincing evidence of the
wholesale destruction of an entire species, one that was largely avoidable.
The destruction had come about for many reasons, but none was more critical than the demand for tuna—especially the demand for *toro*—created by
chefs.

Not so long ago, the bluefin's story was one of unfathomable abundance.
That changed with the international trade in seafood, made possible by advances in refrigerated air cargo. Once the Japanese (with the support of their
booming 1980s economy) could reach across the world to satisfy their insatiable appetite for tuna, a fishing bonanza followed. The American sushi
craze brought even greater demand, abetted by advances in fishing and

distribution. Atlantic tuna populations dropped by up to 90 percent. Safina's evidence was overwhelming: if we continued plundering the ocean for bluefin tuna, there would be nothing left within a generation. I knew all of that—or at least enough to know better—and yet I had gone ahead and served it in my kitchen.

—

I rushed through the final courses as Bob relayed their waning enthusiasm ("I'm not going to lie to you, Chef: cat's got their tongue"), and when it was finally over I invited the table back to see the kitchen. The only one to accept was Caroline Bates. She had barely arrived at the expediting table when she said, "I'm shocked you serve bluefin."

I had anticipated the charge and planned on simply apologizing, explaining the phone call from my supplier, the craziness of the service, and my lapse in judgment. I planned on being overly contrite.

"It was *local* tuna," I blurted out instead. Odd as it was to justify serving a fish that was near extinction, I dug in. "Off Long Island," I said, with an air of *You know, Caroline, I'm not sure you're aware* . . .

She looked at me, puzzled. "What do you mean?"

"Bluefin are headed south down the Atlantic," I said. (Which was true.) "These large schools become available for local fishermen." (Which was mostly true.) "See, there are certain times of the year, when large schools are running, when it's acceptable to catch them." (Mostly not true.)

"Really?" She looked skeptical now. "I don't think that's true."

"Yeah, it's true. There's an ocean conservationist by the name of Carl Safina, who's a renowned expert on tuna. He told me," I explained. (Not true.)

"Carl?" she said, with a laugh. "I just spent a few days with Carl for a major story I'm working on about chefs and seafood sustainability. I really don't think he said that."

I like to think I remained admirably steadfast as Bates stood at the other side of the expediter's table and shook her head. What was I clinging to? I think it was that Safina had argued, quite convincingly, that if a fishery is well managed and regulated, one ought to be supporting the small-scale fishermen whose livelihoods depend on catching within sanctioned quotas. But he wasn't referring to tuna.

I stood silently, replaying the words "a major story I'm working on about chefs and seafood sustainability." There was the usual kitchen commotion around us, but with the added sting of the expediter yelling for the waiters to "pick up the tuna belly for table 41! Pick up the *tuna*!" followed by Bates's exasperation as the glistening pieces of bluefin—the majestic, hauntingly delicious, and nearly extinct fish—were whisked to the dining room in front of her.

—⁓—

I spent much of that summer marinating in fear of the *Gourmet* article. It was cold comfort to know of other New York chefs serving bluefin, or that nearly every good sushi restaurant in America goes through copious amounts of *toro*. They were not part of a nonprofit education center devoted to the future of sustainably produced food.

The article ran that fall under the title "Sea Change," and while Bates did not name me, she did mention bluefin as every chef's "guilty pleasure," admitting that she, too, used to indulge but had since changed her mind (just like me!) after reading Carl Safina's *Song for the Blue Ocean*.

I went back and reread parts of Safina's book after the article came out, stopping at a section very near the end that, nearly a decade earlier, I had underlined extensively (and forgotten about completely). Safina makes reference to Aldo Leopold, whose writing had helped me understand the *dehesa* and *tierra*, the Spanish word for "land" (meaning more than just the ground we stand on). Back when Leopold defined this philosophy as the "land ethic"—what he called land and we call the environment—it was a novel idea. Over the next half century, it became the very heart of conventional environmental thinking, and Safina called for extending it even further, to "below the high-tide line." Explaining this new "sea ethic," he writes, "Simply by offering the sea's creatures membership in our own extended family of life we can broaden ourselves without simplifying or patronizing them. With such a mental gesture—merely a new *self*-concept—we may complete the approach to living on Earth that began with the land ethic."

Safina ends *Song for the Blue Ocean* with a thought that I also underlined: "The promise: that any honest inquiry into the reality of nature also yields insights about ourselves and the dramatic context of the human spirit."

He was right. After some honest inquiry into the reality of nature, I *did* gain some insight about myself, forced on me as it were, and it wasn't flattering. Even though Safina's writing had been affecting and instructive to me as a young cook—introducing me to the power of the chef to create demand for certain foods and demonstrating how profoundly these demands shape ecologies—I had become, somehow, the very person Safina warned against: the consumer who considers the ocean "other."

Safina calls us "soft vessels of seawater"—and he's got a point: water makes up more than two-thirds of the human body, the same proportion it covers on the earth's surface. "We are," he writes, "wrapped around an ocean within."

In the landlocked streets of Greenwich Village and the wooded pastures of Stone Barns, I often treated the ocean and the plight of its inhabitants as separate, both from myself and from our work with sustainable agriculture. Caroline Bates had given me a free pass by not calling me out in her article. I promised myself I would make up for years of ignorance with some honest inquiry into the state of the oceans.

CHAPTER 15

IN 1883, the British biologist Thomas Henry Huxley, one of Darwin's earliest and most vociferous defenders, wrote that "nothing we do seriously affects the number of the fish," a striking statement so high-handed it seems nature was just itching to slap him down.

It's striking, too, because it actually wasn't that long ago. Which is the point, really: so much has happened to two-thirds of our earth in such a short period of time that only the greatest of exaggerators could have predicted the extent of the damage. The bluefin's decimation is just part of the overall decline of fisheries around the world.

There are three stories to tell about the decline.

The fishing-industry story goes like this: over the past sixty years, we started taking too many fish from the sea, more than quadrupling our annual haul, from nineteen million tons in 1950 to eighty-seven million tons in 2005. Simply stated, we are taking fish from the ocean faster than they can reproduce. The result is that the commercial fish catch has been declining by 500,000 tons per year since 1988. More than 85 percent of the world's fish stocks are now reported as fully exploited, overexploited, depleted, or recovering from depletion.

The origins of the most serious damage coincided with the industrialization of American agriculture at the end of World War II. Which is not to say that prior to World War II the oceans were universally respected and well managed; just as destructive farming practices have been a part of agriculture since ancient times, fishermen have been depleting certain populations of fish for ages. But it is to say that before World War II, conquering the seas for the sake of sating our appetites was largely impossible, or at least limited to specific species and geographic locations.

In the same way that munitions factories built during WWII were later converted for the large-scale production of pesticides and fertilizers, paving the way for monocultures—Michael Pollan's "original sin" of agriculture—wartime technologies also radically changed the nature of the oceans. The equivalent sin of modern ocean fishing might be the use of sonar detection equipment (developed to locate enemies at sea) in the hunting of fish. This technology helped easily identify and track large schools of fish, making catches instantly more plentiful and profitable. Safina has written that before these advances, fish catches were naturally constrained by certain stretches of the ocean being "too far" and "too deep." Along with radar came more powerful vessels, able to fish in what had been impossible-to-reach areas, and better refrigeration on board to preserve the day's catch.

With unhindered access, the fishing industry conceived a seemingly endless number of ways to slice and dice the ocean more efficiently. Some of the advances, as sometimes happens in agriculture, were reasonable within limits. But limits aren't something the fishing industry (or industrial agriculture) knows much about. Add technological know-how to a conquering mind-set, and reasonable ideas can easily be taken to unreasonable extremes.

Trawling, for example, a form of net fishing that in one form or another has been around since the 1300s, turned particularly lethal. Today bottom drawlers drag nets (some as large as football fields) that scrape, scour, flatten, and otherwise upset the diverse and vital community on the ocean floor. Much of what gets disturbed—and, in so many cases, destroyed—during routine bottom trawling decreases the ocean's regenerative capacity. It's a scorched-earth policy for the seafloor.

"Imagine what people would say if a band of hunters strung a mile of net between two immense all-terrain vehicles and dragged it at speed across the plains of Africa," Charles Clover writes in *The End of the Line*. "This fantastical assemblage . . . would scoop up everything in its way: predators such as lions and cheetahs, lumbering endangered herbivores such as rhinos and elephants, herds of impala and wildebeest, family groups of warthogs and wild dogs. . . . The effect of dragging a huge iron bar across the savannah is to break off every outcrop and uproot every tree, bush, and flowering plant, stirring columns of birds into the air. Left behind is a strangely bedraggled landscape resembling a harrowed field." We clear-cut the oceans as we once plowed up the prairies.

While bottom trawling destroys underwater habitats, it's led to something equally troubling: the enormous unintended catch these boats bring aboard and then discard. Estimates of "bycatch" vary, but the number could total between eighteen and forty million tons—one-quarter of all the fish caught at sea. Why throw away a quarter of your fish? Because no one will buy them. There's limited space and time to process fish on the boats (and, depending on government regulation, limited quotas, too); bringing aboard what can't be sold would mean leaving behind the fish that can. It's cheaper to throw the undesirable dead or dying back into the sea or process them into fishmeal.

Several years ago, reports of dolphins trapped and drowned in tuna nets raised international concern about bycatch. The alarming news, made all the more alarming by the universal appeal of the bycatch in question (dolphins are more photogenic than, say, sea cucumbers), led to improvements in dolphin release procedures and helped initiate other advances to reduce the carnage imposed by large-scale fishing. But the problem of bycatch is not going away—not until destructive fishing techniques are outlawed and not until a market is created for these underutilized species of fish.

Which brings us to the environmental story: Once upon a time, the conventional wisdom held by Huxley—and most everyone else—was that the ocean was so vast and so resilient, it was impossible to take too many fish. Not true, it turns out. It was also thought impossible to overburden the ocean with our waste. That was never true, either. The problems with the oceans, in other words, aren't just about what we take out. They're about what we put in, too.

The fertilizers and pesticides that feed our monocultures end up in the ocean. So do the chemicals used for maintaining places like golf courses— monocultures with eighteen holes—and backyards. Many of the toxic materials we use on land eventually leach into the sea.

This has led to a whole host of problems, including the appearance of more than four hundred dead zones worldwide, from the Chesapeake Bay to the Baltic Sea. These uncontrolled blooms of algae hungrily feed on the excess nitrogen and phosphorus in the water—extreme examples of the same trend Klaas saw close to home with the effects of nitrogen leaching on Seneca Lake. Their eventual death and decomposition deplete the water of oxygen and slowly choke aquatic life.

The most notorious of these is the eight-thousand-square-mile dead zone in the Gulf of Mexico. The location is no coincidence—positioned at the mouth of the Mississippi River, the Gulf corrals all of the chemical runoff from our nation's Grain Belt monocultures. Each summer, the deadly algae bloom expands to fill an area the size of New Jersey. Most fish and shrimp, sensing the change in oxygen levels, swim to safer waters, leaving the area virtually deserted and paralyzing the local fishing industry. Less mobile creatures—crabs and mussels, for instance—aren't so lucky. University of Louisiana marine ecologist Nancy Rabalais, who has been studying the area for almost thirty years, once described her experience scuba diving in the hypoxic zone. "You don't see any fish," she said, just "decomposing bodies lying in sediment."

The threat to marine life is about more than what we put into the ocean; it's about what we put into the sky, as well. This is another part of the environmental story, the climate change part, which, from a biological point of view, has only just begun.

One way to detect it is through the ocean's primordial lead actor: phytoplankton. Phytoplankton are effectively floating microscopic plants, as opposed to zooplankton, which are microscopic animals. Too small to see, they make their presence known in clusters; the greener the waters, the more phytoplankton are present. They're a real gift to marine life, the ocean's answer to soils rich with microorganisms. The tangy smell we associate with the sea, and the seafood flavor we covet, really come from sulfurous gases produced by phytoplankton. In fact, the entire marine food web begins with phytoplankton, upon which all other organisms in the ocean depend either directly or indirectly. Phytoplankton are the engine for all underwater life, sustaining the microscopic organisms (zooplankton), the supersize ones (whales), and everything in between (herbivorous fish eat the phytoplankton, carnivorous fish eat the herbivores, and so on up the food chain).

Since plant life requires certain conditions for growth, phytoplankton are also very good indicators of changes in the environment. Which is why the grim predictions for their future are so troubling. A recent study suggests that global warming has precipitated a 40 percent decline in phytoplankton since 1950, the implications of which are staggering. Scientists are predicting that this will prove to be the single biggest change to the earth in modern times.

Evidence shows that when phytoplankton populations plummet, as they do during El Niño climate cycles, seabirds and marine mammals starve and die in huge numbers. But phytoplankton's role extends well beyond the ocean. The sulfurous gases they release are not just responsible for the sea's signature aroma; they also help with the formation of clouds, which means less solar radiation. And much of the oxygen we breathe is made by plankton. Like trees and grass, phytoplankton use photosynthesis to capture the sun's energy and convert it into chemical energy, releasing oxygen in the process. In fact, plankton are responsible for 50 percent of oxygen production on earth. And they play a crucial role in the carbon cycle, which determines how much carbon dioxide—the most dangerous greenhouse gas—ends up in the atmosphere.

Scientists often refer to the ocean as a giant sink, one that takes carbon dioxide from the atmosphere and deposits the carbon in long-term storage that can remain undisturbed for thousands of years. The services of the ocean sink are performed largely by phytoplankton. A 40 percent decline in phytoplankton makes the destruction of the rainforests seem almost insignificant by comparison. If there is less phytoplankton to store carbon, there's less of a buffer against global warming. More than that, it increases the inevitability of a disastrous positive feedback cycle in which warmer water supports fewer phytoplankton, which then take up less CO_2 from the atmosphere, which causes the surface water to warm even more due to the greenhouse effect.* And on it goes.

⚊

Chefs have a story in here, too. That's because a large part (albeit only a part) of the increase in fish consumption in this country since the 1970s is the result of chefs. One-third of the seafood eaten in the United States is ordered in restaurants—two-thirds if you measure in total expenditures rather than volume. Chefs control what gets taken from the sea.

And we take the wrong fish. We demand those large fish—the salmon, halibut, and cod—at the highest trophic levels. Trophic levels are a way to measure the position of the fish in the food chain. The larger and more carnivorous, the higher the level, and the more exhaustive they are to the environment. They're like the Americans of the sea—steak eaters and

---

* Rising CO_2 levels pose other, more direct dangers to certain marine life. Once dissolved in the ocean, carbon dioxide forms carbonic acid, which, accumulated over time, lowers the pH of the ocean. Today this acidification is happening at an unprecedented rate. For corals, crustaceans, and mollusks—any organism that relies on calcium to form its skeletal structure—it means the water will become increasingly corrosive and untenable.

three-car-garage owners. And we can't seem to get enough of them. The Northwest Fisheries Science Center recently published a study of cookbooks over the past 125 years, finding a gradual rise in the trophic levels of the fish used in recipes.

You might say chefs help create their own positive feedback loop: by cooking with these fish, we advertise their virtues and make them more popular, which increases demand and drives up prices. Higher prices make these fish seem more worthy to diners. Demand increases further, which puts pressure on chefs to offer more of the same on their menus. It's little wonder that many of these species—salmon, halibut, swordfish, cod, grouper, skate, flounder, and of course tuna—have declined by 90 percent in just the past few decades.

"Does this matter?" asks Clover in *The End of the Line*, referring to the chef's influence on depleted stocks of fish.

> Yes, I think it does, because the attitudes of the great are the stuff of fashion. Why should the leaders of chemical businesses be held responsible for polluting the marine environment with a few grams of effluent, which is sublethal to marine species, while celebrity chefs are turning out dead, endangered fish at several dozen tables a night without enduring a syllable of criticism?

The irony here is that there's nothing lowly about a sardine (except for its trophic level and price). In fact, it is among the most delicious fish in the sea. Just as chefs will tell you a chicken gizzard has more flavor than the breast, a very fresh sardine is more flavorful than a slab of tuna.

So why do sardines sell about as well as chicken gizzards on most menus? Blame the long-standing conventions of the American dinner plate. A thick tuna loin seared on a grill—a tuna "steak"—is coveted, with all the comforting connotations of beef.

But blame chefs, too, because after all, we are the supposed arbiters of taste. Identifying the best ingredients is an essential part of the cooking craft. Of all the complicated problems confronting the oceans, overfishing, especially of the higher-trophic fish, is not only the most immediate threat but also perhaps the easiest one to remedy. Inasmuch as the industry story of the depleted oceans is about greed—or the lack of what Carl Safina called the "sea ethic"—the chef's story of the sea is about a lack of imagination.

CHAPTER 16

I'M OFTEN ASKED about my favorite meal, one meal that stands above all the others, and I struggle to answer, because I find myself editing my reply based on who's asking the question. How do you rank my aunt's double-boiled scrambled eggs when I was sick with strep—the barely coagulated eggs sliding down my throat and relieving my pain—against, say, a parade of courses at Alain Ducasse's Le Louis XV, in Monaco, after a grueling yearlong internship at a Paris restaurant—one dish after the other so delicious I ended up weeping alone at the table? And where to place Eduardo Sousa's natural foie gras, a liver that changed the way I thought about food— not despite but *because* everyone in the world believed it impossible to produce?

The answer is that you cannot compare the greatest of meals. There is no one great meal, no perfect standard against which you measure every other meal. But you can think of certain meals as game changers, meals so profoundly original and singular that they forever transform how you think about food. Here was one such meal.

On our way to a second visit with Eduardo, Lisa and I made a pilgrimage of sorts to Aponiente, a tiny thirty-seat restaurant in the town of El Puerto de Santa María, at the far southwestern tip of the Iberian Peninsula, not far from the Strait of Gibraltar. The restaurant belongs to Ángel León, a man frequently referred to as the "Chef of the Sea"—an epithet he both promotes and laughs off.

I had met Ángel briefly once before, at a talk he gave in New York to a slack-jawed group of cynical city chefs. He was renowned only in the culinary world, the kind of chef who fearlessly broke boundaries—not with wild juxtapositions of different foods or with chemical manipulations, but by looking to nature, and the sea in particular, to define his cuisine. The results were astonishing. Instead of butter, for example, he used a puree of fish eyeballs (detritus for most chefs) to thicken fish sauces, giving his dishes an added boost of ocean flavor. And then there was his preparation for "stone soup," made with algae and weeds from stones he'd plucked from the ocean floor.

At the talk that day, Ángel showed off his latest invention, a mixture of dried algae that looked like sand. He used it to clarify a broth without the addition of heat and—even more amazing—without removing any of the flavor. Over the course of a two-hour demonstration, he rambled a bit about the plight of the oceans and gave what seemed like a loopy defense of a certain kind of tuna fishing, but when you are in the presence of Ángel's extraordinary creations, you quickly adjust to the fervent and poetic way he experiences the world around him.

Having never actually eaten his food, I arrived at Aponiente with enormous expectations. I came not only because of my own curiosity about Ángel and his work, but also because of Caroline Bates, of *Gourmet*. What better way to start righting the wrong of that bluefin than to learn about cooking fish from a chef who actually understood the ecology of the oceans? Ángel surprised me by sitting at the table and joining us for lunch.

I should pause here and say: this never happens. Chefs don't eat in their dining rooms, just as conductors don't sit with their audiences during concerts. I was flattered that Ángel would join me, until I discovered there were no other reservations for lunch. Not one other table. El Puerto de Santa María, like many southern coastal towns, is a haven for summer vacationers.

The population swells tenfold in July and August, which is when Aponiente does 80 percent of its business. The rest of the year—especially in March, when I was there—the restaurant loses money.

"I learned everything I know about fish from my father," Ángel said, toying with his fork as we waited for our server to arrive. (As he does not speak English, Lisa became the bridge once again.)

"We went out fishing one day, just the two of us, and we found this great place to fish for grouper—it was so great, and we caught a ton of grouper, like five fish that same day. It was a huge haul." He took a long, slow drag of his cigarette, drawing out the sweetness of the memory. "But it wasn't an accident, this place we found. Whenever we caught a fish, the first thing my dad made me do was open up the belly. If, inside, we saw razor clams and certain kinds of shellfish, we would take the boat to the area we knew had a lot of razor clams and shellfish and we would start fishing there. I was just a kid, eight or maybe nine, so I didn't pay any attention. I just thought, *Well, this is how you know where to go and fish.* But it was really investigative work—you know, Quincy-like—and I learned to love it, to know my fish. It informed me as a chef so much because it has all these culinary implications. You're paying attention to what you eat eats." He paused to extinguish his cigarette and waved his hand in the air, signaling for the food to start coming out of the kitchen.

(A few years after this meal, I met Ángel's younger brother, Carlos. He described their relationship as exceptionally close, except for when they fished with their father. "We would fish together, the three of us, and Ángel wouldn't let me touch any fish when it came out of the water," Carlos said. "Nothing. He'd scream: 'Don't touch!' He didn't want me to tamper with the evidence, you know? He was afraid even my fingerprints would make the fish less pure or something. And then he'd go in the corner of the boat and dissect the fish, so slowly and carefully. He'd carry it over to me and say, 'You see that?' And I was like, 'See what?' 'The stress,' he'd say. But to me it just looked like a dead fish. He started to see things I couldn't see.")

Ángel sat up and continued. "And so the next day we got up at five o'clock in the morning to go back to fish in the same spot. My dad was so excited, I remember—only, when we got there, a dragnet had been set up and the fish had been caught up in it. These people were taking all the fish! So my father got super pissed off. You didn't want to be around my dad when he was pissed like that—completely . . ." Ángel raised his eyebrows and whistled at the memory. "So he took out his big knife, and he just started cutting the nets. *Chhh, chhh, chhh.*" Ángel cut through the air with his fist, making vertical slashes. "The people that had done it were in a boat with binoculars and could see him. They were in a pretty good-sized boat with a large motor in the back. And they started chasing us. My dad was calm, but he had us rowing hard until we got close to land. When my mom found out, she was furious and wouldn't let my father bring me fishing for a while. She was sure he would get me killed. My dad only said, 'They were taking too much,' and that, at least, I understood."

Ángel calls himself a sea-ologist. He is in his thirties and stocky, with a thick neck and dark, deep-set eyes. His manner suggests he's not easy to get to know, and that if you try, he may at any minute blow up at you.

After a glass of fino, warm bread was served. It was dark green and smelled overwhelmingly of the sea. "Plankton bread," said the server, but he didn't have to. I had heard about Ángel's signature bread, with its homemade brew of phytoplankton, which Ángel had a laboratory grow for him. "You mix the yeast with the plankton," he said, "and it gives you a 70 percent better rise in the dough."

He told me that when he opened Aponiente, he initially didn't want to serve bread. ("What the fuck for?") He said it had no meaning for him in the context of what he was trying to express. But when customers started demanding it, he relented. "I said, 'Okay, fine, bread—but you'll have to taste it with the sea, because the first thing I want you to taste here is the expression of the sea.'"

And the second thing, too. I was served a single clam—an unnamed variety that Ángel explained was quite unpopular because it supposedly had very little flavor—poached so lightly in its own juices it appeared raw, sitting in phytoplankton sauce. I mopped up the puddle of phytoplankton with the warm phytoplankton bread and breathed in through my nose, smelling the sea.

"It's a very elemental plate—humble, but at the same time it's the greatest thing that you could eat. You've got the sea, and the primary ingredient of the sea, the origin of life," he said. And then he paused. "The beginning of every meal should start with the origin of life, don't you think? I feel very lucky to be creating cuisine with the origin of life."

I asked him how he came to make the origin of life in a laboratory. "To me it was always like a *Star Wars* adventure," he said. "I had this infatuation with being able to use the primordial form in my cooking. But nobody else really wanted to figure it out with me, so I stopped talking about it. I didn't stop dreaming about it."

Eventually he drummed up the nerve to approach the University of Cádiz, where he learned about a kind of netting they use to harvest phytoplankton in the Strait of Gibraltar in order to test for pollution. "I thought, *Okay, not so hard*. I will finally have plankton for my kitchen." Ángel took a boat and dragged the net for four hours. In the end, he collected less than two grams, enough for a small loaf of his bread. He decided to test it and he learned that nearly the entire periodic table was present in those two grams.

"You know what? That confirmed for me that I was right to be so excited about it. The itch wasn't going to go away. Lucky for me, I live near the sea, surrounded by people who relate to the sea." Working with a group of scientists, Ángel and the team created a kind of marine garden using special lighting and heavily oxygenated water to grow uncontaminated, superconcentrated plankton. Ángel could now harvest twenty kilos every five months.

As our plates were cleared, I was amazed at how long the smell of phytoplankton lingered. Ángel nodded, pleased. "The taste, the aroma, it stays with you. That's what I want. I want it to stay with you for the whole meal. It's almost like I can take fish out of the equation—I can go straight to the source. I can say to my diners: if there's no phytoplankton, there's no life."

———

The next course, horse mackerel with wasabi sesame seeds and lemon caviar, was rolled in nori like maki sushi. The mackerel had been deboned and pressed into a medallion shape, the seaweed resembling the skin of the fish. Ángel told me the fish had been sold to him that morning just off the boat, but that the lot he bought was badly bruised. For a chef so obsessed with, and knowledgeable about, great seafood, I was surprised to learn he went ahead and purchased the damaged goods.

"Of course I did," he said. "The fisherman came to me because he knew I wanted it. And why not? Fish are bruised all the time, just like us, but they're no worse off. You know who's worse off? Fishermen, that's who."

By pressing the mackerel together and presenting it like a sushi roll, Ángel hoped to create something exalted out of what might be thrown away. He explained that he had every intention of creating a market for what the fishermen would otherwise treat as a loss.

When I asked Ángel, gently, if he saw a contradiction in buying damaged goods when his restaurant charges so much money for a meal, he was quick with a response. "Isn't this what it means to be a chef? To use what is merely half-usable and make it delicious?"

The next course was tonaso, a fish I had never heard of, and one that in Ángel's parlance was less than half-usable. It was trash fish, literally, a frequent bycatch of shark fishing. Ángel explained that it's usually either ground up into fishmeal or, in the case of the fishermen he deals with, thrown back into the sea. But because he has cultivated a relationship with these fishermen, they humor Ángel by keeping a handful of the fish that are in better condition and delivering them to his door.

Steamed and thinly sliced, the fish had the consistency of tofu, almost custard-like. Ángel finished the dish with fermented black garlic and an intensely rich reduction of infused shrimp shells and fish bones. There was no butter and no oil in the sauce, and I guessed that Ángel had thickened it with his famous eyeball-puree technique, but he shook his head and said its sheen and powerful aroma came from a long, slow reduction. It was the kind of sauce that reaches out and pulls you in long before you taste it, not unlike the sauce Palladin had prepared in Los Angeles when I was a young line cook. The tonaso itself was rather plain, but, again like tofu, it was a perfect vector for tasting the sauce. In the same way that Palladin used sauces to elevate discarded and underappreciated parts of an animal, Ángel's mastery of the craft helped highlight an underappreciated fish.

Ángel discovered tonaso, and many of the other fish on his menu that I didn't recognize, by getting onto the larger boats that supplanted the small-scale fishing vessels he grew up knowing so well. These weren't the largest of the industrial fleet boats, the bottom trawlers that wreak havoc on the ocean's floor. Their methods were slightly more humane. Nevertheless, in them Ángel saw a level of destruction and wastefulness that haunts him to this day.

"I remember how excited I was to finally get on my first boat," he told me as the empty plates of tonaso were cleared. "I had tried for months, but these fishermen don't want to fish with strangers. I kept asking and pleading, and then finally I was allowed on. And what I saw from the second day horrified me. Can you guess out of every tonne of fish caught how much is kept on board?" He didn't wait for me to guess. "Six hundred kilos! The rest—either dead or damaged—are dumped right back in." He lit another cigarette. "I decided, at the moment, that whatever I ended up doing with my life, even if it was on the smallest scale, it would have to include being the caretaker for the other four hundred kilos."

He convinced the fishing captain to hire him for a stint as the boat's chef. One day he asked if he might cook with some of the catch they were going to discard overboard. "They were really impressed with how delicious it all was, but not that surprised—they'd been saying the same thing, only no one would buy it." Ángel was hired as a chef on different boats over several years, learning more about the industry.

It's tempting to look at Ángel's motivations through the lens of that young boy who witnessed his father risking both their lives to free fish from human greed; instead of direct confrontation, Ángel infiltrates the boats and changes the crewmembers' attitudes from within. But that's a thin read of his philosophy. Born and raised in El Puerto de Santa María, Ángel is the fishermen's greatest defender.

"I'm much more pro-fisherman than I am fish," he told me. When I asked if he had a guiding principle for his work, he waved his hand across the table. "It comes from a feeling," he said, and then, as if to correct a drift toward self-involvement, he added, "It comes from watching fishermen cry as they throw fish back into the sea because they know they can't sell it."

Ángel has become something of a hero to the fishermen. After Aponiente first opened, word spread about a townie turned world-famous chef willing to pay for damaged and unwanted fish, for fish so obscure they were nameless. That was when Ángel set about convincing diners to pay for these fish, which has proven to be more difficult.

"Because of the way we're socialized about fish eating, we want to eat glamorous fish—the ones with names," he said. "They might not even be the real names of the fish—they're names that were created for marketing. Oceans cover 70 percent of the planet, and yet we eat like there are only about twenty kinds of fish out there. I want to change that."

I headed to the bathroom before the next course and came across a Photoshopped picture of Ángel, his hulking frame emerging, merman-like, from the body of a squid. Ángel was smiling joyfully. Just opposite the photo, to the left of the entrance of the kitchen, was a small silver plaque. It read, WHEN THERE'S A CAPTAIN, THE SAILORS ARE NOT IN CHARGE. You could interpret this as Ángel's credo on the hierarchy of the kitchen. The chef is in control; the lowly line cooks are there to obey. But perhaps Ángel wants it applied to the diners as well. And why not? He placed it at the entrance to the kitchen, for everyone to read.

It occurred to me that Ángel saw his diners as sailors, along for a ride. Eating at Aponiente means a kind of surrender to the sea, the way I'd hoped that, in doing away with menus, we had made eating at Blue Hill at Stone Barns feel like a kind of surrender to what a landscape can provide. Just as nouvelle cuisine chefs broke apart the traditional recipes and conventions of fine dining—the "it is prohibited to prohibit" school of cooking—Ángel was doing the same through the ocean's vast offerings. As a sailor on his boat, you are prohibited from expecting one of the twenty fish you associate with the sea.

The waiter announced the next course as "a little nose-to-tail eating." It was three small prawns, floating in a bisque of shellfish, with the shells of the prawns lightly smoked and then fried. A phytoplankton cracker topped with a spoonful of aioli accented the side of the bowl.

This was a calculated, ironic dish—"nose to tail" referring to eating all the parts of an animal, a buzz phrase for sustainable cooking—and I smiled at the oddity of it. The prawns, normally hulking in size, were small and, to be frank about it, unattractive, as were the fried shells (what I surmised were the "tail" equivalent). It looked more like an attempt at a political statement than a tasty dish.

But then Ángel started talking. The prawns were in fact bycatch, decapitated by the netting process. Ángel said he refused to cook with head-on shrimp, the way it's traditionally served in Spain, the head supposedly being one of the best indicators of quality. His own research with the laboratory at the University of Cádiz confirmed what he'd suspected. More than 80 percent of the shellfish in Spain, he learned, comes with boric acid on it. The chemical preserves the bright red color of the shrimp, providing the illusion of freshness.

"Who needs the head if it's marinated in boric acid?" he said to me. Instead, for a fraction of the price, he buys decapitated shrimp, which would otherwise, like the damaged mackerel, be ground for fishmeal.

"Most of the ingredients I cook with are too ugly to show in their entirety. I've stopped caring about that. I still want them to be pretty and look nice, but it's not important to me that a dish be more beautiful than it tastes."

We both ate the prawns. "Delicious, no?" he said. And they were. "Every nine days or so, I get a delivery of decapitated shrimp and a mixed bag of other shellfish from one particular boat that's been supplying me since I opened. A lot of it we can't use, it's really beat-up, but those go into the bisque."

"Bisque" doesn't do justice to the sophisticated soup. Actually, I don't know that "soup" does, either. "Velouté" comes to mind, because it was thick and richly satisfying in a way that reminds me of what the best veloutés are all about. It was redolent of Ángel's skills as a chef, coaxing flavors out of seafood that I didn't think possible to express, and creating something smooth and unctuous without butter or cream.

Again I asked Ángel if the broth was thickened with his famous fish eyeball puree. "You know, that idea also came from the laboratory. I learned fish eyes are about 67 percent protein. So, yes, I used it as a thickener," he said, in a tone that suggested *If you knew what I knew, wouldn't you do the same thing?* "It got a lot of media attention, which was nice, but I stopped doing it, because in order for it to work, you really had to use the eyes within the first few hours after the fish were caught. When the delivery arrived, all the cooks were racing for the eyeballs. It was ridiculous," he said, and added, "Look, in the end I'm a pragmatist." His bisque is thickened now with the protein from the bycatch, which he said is just as delicious.

Ángel told me he used to dream of dishes like that soup. "I was always given to fantasy—I had a very strong fantasy life as a kid. I was in my head so much of the time, imagining things, then getting into a lot of trouble. I caused my parents a lot of hardship. I didn't do well in school, for example, because I hated sitting still—actually, I just really *couldn't* sit still. My father would tie me to the dining table chair in exasperation and say, 'Okay, fine, you're not going to study, but you're going to sit still.'"

Then one day, when he was ten years old, he set up a *chiringuito*, a beachside clam shack, outside his home. He cooked with different fish and shellfish, selling them to anyone who passed by. Ángel's version of a lemonade stand.

"That's how the world of cooking saved me," he said. "It gave me a place where I could put that imagination. For a little while, anyway."

It wasn't until he was twenty years old that he was diagnosed with attention-deficit hyperactivity disorder (ADHD). "Things got a lot better after that. My parents, they're both pathologists, which means they needed empirical evidence to prove I wasn't just an asshole. And it worked. They finally understood what it was with me, and I felt liberated to be myself, because I wasn't acting. This is who I was."

—◦—

The last course, arresting in its simplicity, was sea bream: two small fillets, with skin so perfectly grill-marked it looked as if it had been branded, and a small pool of phytoplankton cream. The waiter said the bream had been "perfumed with olive oil."

Sea bream is a popular fish in Europe, sought after for its mild, white meat. It is one of the few species not threatened by overfishing, and when I suggested that this might be the reason it was on the menu, Ángel shrugged. "Bream remind me of my dad," he said. "We used to fish for them. They're a smart fish, very sensitive. So when we fished for bream, my father had all these rules, like you couldn't talk on the boat—you couldn't make a sound, actually—and when you cast your line, it had to go out 150 meters, not a meter less. He was so particular about it because he knew the bream were just too smart."

I asked him about the perfumed bream, which tasted like it had once swum in an ocean of olive oil. He explained that the bream were grilled over olive-pit charcoal. For Spaniards, olives are just behind ham as one of the most basic products of the culture. "And do you know what that means?" he asked, again without allowing me to hazard a guess. "It means a table full of pits."

Ángel carbonizes olive pits like you might carbonize wood to make charcoal. He said he's perfected the charcoal to the point that it burns much hotter than wood—he usually brings it to 750 to 1,000 degrees Celsius, using a hair dryer to activate the heat. The fish cooks quickly. "You want crackling skin, but you also want the fish to gently confit," he said, which is why for the bream he only brings the temperature to 75 degrees. "It's a tricky balance, infusing the flavor of the delicate oil and the flavor of the smoke."

Ángel sat back as the plates were cleared and lit a final cigarette. I asked him if eating sea bream reminded him of fishing with his dad. "Everything I'm doing lately is reminding me of my dad," he said. "My very first memory of my dad is a heroic one. My father is deaf in one ear, and so as a result of that he never got seasick. We would go out in horrible seas, and my father would be fine. I remember one time in particular, a really rough trip out at sea, when everyone on the boat—there must have been ten of us—we were all throwing up all over the place. Nonstop vomiting. I looked over at my dad, and he was slowly sipping his beer and smiling at the sea. This happened several times when I was growing up. As a boy I didn't understand why that was. I didn't know that liquid in eardrums controls the sense of balance. All I thought was, *This guy is Superman*."

Ángel sat up suddenly, checking his watch. "Anyway, my point is that my dad got me thinking anything is possible."

It was nearly 5 p.m. when I said goodbye to Ángel outside his kitchen. He was anxious to get started on the evening's menu, and I didn't want to delay him. We promised to meet again soon.

As I waited for my bag, I walked back to the picture of Ángel as a squid. There was something revealing about the image; I thought about it long after I returned to Stone Barns. Visit a seafood restaurant anywhere in America, whether there's white linen or checkered plastic covering the tables, and undoubtedly you'll find a picture of the chef or owner posing alongside a prized catch—a swordfish, or an impossibly large striped bass. These triumphant photos smack of hubris, harking back to the days when the seas were full and the taking was good.

But Ángel's portrait—if a little cheesy—is quite a bit different, and the effect is subtle but significant. He's not lording over the giant squid (or any of the hundreds, if not thousands, of fish he's caught in his lifetime); he's emerging from its core, literally at one with the squid. The image is neither conquering nor celebratory. It's humble in the same way Ángel's cuisine is infused with humility, bringing to mind Klaas's unassuming air, or Eduardo's quiet respect for the *dehesa*. You look at the photo and see great reverence mixed with high-octane enthusiasm. The picture, like his food, tells you he speaks for the fish.

CHAPTER 17

Not long after my transformative meal at Aponiente, I was back at Blue Hill at Stone Barns inspecting a fish delivery from our main purveyor. One of the drivers, Howard, always backs the van close to the receiving entrance and, as is my habit, I checked to see the fish awaiting delivery to other restaurants. We were the last stop that day; only a large tub of cod heads and bones remained, piled in the corner.

"Headed to Chinatown," he said with a laugh, anticipating my question. I asked if he planned on selling them. "Sell the heads? Hell, no, I throw them away myself."

I took them all, in an impulsive nod to Ángel. We've been serving steamed cod head ever since, one to share for a table of two. Some diners interpret it as a political statement, a way of saying there's no more cod left in the sea. Some see it as a celebration of the fish. Still others are offended by the offering, if for no other reason than because of the cost of the meal. "An insult" was a comment left by one diner. Having now served a lot of fish heads—because they're delicious, full of collagen-rich meat that's nearly impossible to overcook—I'm reminded, with almost every delivery, that the real insult is throwing this large and delicious part of the animal away.

Americans consume seafood much as we consume meat. We'll order bass or salmon or cod in a restaurant, just as we might order chicken or pork, and what we expect to eat (and are invariably served) is a seven-ounce chunk of the fillet. We eat high on the *cod*, in other words, as guilty of wasteful and blind consumption as eating high on the hog.

Both luxuries are sustained only by illusion. Industrial efficiency, fueled by grain subsidized with our tax dollars, has ensured a seemingly inexhaustible amount of meat in the United States, allowing us to conveniently discard the lesser cuts (or process them into chicken fingers and dog food). Likewise, our profligate seafood-eating ways are fueled by a seemingly endless supply, and that's because fish are now farmed like any other protein.

—◦—

The business of fish farming is nothing new. In China, it's been around since the fifth century B.C. But in the past fifty years, the industry has experienced an unprecedented boom. The 8.8% percent annual rate of growth, a clip the industry has averaged since the 1980s, is just enough to keep up with the world's expanding appetite. (Production will somehow have to double by the middle of the century to meet demand.) More and more, experts believe it's the future of fish eating. In fact, recent estimates suggest that by 2018, the seafood on our plates will more likely be farmed than fished.

There are plenty of arguments against aquaculture. For one thing, most aquaculture operations are located near the shore. Keeping your fish in calmer and more accessible waters makes good economic sense, but in most cases it does not make good ecological sense. Since the shoreline comprises a complex web of life—a vibrant edge effect—you're inserting the equivalent of a large monoculture into a delicate cradle of diversity. With it come all of monoculture's attendant needs, like antibiotics to stave off disease—most fish farms, like animal feedlots, require a near constant supply—and all of its destructive capacity, too, like fouling coastlines with concentrated waste.

Add to this the most compelling argument against aquaculture: it's inefficient. To get a fish to market weight quickly enough to clear a profit, you usually have to feed it somewhere from two to five times its weight in wild fish. This *feed conversion ratio* (also known as FCR—the ratio of feed consumed versus pound of weight gained) means that in order to farm fish, you have to deplete wild fish stocks. Which means you're still banking on the

ocean's productive capacity—borrowing from Peter to pay Paul. More recently, fish farms have begun substituting grains and oilseeds in their feed
as fishmeal has become more expensive. But the rising cost of grain and competition from livestock producers (to say nothing of our flawed system of
agriculture) suggests that these alternatives are no more sustainable in the
long run.

Though chefs serve farm-raised fish all the time—they're cheaper, they're
abundant, and they're consistent—I didn't know of any well-respected chefs
promoting farm-raised fish. Farmed fish have an unsavory reputation among
chefs for the same reason most musicians don't talk up the wonders of
computer-generated sound effects: we prefer the real thing.

FALLING IN LOVE WITH A FISH

Not long after our meal at Aponiente, Lisa called to tell me she'd recently
returned from a food conference where Dani García, a young chef from
Marbella, Spain, had spoken about a farm-raised sea bass so delicious it
rivaled any wild fish he'd ever served. I told her he most likely didn't mean
it, discouraging her from learning more.

But Lisa, who knows a good story when she hears one, ignored my advice
and forged ahead. The next time we talked, she was exploding with enthusiasm: this farm, Veta la Palma, was a model of sustainable production, she
said. It raised fish in a way that didn't harm the environment or the fish themselves. She claimed the fish were delicious and virtuous.

After months of repeated attempts, Veta la Palma was finally allowing
her to tour the farm. The visit was to coincide with another trip I was making
to Spain for a chef's conference and then, afterward, to see Eduardo and the
geese again—both within a short driving distance from Veta la Palma. I
tried to get out of it, but since Lisa was my liaison with Eduardo (he answered her calls, not mine), I eventually relented and offered to drive out to
visit the fish farm with her before seeing Eduardo.

The conference was in Seville, and the night after my presentation, we went with a group to a new restaurant in the center of town. Lisa pointed out a sea bass from Veta la Palma on the menu. Since we were headed to the farm the next day, I decided to order it.

I remember when the waiter set the bass down in front of me. The flesh, a pristine white, shimmered against the backdrop of a dark green herb sauce. The skin was cobalt black. Seared and roasted, it cracked apart with the tap of my fork, and right away the brilliant white flesh revealed that the fish had been overcooked. There was almost no moisture, for one thing, and the proteins had coagulated, so that the bass felt to the touch of my fork more like a tensed biceps. It had either been roasted at too high a temperature or for too long, or, from the looks of it, both.

Cutting off a piece of skin with my fork, I opted to taste it first, which was odd. I don't like fish skin. I don't like the acrid, tar-like taste, and I don't believe you need a crisped skin to counter the softness of the flesh. We almost never cook the skin at Blue Hill. But the skin on the sea bass was wafer-like—delicate, crisp. Quick, flanking bites of the fish's perimeter were excellent, too, and soon I found myself pushing the remaining fish into my mouth like a log into a chipper.

The fish was incredible. Even overcooked and tough—even D.O.A. ("dead on arrival"), as line cooks like to say when a fillet has seen too much heat—it made my mouth water. It was so richly flavored, you'd be forgiven for comparing it to a slowly cooked shoulder of lamb or a braised beef short rib. I'd never known bass could be so delicious.

———

We arrived at Veta la Palma the next morning. Miguel Medialdea, Veta la Palma's biologist, met us at a bar in the town of Isla Mayor, just at the outskirts of the 62,000-acre Doñana National Park.

Miguel was wearing jeans, a flannel shirt with rolled-up sleeves, and worn-in work boots that made him look like what I'd imagined Eduardo would look like: a farmer. He had the watchful reticence and physical bearing of a cattle rancher. Standing in an empty bar on the main street of the town, I almost expected tumbleweeds to blow through.

We ordered a round of espressos, and Miguel repeatedly welcomed us, admitting that we were among the first visitors he'd toured through the property. The company that owns Veta la Palma, Pesquerías Isla Mayor, almost never allowed visitors (though they probably had never encountered a visitor as tenacious as Lisa). I got the sense that Miguel was even happier to see us than we were to see the farm.

Downing his espresso like a shot of tequila, he began with a brief history. From the 1960s to the 1980s, Miguel said, the land was in the hands of an Argentinean conglomerate. The company rebuilt and expanded a series of canals the British had originally constructed in the 1920s. They strengthened flood fences and extended drainage systems, eventually draining the marshland and creating grazing land for a large cattle operation. By almost every measure, the plan failed. Economically, it never made much money. Environmentally, it was a disaster. Birdlife, for example, plummeted by 90 percent—and for an area at the tip of southern Spain, the final layover along the migratory route to Africa, this was a lot of birds. Political problems developed as the Argentineans' relationship with the Spanish government became fraught. In 1982, fearing a government takeover, the Argentineans opted for what amounted to a fire sale. Veta la Palma was born.

Now at the driver's seat of a minivan and taking us around the outskirts of the farm, Miguel described what happened next. "We used the same channels built originally to empty water into the Atlantic, and reversed the flow," he said, circling his hand counterclockwise to show the simplicity of the idea. They flooded the canals with estuary water instead of pushing it out, filling the forty-five ponds and creating an eight-thousand-acre fish farm.

Working with the Spanish government, which by 1989 controlled Doñana National Park, the company went about integrating the conservation ethic of the park with the economic activity of the farm.

"It's the idea of utilizing, or using, in order to conserve," Miguel said as we sped along a dirt road, the flat marshland intercut with long canals on either side of us. "We like to say we're a national park with a belt of human activities."

As Lisa spoke to Miguel in Spanish, I sat in the back of the van and stared out at the endless expanse of wetland. A burnished tobacco light washed over the scene. In the distance, several cars were caravanning along another long stretch of dirt road, kicking up dust, their windows reflecting the southern Spanish sun as they moved across the landscape. It was like being in the middle of a desert, except that we were surrounded by lush grass and plant life, and wide canals corralled water everywhere around us.

And yet the placidity of the landscape tells only part of the story. There's another angle, or maybe another story entirely, about the wetlands. It's a story that looks at Veta la Palma as yet another edge, a place where two ecosystems—land and sea—meet, and where life flourishes as a result. That calm surface is a front for the furious activity taking place just below.

We usually think of a beach as the demarcation line between land and sea, the land gradually handing itself over to the waves. But the neat divide of an ocean tide line is deceptive. Land doesn't give up easily, as anyone who snorkels or dives knows well. The continental shelf continues for many miles underwater, sloping down along the way until, finally, it ends.

The ocean's edge is "the primeval meeting place of the elements of earth and water, a place of compromise and conflict and eternal change." Those are the words of Rachel Carson, an English major turned marine zoologist and the author of *Silent Spring*. While Carson is widely credited with winning the ban on DDT and helping ignite the environmental movement in this country, she is less well-known for what occupied most of her life: the oceans. Before publishing *Silent Spring*, Carson wrote three books on the subject, all best sellers, including *The Edge of the Sea*, in which she takes readers to the shoreline and just beyond.

Carson owned a home on the coast of Maine, which served as both an observatory for her fieldwork and a place to write. In her introduction to *The Edge of the Sea*, oceanographer Sue Hubbell explains that for many years Carson worked on a field guide of what could be found along the coast. Except it didn't get very far. Carson struggled. She complained of writer's block in letters to her editor. Realizing she was writing "the wrong kind of book"—that readers needed to feel an emotional connection if they were going to be compelled to protect something—she abandoned the field guide project and decided instead to focus on the interactions between the organisms.

"She soon realized," Hubbell writes, "that it was more interesting to write about the relationships."

Veta la Palma is a unique lesson in these relationships. Several miles inland, the argument between sea and shore is already under way. Exploding with life, it's the edge effect in action. Miguel's job is to create conditions that find advantage in this dynamic ecosystem.

⁓

We stopped to walk along a canal and take a closer look at one of the ponds. Miguel navigated the marshland with Crocodile Dundee–like ease, wading through the thick vegetation as if taking a stroll on a neighbor's lawn. In between pointing to rare birds and aquatic plants, he explained how Veta la Palma's natural biomass—the greenery I was slogging through and all the greenery I couldn't see in the water—determined the health of the aquaculture system. Which in turn determined the number and the quality of the fish they raised. If the production were too high, the density of the natural feed would plummet.

"Natural feed?" I asked. "Like what?"

"Well, like the phytoplankton."

"But bass don't eat phytoplankton, do they?" I asked.

"No, but they're eating the shrimp. And the shrimp eat the plankton." We crouched beside a pond filled with young shrimp. He wanted me to taste them. They were so tiny—each one no larger than a grain of rice—that I popped several in my mouth to get a sense of their taste. They turned out to be rich and sweet, with that same fullness of flavor I'd found in my bass the night before.

I made sure Miguel could hear my hum of appreciation. "They're incredible, Miguel, really. But what are *you* feeding the bass? Fishmeal?"

"Dan," he said patiently, "we don't feed them, not at this time of year, at least. The natural productivity of the farm is so high during most of the year, we're not feeding the bass." He paused. "In August and September, it is very dry, and the productivity is low, so yes, of course, we feed a supplement that is available to them—it's a self-feeder, really. They come to eat as much as they want, but they have to work to get it.

"It's a coevolution between nature and the farm's productive capacity," he continued. "We're in this thing together. The response of nature has been much stronger than we thought. We're good partners."

He stopped at the edge of one of the canals and, looking at Lisa—who somehow seemed to understand the farm intuitively, almost as if she had designed it herself—drew an x in the dirt with a stick. "We are here," he declared. He outlined a rough approximation of Spain, which in his artistic representation resembled a human heart. Tracing the Guadalquivir River like a major artery, he started at the southeast of the country and wound his way to the southwestern corner, where we were.

"The Guadalquivir River is our lifeline; it runs through us, and then it gets flushed out here," he said, pointing to a patch of dirt and marking it with an A for the Atlantic. "The ocean water flows into the system here." He dragged his stick from the Atlantic to his original x, the spot where we were standing. "And the pumping station, located here"—Miguel x'd another spot close by—"this redistributes the water throughout our farm."

Because the water in the farm comes both from the ocean, flushed in at high tide and pumped through the system, and from the Guadalquivir River, the resulting estuary is made up of brackish water—a mix of salt and freshwater. It's teeming with microalgae and those tiny, translucent shrimp, which provide food for the fish they raise. And the species they raise are native to the estuary: twelve hundred tons of grey mullet, shrimp, eel, sole, sea bream, and that bass I had fallen for at the restaurant. Miguel told me that more than half of their production was sea bass.

The enormous scale of the property means there's no overcrowding of fish. They suffer none of the problems—injury, disease, parasites—usually associated with farming (Veta la Palma loses 1 percent of its fish to disease; the industry average is more than 10 percent), and the ingenious web of canals provides a filtration system against pollution.

In a vain attempt to impress Miguel, and in the hope of getting what I really wanted—an answer to why the bass had been so spectacular—I casually asked how long it took for the bass to reach market weight. Miguel pointed to a group of men dragging a small net through one of the ponds. As they lifted it out of the water, I saw three large, muscular bass caught up in the netting, flapping powerfully in a vain attempt to escape. I was amazed at the size of the bass and, forgetting my thought, remarked at how beautiful they looked.

"Thirty months," Miguel muttered, seemingly to no one in particular.

"Thirty months!" I said. "It takes two and a half years to raise . . . a bass?"

"Yes, that's the average, which is more than twice the aquaculture average." I asked how the company could make money.

"So far there's profit, enough to keep us working at an optimum, not a maximum." He paused in deference to the ritual taking place in front of us. One of the men, emerging from the pond with the netted fish, dunked the bass headfirst into a plastic bucket of ice mixed with seawater. The bass flapped wildly, thrashing in the cold water, but quickly calmed and in just a few seconds seemed to fall asleep.

"This we've found to be the most humane slaughter," Miguel said. "They lose consciousness without a struggle. We've found a large correlation between this kind of slaughter and the best-tasting fish." Lisa smiled, probably remembering Eduardo and his method for gently gassing his geese.

"In the end I'm very lucky," Miguel said, dialing back to his point about the company's executives. "They know that you can't, and they don't want to, surpass nature's capabilities. So we keep production low, and we don't surpass the limits of the natural ecosystem."

"But Miguel," I said, looking around at the endless miles of canals, "This isn't really *natural.*" It might have been my abrupt chef-speak kicking in, but I sounded more cynical than I felt.

"It's a healthy artificial system. Yes, artificial. But what's natural anymore?"

A LESSON IN RELATIONSHIPS

Back in the van, still marveling at the vastness of the farm surrounding us, I commented again on the beauty of the landscape. Miguel nodded.

Miguel has that California habit of speaking slowly, letting his words hang in the air. You feel his intensity, but it's beneath the surface, coiled and ready for action. Several moments passed, and I thought that was that. Then he said, "Some days, driving around the property like we are right now, I'll look out the window and I swear I see a zebra, or an elephant, and I have to shake my head to stop my dreaming. It's really powerful. What's the saying? 'If you drink once from the African soil you will do it again and again'?" He told me he'd studied in Tanzania for many years.

I asked what kind of aquaculture there was to study in Tanzania. "No, not fish. I studied the grouping patterns of giraffes in the Mikumi National Park."

"Giraffes?"

He nodded. "The giraffe is not a very well-known animal, from a scientific point of view, and yet it is one of the most beautiful and most handsome animals. I fell for them. And over my long observation, I studied how members of the herd hardly interacted. How could this be? Giraffes live in herds, move in herds, feed and sleep in herds, but they don't really mix with each other, they don't socialize much. So the question is, why? Is it a defensive behavior? Or a relic behavior, still maintained because it's not expensive, energywise?"

He stopped speaking, and for a few moments all I heard was the kicking of gravel underneath the van as we sped alongside yet another canal. He appeared to be daydreaming. "How did you become such a fish expert?" I asked.

"Fish?" Miguel looked at me in the rearview mirror, genuinely perplexed. "When I came here, I didn't know anything about fish. I was hired because I'm an expert in relationships."

———

As impressed as I was with Miguel's philosophy, I couldn't help asking the pragmatic question: how did this kind of place decide it was doing well? I thought of Eduardo, and how he believed the size of the livers determined the extent of nature's gift.

"Miguel," I asked, "for a place that is so natural, how do you measure success?"

Miguel nodded as if expecting the question, and in a perfectly (almost unbelievably) orchestrated act of good timing, he pulled alongside a shallow levee. Thousands of pink flamingos stretched before us, a pink carpet as far as I could see.

"That's success," he said. "Look at their bellies." He pointed. "They're feasting."

I was totally confused. "Feasting? Aren't they feasting . . . on your fish?"

"Yesss!" he said, as proud as I'd heard him all day. Lisa and I laughed, but he ignored us, looking out at the flamingos. "There are thirty thousand flamingos. Overall, we lose 20 percent of the fish eggs and the baby fish to the birds here."

"But, Miguel, isn't a thriving bird population the last thing you want on a fish farm?"

He shook his head slowly, with the same calm acceptance Eduardo had shown in the face of losing half of his goose eggs to hawks. "We're farming extensively, not intensively," he said. "This is the ecological network. The flamingos eat the shrimp, the shrimp eat the phytoplankton. So the pinker the bellies, the better the system." The quality of the relationships matters more than the quantity of the catch.

Those flamingos are only a small representation of the thriving birdlife. There are now around half a million birds at Veta la Palma—more than 250 species, compared with fewer than 50 in 1982, when the canals were flushing out water and creating grassland for the Argentinean beef cattle. Miguel even created Lucio del Bocón, a 740-acre lagoon set aside as a bird refuge—no fishing allowed. It has become the most important private estate for aquatic birds in all of Europe.

Technically I was correct: thirty thousand hungry flamingos—one of the largest populations of greater flamingos in the world—are the last thing you want on a fish farm. But Veta la Palma isn't just a fish farm. Miguel meant "extensive" in the broadest sense of the word. I came to understand it in two ways.

The first is practical. The fish's excrement produces nitrogen. The phytoplankton and zooplankton and the micro-invertebrates feed off this nitrogen, and in turn become food for the fish and for filter-feeding birds like flamingos. Because the system is so healthy, there's more than enough feed for fish and birds. In fact, without the birds, problems would inevitably arise. Much in the way that the Mississippi drains excess nitrogen from corn production into the Gulf (causing irreparable damage in algae-blooming dead zones), the mighty Guadalquivir River carries nitrogen runoff from the heart of Spain into the Atlantic, passing Veta la Palma on the way. Filter-feeding birds scoop up the excess nutrients, balancing the system and keeping it pure.

When I pressed him about the role of all the other birds, filter-feeding ones aside, there was even more humility in Miguel's voice. "Ninety percent of the things that happen between the species at Veta la Palma we can't see," he said. "But I am absolutely sure they are allies of the system."

The bottom line is, you must embrace life, which is to say *all* life, not just what you're trying to grow, or, as Klaas would argue about healthy soil, what you can actually see.

A BIRD'S-EYE VIEW

A stock market index, if you know how to read it, provides small indicators for the direction of the market's overall resiliency. There are equivalents in agriculture. Klaas's weeds, for instance, are like an index to soil health. Knowing what the weeds are saying is the first step in correcting an imbalance, and perhaps the first step to better food as well.

Could birds serve the same purpose? Before meeting Miguel, I would have said no. My earliest memory of any kind of bird consciousness is eating dinner on a porch in Cornwall, Connecticut, at the home of my childhood friend Jon Ellis. Jon's father, an avid gardener, had memorized a catalog of birdsongs. During dinner, whenever a bird chirped or whistled, Mr. Ellis would raise an eyebrow or hold up a finger and announce the species. "Warbler," he would say periodically throughout the meal, or "blue jay." It didn't matter if we were in the middle of a heated debate or in the silent spaces between bites of food: if there was a song, Mr. Ellis identified the bird. More than anything at the time, I thought this was funny. Later, Jon would use the technique to impress girls (though he made up the names of the birds). But looking back, I realize Mr. Ellis had a point to make. He wanted to show us that knowing about the natural world is a more enjoyable way to be in the world.

Like most chefs, I've always thought of myself as concerned about the environment. Carlo Petrini, the founder of Slow Food, once said, "A gastronome who is not an environmentalist is stupid," and I think he's right. Our first role as chefs is to identify the best ingredients. We know that delicious ingredients come from good farms, which, almost by definition, means farms that promote healthy environments. How could they not? A badly managed farm will not produce consistently good food. That makes chefs gastronomes and, by definition, environmentalists, too.

But a place like Veta la Palma broadens the definition of "environmentalist" (and "chef") even further. It makes you realize that healthful ecologies are determined in large part by the ecologies that surround them. How is your land healthy if the water feeding it is not clean? Every farm is intimately linked to the larger ecosystem—to what Leopold called "the land" and what we call the environment. When you farm "extensively," you're taking in the world.

Birds are the bellwether of that environment. In the same way that phytoplankton can tell us about the state of the oceans and the climate, and weeds, as Klaas showed me, can tell us about the condition of the soil, birds can tell us about the state of . . . just about everything. They live everywhere, after all—in farmland and meadows, in forests, and in cities. And also in between. Few species thrive in such disparate ecologies, which makes birds perhaps the most sensitive environmental barometers we have. And birds, by almost every measure, are in great decline.

Today, birds' biggest threat, especially in Western countries, isn't from hunting or predation. Their greatest foe is unrelenting, intensive agriculture. Fertilizers, pesticides, modern seed varieties, and mechanization have radically transformed the landscape, which means birds have less food (there are fewer insects and seeds from weeds), and their living and nesting habitats have become smaller and scarcer (natural preserves for animals and uncultivated land have disappeared, crops are harvested in rapid succession, and birds have nowhere to move).

Since 1980, bird populations on farms in Europe alone have decreased by 50 percent. And though there's been some improvement among the most endangered species, especially in North America, the outlook for the future is grim: more intensive agriculture, more habitat loss, fewer birds.*

Grimmer still is the challenge birds face due to overfishing. Studies have shown that the health of seabird populations is tied to the abundance of prey—an obvious connection—but the recent damage is difficult to comprehend: almost half of the world's seabird populations are in decline.

The destruction of animal populations, especially bird populations, was happening long before industrialized agriculture. You don't need technology to destroy nature after all (as the history of fishing demonstrates). But nature writer Colin Tudge sees the destruction from a historical perspective. He calculates that there were once 150,000 species of birds and that of those, 139,500 went extinct over the course of 140 million years. That's an average of one species of bird lost every thousand years. "In contrast," Tudge writes, "modern records show that the world has lost at least eighty species of birds in the past 400 years—one every five years."

Tudge's predictions are substantiated by other scientists like Miguel Ferrer, a leading Spanish ornithologist who believes global climate change will wreak havoc on flight patterns. Ferrer estimates that twenty billion birds

---

* Rachel Carson was an ornithologist before she became America's most famous environmentalist. She took frequent birding trips with her friend Roger Tory Peterson, who revolutionized bird watching with *A Field Guide to the Birds* (1934). Those birding experiences proved useful for Carson in 1958, when she received a letter from the owner of a bird sanctuary in Duxbury, Massachusetts. The letter described the bird mortality rates after a particularly strong DDT spraying. For the first time, Carson saw evidence to support what was then only a hunch—that the casual, indiscriminate spraying of pesticides and herbicides would impose long-term damage on the environment as well as incalculable damage to animals and humans. She pursued the issue, amassing more data, connecting the dots, and ultimately presenting her findings in her revolutionary book *Silent Spring*.

have already changed their migrating habits due to climate change, a trend that "has a knock-on effect on almost everything they do, from breeding habits to feeding habits to their genetic diversity, which in turn affects other organisms in their food chain." Eduardo's geese and their waning instinct to gorge in the fall are just one example.

Jonathan Rosen, in his book *The Life of the Skies*, reminds us that there is no such thing as a single farm. As all bird-watchers in America know, most migratory birds need Central and South American forests to migrate to in the winter, woods and fields (and farms) to land on somewhere in the north, and, still farther north, regions hospitable to nesting in:

> The birds of Walden, local as they seemed to Thoreau, might have flown a thousand miles or more to get there. They are like a story told by one part of the world about another part of the world. Which is why backyard birding is a kind of misnomer after all. Birds are like castles in the air that Thoreau said we must now put foundations under. This is how birdwatching, which grows out of books but can never be satisfied by books, creates environmentalists. If we don't shore up the earth, the skies will be empty.

Our plates may be a little emptier, too. I won't ever fully understand birds' relationship to good food. But it is clear, as Miguel said, that they are allies of the system.

If you want evidence, just look at the pink bellies of those thirty thousand flamingos, who, it turns out, shouldn't even be at Veta la Palma in the first place. They brood in Málaga, in the town of Fuente de Piedra, 100 miles from the farm, because there they find the right soil to build their nests. Every morning they fly to Veta la Palma, and every evening they make the journey home to Málaga, back to their newborns. The males go together one day, and the females make the flight the next.

How are they able to find the farm? Miguel told me that scientists have studied the flight pattern and what they've discovered is that the flamingos follow the yellow lines of Highway A-92, the most direct route connecting Veta la Palma and Málaga.

Moved, I asked Miguel why the flamingos would fly hundreds of miles each day. "Miguel, do they do this . . . for the children?"

He looked at me, confused. The answer was apparently obvious. "They do it because the food's better."

CHAPTER 18

NEARLY TWENTY YEARS AGO, on a busy early summer afternoon, David Bouley appeared in the kitchen of Bouley restaurant with a large UPS delivery under his arm. He was excited. Brian Bistrong, one of the sous chefs at the time, described the scene to me not long after I started working there: "He was smiling from one end of the kitchen to the other, talking a mile a minute."

Bouley enjoyed lecturing his cooks. I remember these sessions fondly. They often happened over black spicy shrimp and peanut noodles in Chinatown while waiting for the Fulton Fish Market to open. It was a kind of culinary catechism—a mixture of anecdotes and old kitchen wisdom—meant to inspire and instruct. Which was why, as he removed a bundle of newspaper-wrapped fish from the box, a tight circle quickly formed around him.

Bouley introduced the fish. They were Copper River salmon from Alaska, prized for their rich, fatty flesh and famous for traveling thousands of miles in the ocean, only to return to their birthplace along the river. By the time they enter fresh river water, they don't eat at all, focused as they are only on reaching their home to spawn. Bouley explained that the flavor of Copper River salmon depends on the fat, and the amount of fat depends on when they're caught. The best time to get them is just as they return to the river, when their bodies, like Eduardo's geese in late fall, are saturated with fat (the better to fuel their arduous swim upriver). Bouley had arranged for the salmon to be shipped overnight, something that was rarely done back then.

"By the time the fish got to us, it had been out of the water for like sixteen hours," Brian told me.

Having unwrapped the salmon, Bouley laid the still-shimmering, clear-eyed fish in a row on the table. The cooks went back to work, but he stood over the fish, rocking on his heels. Bouley had seen a lot of incredible fish in his lifetime—he'd worked across from a docking port in Hyannis, chartered planes from Maine for fish just out of the water, and at this point in his career had the best seafood in the world delivered to his kitchen door. Still, he examined the salmon as if they were ancient scrolls.

Brian described Bouley's lips moving silently, his body a kind of divining rod. "We just looked on like it was the most amazing thing," he told me. "Because it was."

FRESH FISH

Chefs obsess about fresh fish. Fish have twenty-four charms, so the saying goes, and they lose one every hour. Which is why it's difficult to comprehend, even for those chefs old enough to remember, that distinguishing the very freshest of fish is a new idea in America. By definition, fresh fish, which is to say seafood that isn't dried or smoked or frozen, has been available to chefs—and, to a lesser extent, to the general public—in many parts of the country for more than fifty years. But by today's standards it wasn't particularly fresh-*tasting*.

"I could always choose from a very large selection," Alain Sailhac, a four-star chef in the 1970s and now the executive vice president of the International Culinary Center, told me. "This was never the problem. The problem was there were no standards. I'd order, let's say, snapper, and some days the snapper was very fresh, very delicious, and then other days, oh my God, you wouldn't have believed it, so mediocre. You never knew what would arrive."

The widespread availability of superior-quality seafood today makes it easy to forget how impossible it was to obtain only forty years ago. Large distribution networks didn't exist, overnight options weren't available, and Americans ate less seafood. (Today our annual consumption averages about fifteen pounds per person—over 20 percent more than forty years ago.) It's a chicken-and-egg question: was there less seafood because there was less demand, or was there less demand because seafood didn't taste all that fresh?

How to define "fresh" was at least part of the problem. Fish was fish. The industry didn't differentiate gradations of quality. Dave Samuels, the fourth-generation owner of Blue Ribbon Seafood, a highly regarded supplier to high-end restaurants, told me that until the early 1980s, even the fishing industry didn't really grade fish. "There was always 'top of the line'—the stuff they caught last on boats before they headed in to dock. They got a couple cents more for this, but by and large these guys were catching cod and flounder and all the rest of it and just figuring out how to get it into a can."

In 1986, the Portland Fish Exchange, in Portland, Maine, became America's first display auction of fresh fish and seafood. Before this, the large fish distribution centers—Gloucester and New Bedford, Massachusetts, to name two of the most famous—held blind auctions. Fish buyers simply bought fish in bulk.

"Picture the Pit in Chicago," Samuels said, referring to the famous Chicago trading floor. "They were bidding on a commodity, that's all."

Few seafood distributors have the historical perspective of Dave Samuels, but most agree that over the past few decades the industry has undergone the greatest transition since the advent of longline fishing. "My childhood is one blur of my father saying, 'How am I going to get rid of all this fish?'" Samuels told me. "Which is pretty much all I was saying when I took over in the early '80s. And then, poof, the world turned upside down. All of a sudden people were willing to pay more for better quality. Now all I think about is how to service the unbelievable demand for quality. We went to bed in the Dark Ages and woke up in the Enlightenment. The whole market changed."

There were many reasons for the change—better fishing technology, an increasing awareness of seafood's health benefits, and greater exposure to cultures and cuisines that celebrate fresh fish. But among all these influences, nothing compares to the work of just a few chefs who very nearly alone up-ended the seafood industry, perhaps none more completely than Gilbert Le Coze of Le Bernardin, in New York.

Le Coze, like Jean-Louis Palladin, trained under the great pioneers of nouvelle cuisine, chefs like Paul Bocuse, Alain Chapel, and the Troisgros brothers, who stressed lightness and simplicity—modernizing classic French gastronomy in the process—and successfully established direct connections with farmers and foragers. And, like Palladin, Le Coze's most enduring leg-acy was to embrace America's natural resources—to see the potential in what the American landscape could offer, which in turn dramatically trans-formed the supply. Palladin and Le Coze defined what they were looking for and then found it. For Le Coze, that meant going to the fish.

"I'll never forget the first morning Gilbert came to the Fulton Fish Mar-ket," Samuels told me. "He was walking through the stalls, and his eyes were bulging out of his head—a kid in the biggest candy store he'd ever seen." It was 1986, and Le Coze and his sister, Maguy, had just opened Le Bernardin, a seafood-only restaurant they transplanted from Paris. Le Coze's knowl-edge of seafood dated back to his days watching his fisherman grandfather and working in his father's small hotel restaurant on the coast of Brittany.

"Finally he gets to my stall," Samuels remembers, "and we're being intro-duced, and he totally ignores me, sticking his nose right into a large bin of skate. He dove right in. The skate had just gotten off the boat, so fresh it's still covered in slime, but he looks it over and says it's crap, it's no good. He starts shaking his head no, and I was like, *Who the fuck's this guy?*"

After a few minutes of inspecting more fish, Le Coze came to another bin of skate, caught by a different fisherman but looking identical to the first, and again stuck his nose close to the surface. He raised his head with a large smile and began waving his hands. "He thought the second bin of skate was in-credible. Why? I have no idea why. He didn't speak a word of English. But he was so emphatic. He was smiling and laughing and asking me if I could get him more of this skate here, in the second bin. I told him sure, whatever he wanted, but I had no idea if I could get more skate like the second bin, because I had no idea what he was talking about. It was skate, for God's sake! I was selling it for pennies a pound. It seems crazy to say now, but until that moment I had never considered there could be a difference."

Soon Le Coze was visiting the fish market after every dinner service. "I remember one day he was bouncing all around," Maguy Le Coze told me. "He said, 'Maguy, we will not have to ship fish from Europe!' I thought he was talking crazy. *Every* chef in New York imported from Europe. It went without saying. If you didn't import your fish, you didn't serve fish. But this is what he wanted. We opened the doors with monkfish, skate, and sea urchin—no one knew what they were." (Maguy became the intermediary. "When we opened, my brother wanted two things from me: to keep the reservation book at eighty guests per night—not one more—and to work the room to convince guests to eat rare fish," she said. Her confidence, and her salty charm, won over a skeptical New York public.)

Not long after Le Bernardin opened, Le Coze was arguing that the selection of fish was better in America. From a biological point of view, he was right. The continental shelf stretches the length of the Eastern seaboard, much of it over 60 miles wide: one continuous edge effect. The number of habitats is enormous. Compare this with France, or even the West Coast of the United States, where the shelf suddenly drops off into the Pacific Ocean, and you can understand Le Coze's amazement. He'd never seen so many fish.

Le Coze purchased seafood no other chefs thought to buy. He's credited with creating the market for black sea bass and monkfish—his personal favorites—as well as skate and sea urchin. In those days, these fish were considered too lowly and flavorless for refined cuisine. So was tuna. As Samuels explained, in the 1960s and '70s, bluefin tuna was hacked up with an ax and canned for tuna fish or made into cat food—at least until Japanese sushi chefs came to New York and began paying more for tuna that was killed and processed the right way. Le Coze was the first high-end French chef to offer raw tuna on his menu. He is the father of the tartare and carpaccio craze that spread around the country in the 1980s and '90s.

"It's a funny thing," Bill Telepan, the thoughtful New York chef and one-time protégé of Le Coze, said to me recently. "People assume Le Coze was influenced by sushi. It's the other way around. He influenced sushi—at least its becoming so quickly accepted. There were sushi restaurants all over New York in the '80s, but they never really caught on until Le Coze came along and devoted an entire section of his menu to 'simply raw.' Chefs around the country were suddenly serving raw fish. Sushi's success is a mainstream progression of what Le Bernardin introduced."

In the beginning, the chef's demands put some suppliers off. "Very little was good enough for him," Samuels said. "We didn't really need his business, so it was easy to laugh him off. But you know what? He knew more than we did." So Samuels began calling his fishermen, explaining what Le Coze wanted them to do with the fish on the boats—how to bleed fluke, for instance, and how to chill bass and monkfish. He had specific instructions for gutting fish. He knew what time of year he wanted which fish and explained what kind of fat he was looking for. The fishermen started to care, because a market developed that paid them to care.

Samuels's most distinct memory of Le Coze is from his first dinner at Le Bernardin. He sat down to a glass of Dom Pérignon and never received a menu. Instead Le Coze sent out a parade of different fish courses. Samuels can still remember the cod course. "It was our fish, which my grandfather and father sold, but it was like I was meeting it for the first time. It was so white. It was moist and delicate. I took a few bites and turned to my wife. I said to her, 'This guy is going to change the world.'"

⸻

By the late 1980s, Jean-Georges Vongerichten and David Bouley had followed Le Coze to the market, seeking the same quality and further expanding the supply. They became loyal customers of Dave Samuels.

"These guys took the football and ran with it," Samuels says. "Especially Bouley. He was such a fanatic about knowing where the fish were being caught—not whether it was Atlantic halibut or whatever, but *where* in the Atlantic. His influence on other chefs is enormous." Bouley learned about how temperature, ocean currents, and time of year affected flavor. And he remembered every detail. "You know those people who can't remember your name but recall the lyrics to every song they've ever heard, even if they've only heard it once?" Samuels asked me. "Bouley was like that. He never forgot a fish."

Bouley would soon make his own connections with fishermen, sending waiters and cooks to the docks to buy from them directly. Fishermen came to anticipate Bouley's little red pickup truck full of cooks snoozing as they waited for the boats to land. And Bouley began identifying where the fish were coming from on his menu. You didn't choose between salmon or squid—you chose between Copper River salmon and Rhode Island squid.

"I'd always been a fishmonger, just trying to get my catch to market before it went bad," one fisherman on Cape Cod told me. "Then Bouley shows up and does a little freak-out over my stuff, and suddenly his cooks are coming *to me*. Other chefs start calling *me*. One day I see the menu and it says, 'Cod, line-caught by a Chatham day-boat fisherman' or something like that. Hell, that was me—the fisherman. I felt like a prostitute who suddenly gets called a *lady of the night*."

Highlighting the provenance of each fish, Bouley elevated their stature (and their fishermen's) in the same way Alice Waters elevated Mas Masumoto's peach.

That night in 1992, after Bouley brought in the Copper River salmon, he left one whole so the guests could see it themselves. It was walked around the dining room on a silver tray. The waiters spoke about how the fish had been caught when their fat and flavor content were at their highest.

According to Brian Bistrong, two orders of the salmon were among the first of the night.

Bouley himself cooked the fish. The restaurant wasn't yet busy, and Brian remembers the cooks gathering around again as the chef plated the dish. He had seared the salmon with the skin on, then cooked it gently in a low-temperature oven so the flesh stayed red and moist. He served it with creamy watercress rice and a sauce made from a puree of peas and pea juice. The cooks nodded and told the chef it was a nice dish, and apparently Bouley seemed particularly pleased.

Five minutes after serving the plates, the waiter returned with the salmon, uneaten. He stood at the expediter's pass until he could get Bouley's attention. The kitchen was suddenly quiet. The waiter said the diners didn't care for the salmon and instead wanted the shrimp appetizer.

"What do you mean they don't care for it?" Bouley said, stepping closer to the expediter's pass to make sure the salmon was not overcooked. It was perfectly cooked. The waiter was told to go back out to the table to ask. He turned to take the plates to the dishwashing station, but Bouley told him to leave them at the pass. Bouley's mood turned dark. He mumbled to himself and spoke angrily to the cooks. As the dining room filled, finished plates had to be placed around the offending plates of uneaten salmon, which were awkwardly maneuvered to make room.

When the waiter returned and merely repeated the couple's preference for the shrimp, Bouley summoned Dominique Simon, the manager of the restaurant, and told him to find out why they had sent back the salmon.

Several minutes later, he returned with an answer. "Chef," Dominique said, in his heavy French accent. "They think the salmon is not fresh. They say they know fresh fish and that this fish is not fresh. They think it has turned."

"Bouley was busy plating a few of the other orders, but I could see his face was getting red," Brian told me. "Then Dominique says, 'They said they think the fish is rancid.' Bouley kept working while Dominique stood there. It felt like forever."

What happened next is proof of the chef's prerogative, a privilege that couldn't have been contemplated until the nouvelle cuisine chefs of the 1960s and '70s invented it. Bouley turned to Dominique and said, "Tell them their meal is over." Dominique, the embodiment of civility and calm, was momentarily baffled, until Bouley grabbed the table's ticket hanging from the pass and crumpled it in his hands, then returned to cooking. The cooks erupted in cheers.

The diners, a surly couple who had been rude and overly demanding since the moment they'd arrived in the restaurant, were confused when Dominique returned to the table. When the manager asked them to leave, they apparently thought it was a joke. They laughed. They said they merely wanted to try the shrimp. They said their meal had only just begun. "Messieurs-dames," Dominique said, shaking his head, "Chef Bouley has finished."

CHAPTER 19

Six months after our first visit, Lisa and I returned to Veta la Palma, this time with the world's supreme ocean authority: Carl Safina, author of *Song for the Blue Ocean*.

I'd been in touch with Carl after reaching out to several marine and aquaculture experts about the farm. I wanted to introduce Veta la Palma's fish to U.S. chefs, and though I knew I wasn't wrong about the brilliance of the fish themselves, I didn't want to be wrong about the brilliance of the operation. I spoke to Lisa, who was planning a story on Veta la Palma for *Time*. She, too, wanted verification. Despite my best efforts on the phone and in e-mail, Carl was not easily convinced that a fish farm in southern Spain was (a) truly sustainable and (b) truly worth his time. On a whim, I challenged him to join us on a visit and see the place for himself.

When we parked our rental car at the front gates of Veta la Palma, I was feeling proud of myself. I'd pulled a nice little coup: I'd brought the leading expert on sustainable fishing to the leading European model of sustainable aquaculture. Or so I hoped. It wasn't until we entered the farm that I began to worry I'd made a mistake.

I stepped forward to hug Miguel, a little awkwardly. I want to say my embrace was out of appreciation for his tour six months earlier and our continued correspondence. But I found myself hugging Miguel out of anxiety. What if Carl, standing in Veta la Palma's greeting room, as we now were, examining a twelve-foot-long replica of the farm and the national park that

surrounded it—the seemingly endless, intersecting canals, the rich marshland, the forty-five fish ponds, and the 740-acre lagoon reserved just for the
birds—what if he took one look at the place and skewered it? Was it crazy to
think that his critical eye could find fault where I had found none? No, it
wasn't. The craziness was in persuading the world's foremost ocean conservationist to travel thousands of miles to drop in on a place I knew next to
nothing about. I didn't even know enough to know what I didn't know.

Standing in front of the replica with a long professor's pointer, Miguel
took us on a virtual tour of Veta la Palma, starting with how the water flowed
through the system and out into the Atlantic, and then back in again with the
help of the pumping station. If he was nervous to present his life's work to
Carl Safina, he didn't show it. He sounded thoughtful, well informed, and
passionate. Carl looked bored, and even a little annoyed.

I was sure his mood would change when Miguel described how the canals
had been drained in the '60s to raise beef cattle on the land. Instead Carl politely nodded, as if he were learning about draining a kitchen sink. "Marshes
have been emptied forever and a day," he said. "It was always the fear of disease, especially malaria."

Miguel added details I didn't know, like the fact that they were devoting a
section of the surrounding fields to raising grass-fed beef. I asked him how
the beef cattle benefited the fish, knowing Miguel enough to believe that it
would, and that the connection might impress Carl.

"It's a totally different activity," he said. "There's no relationship with the
fish. But the result is that the northern sector of the property gains in value."

"In other words," Carl said, his tone tinged with derision, "there is no
ecological relationship—it's a financial one."

"Yes, that's right," Miguel said. "It makes the whole operation more viable. Also, raising livestock is a really deep tradition in the area. So it forms
part of the eco-cultural landscape. Because it has such deep roots." Carl nodded. I looked at Miguel.

"For the future," he said hopefully, "we should not consider agriculture and aquaculture as separate entities. We should consider them as part of the same ecosystem, and we will." Though we were only standing over a replica of Veta la Palma, I had the same sense as when I'd looked over the *dehesa* from Placido and Rodrigo's rooftop. Everything was connected.

Carl pointed to one of the bass ponds. "What's your density?" he asked.

"We have low density for the ponds. It's roughly nine pounds of fish for every one cubic meter of water, which enables us to keep free of parasites—"

Carl interjected. "Hang on. That's nothing? That's crammed!" he said, translating the number to twelve cubic feet on a piece of paper. At the sound of the word *crammed*, my heart sank. "That's not a lot of space at all." He outlined twelve cubic feet by walking in a rough approximation of a square, and then stood inside it and scrunched his shoulders to illustrate the confinement. He looked like a caged bird. I cringed.

But Miguel walked over and gently corrected his calculation. One cubic meter was thirty-five cubic feet, not twelve. Miguel outlined the larger square, and Carl looked at it approvingly. "Oh, okay, not bad."

I didn't relax until we were walking the farm. Rather, Carl didn't relax until we were walking the farm, and as a result, I didn't, either. What turned him wasn't the flooded plain—the flat relief of endless pockmarked depressions filled with lush vegetation—nor was it the engineering feat of the pumping station, or the talk of phytoplankton and microinvertebrates. It was the birds. Compared with my last trip, there seemed to be even more birds, thousands more. They crowded the skies, Hitchcock-like.

Every few seconds, Carl craned his neck in another direction. "Wow," he would say. "A redshank." And then, "Holy cow, willow warbler." It was Mr. Ellis with a beard and an ironic glint in his eye. I smiled. Miguel smiled. And then, after we had walked a bit more, Miguel tried bringing our attention back to the fish. Carl kept right on surveying the sky, wide-eyed, overcome with excitement. "Oh, my," he said, ignoring Miguel. "That's a black-shouldered kite."

The scene reminded me of something Jack Algiere had told me in frustration one day. He said visitors to Stone Barns Center's 23,000-square-foot greenhouse arrive and immediately look up at the computerized overhead irrigation, the retractable roof bays, and the sprawling structure. "They look up at the bells and whistles," he said, "but really they should be looking down." He pointed at the soil, and the billions of organisms that thrived beneath the surface. "This tells you everything you need to know."

Carl was looking up because it told him everything he needed to know.

Back in the car, Carl leaned forward on the passenger seat, holding his long-nosed camera at the ready. He asked how the bird population changed during the dry season, from June to September.

"Doñana, the national park, is dry, but Veta la Palma stays flooded the entire year, as"—Miguel added, in a tone that suggested perhaps Carl had forgotten—"it is a fish farm. Which makes Veta la Palma the only place aquatic birds can land and rest." Carl rolled down the window and maneuvered his camera into position. He caught a group of spoonbills nesting nearby.

We were driving on the periphery of the farm, heading toward the pumping station at Carl's request, when we saw tractors harvesting rice on the outskirts of the property. Miguel told us that almost one hundred thousand acres of rice production surround Veta la Palma in the province of Seville. The region is one of the largest rice-producing areas of the world, little of it organic. I wondered aloud how such intensive, conventional rice farming affected Veta la Palma's waters.

"It's a struggle, always," Miguel said. "But what you said before about the bass skin you tasted—how the flavor is so sweet and clean—tells me things are working." He explained that fish skin is a kind of last defense against contamination, evolved to soak up the impurities before they cause harm. "But," he said, "we don't have impurities."

Veta la Palma is continually testing the water for nutrient levels. The water in the canals, after all, is flowing in from the Guadalquivir. It courses through the farm and carries all the things rivers tend to carry these days, including chemical contaminants and pesticide runoff. The system, though, is so healthy—because of the continuous water flow, the plant biomass, the filter-feeding fish and birds, and of course the plankton—that it acts like a sieve. Water flows into the farm, and when it flows out into the Atlantic it's actually cleaner than when it started.

"This is a good example of how a natural system can soak up the artificial management of land even across the street," Carl said. He explained the value of Veta la Palma in the context of estuaries, which are the most productive areas of the oceans—twenty times more so than the open sea. (Of the one hundred or so recognized estuaries in the United States, Carl noted somberly, most, unfortunately, are in severe decline.)

For the first time, I learned the reason for their explosive growth. Phytoplankton, those organisms on which all ocean life depends, can grow only with abundant sunlight. And sunlight is abundant only near the surface of the water, the sea's uppermost layer. This is called the sunlight zone, and it's the habitat for 90 percent of ocean life, just as the topsoil—the outermost layer of the soil—is home to most subterranean life. Which is why the deep sea, from the perspective of producing life, is really more like a desert: no sun, no feed.

Sunlight is not the only thing plankton need. Carl listed their nutrient requirements, like nitrogen and phosphorus. This is where rivers come into play. Rain-drenched, nutrient-rich soils wash into rivers, infusing them with a vitamin pack that flows into estuaries and then eventually gets dispersed in the ocean.

Rivers, in other words, not only connect land and sea; they are the feeding tubes of the sea. For the first time I understood that the Gulf of Mexico—teeming with diversity—owes some of its initial good fortune to Mighty Mississippi. America's heartland always gave up a little of its fecundity every year to feed the system that gives us the shrimp and the bass for our menus.

Only now there is so much fertilizer leaking into the ocean that a large area of the Gulf is choking to death on *too much* nutrition. Carl once wrote that pollution "comes from land. Gravity is the sea's enemy." Gravity is also the sea's lifeline. What suffocates the Gulf is the misuse of what enriches the Gulf.

Carl's call for a "sea ethic" reminds us that we tend to think of farmland management and ocean sustainability as distinct pursuits. But stand in the middle of Veta la Palma, surrounded by nearly one hundred thousand acres of chemical-intensive rice farming, the lifeline of the polluted Guadalquivir coursing through it, and you feel irrefutably—and, if you're Miguel, maddeningly—connected.

The water that flows out of Veta la Palma, cleansed by the health of the system, is dumped into the Atlantic. As Jonathan Rosen said about birds, it is like a story told by one part of the world about another part of the world.

THE HEART IS NOT A PUMP

Rudolf Steiner, the Austrian philosopher and educator, was once presented with a problem. A group of farmers, alarmed by their experience with "mineral manuring," or chemical fertilizers, and worried about how these new practices would affect soil health, asked Steiner for advice. These were prescient farmers; the year was 1924.

In response, Steiner provided a series of lectures and follow-up lessons that became the foundation of what would later be called biodynamic farming. Farms are living organisms, he explained, and they operate in the greater organism of earth. When Steiner said everything was connected, he meant it. Farming in concert with nature also entailed, for him, planting and harvesting according to phases of the moon. This idea, along with instructions like cow-horn clay preparations (pulverize clay, add water, stuff in a cow horn, bury) and discussions of "supersensory consciousness," can make biodynamic devotees appear eccentric and occult. It's all part of a powerful spiritual history of the organic movement, but as one farmer said to me, "Every good organic farmer follows the precepts of biodynamic agriculture, just shorn of the mysticism."

Still, I thought of Steiner as we stood before Veta la Palma's famous pumping station that afternoon. Miguel described the pump as the geographic center of the farm, moving over 250 million gallons of water a day. It has the ability to open, partially open, or close, depending on the water and weather conditions. It was impressive to see up close, and Miguel's enthusiasm was affecting, but I found myself distracted by the meaning of what he described: the pumping system doesn't pump so much as assist the water. When the tide is up, water flows in, and when the tide goes out, the water returns to the river. Because of the force of the tide, this would happen with or without the pump, as it happens in every estuary in the world. The difference is the need to elevate the water into the irrigation canals. From there gravity eventually pulls the water into the fish-rearing ponds. The pumping station's job is to automatically adapt, evaluating the changing levels of water and distributing it accordingly.

"It's a continuous movement, so the pumping station works all day long, year-round," Miguel said.

Which is why Rudolf Steiner came to mind. A student once asked Steiner what he believed the betterment of humanity required. He gave three answers, and it's the last one that stuck with me. Steiner told the student that in order for human beings to improve and make true progress, they needed to understand that the heart is not a pump.

I remember hearing this story many years ago at a lecture given by Sally Fallon Morell, a longtime advocate of traditional diets, and thinking two things. The first was: *Really?* Steiner, the great philosopher king, gets three suggestions to improve humanity and he gives us one that has as much clarity as a haiku? It confirmed for me what many think of Steiner: creative, provocative, and a little loopy.

The second thought was: *Really?* If the heart doesn't act like a pump, what does it do? At the suggestion of Fallon, I read Thomas Cowan's *The Fourfold Path to Healing*. Cowan spent twenty years contemplating the question, and, while his analysis helped me decipher Steiner's statement, it wasn't until I stood at the pumping station—at what Miguel called "the beating heart of Veta la Palma"—that I began to understand its significance. Steiner wrote that science "sees the heart as a pump that pumps blood through the body. Now there is nothing more absurd than believing this, for the heart has nothing to do with pumping the blood."

Cowan argues that Steiner was right. For one thing, when blood enters the heart, it is traveling at the same speed as when it exits. It slows down as it heads to the smaller capillaries to transfer nutrients, then moves to the venous system, a highway of larger and larger veins that eventually lead back to the heart. As it approaches, the blood speeds up again. The heart acts more like a dam at this point, trapping the blood and holding it in its chambers until they're filled, which is when the valves open and the blood is released, resuming the cycle.

As Steiner explained, *"The circulation of the blood is primary.* Through its rhythmic pulsations—its systole and diastole—the heart *responds* to what takes place in the circulation of the blood. *It is the blood that drives the heart and not the other way around."* The heart doesn't pump the blood. The blood pumps the heart.

So what does the heart do? It listens, according to Cowan, which is Steiner's larger point. It's the body's primary sensory organ, and it acts like a conductor, controlling the rhythms of cellular management. A scientist might call this maintaining homeostasis. Either way, the idea is that the heart serves at the pleasure of the cells, not the other way around.

Veta la Palma's pumping station works the same way: the pump activates what the tidal action from the Atlantic and the powerful Guadalquivir River (the largest vein entering the farm, fed by thousands of smaller, capillary-like rivers) demand. It listens as the water "pumps" itself through the system. The pumping station is programmed to react, instead of control, the flow of water. And the distinction is not small.

The first part of the distinction is that Veta la Palma's technological pursuits are in the service of a better-functioning natural system. As Miguel later wrote to me, "It's about technology working, side by side, with ecology. Without this engineering project, we biologists would not be able to guarantee the viability of what we grow, nor its 'added value' (i.e. the birds, since in summer Veta la Palma is the only place there is water). And without our biological and ecological knowledge, the engineers would never have been able to build such a sophisticated hydraulic system."

The second part is, to borrow a phrase from Steiner, spiritual, and that takes more time to decipher. Steiner was one of the earliest writers to question the so-called mechanistic approach to science, which viewed the workings of the environment in separate parts—machinelike more than lifelike. According to Fred Kirschenmann, a sustainable-agriculture writer, theologian, and the president of Stone Barns Center's board, these ideas took root during the scientific revolution of the seventeenth century, led by people like Sir Francis Bacon, who believed you could bend nature to your will, and René Descartes, who saw humans as masters and possessors of nature. That may seem simplistic now, but who can blame them when the presiding wisdom of the day was the theology of the church? These scientists were reactionaries just like Steiner, questioning accepted theological teachings with the intention of breaking things down to their simplest forms. They wanted to show how the world really worked. Rationality and physicality were at the core of their thinking.

The core of Steiner's thinking was that biology is a lot more complicated than that. It is decidedly not linear. He didn't see simple cause and effect applying in a natural system. Just like Miguel, Steiner valued relationships, and in that sense he was a kind of complexity theorist, seeing the workings of nature as continually in flux.

"Steiner said what the science community at the time wasn't ready to hear," Fred told me once. "Nature doesn't allow you to impose one idea, or one solution, because it inevitably changes the game." The only way to comprehend nature was to recognize an inherent spirituality in its workings.

It was the spirituality that always rubbed me the wrong way. And yet, as I stood in the middle of Veta la Palma, with the backdrop of the enormous pumping station, Steiner's message began to seem more right than wrong. Today the mechanistic worldview is mostly old news. We don't, for example, hear talk anymore of the search for the one gene that will cure heart disease. Or any other disease. For much of the past half-century, the prevailing view was exactly that: one gene, one trait. Identify the gene, suppress it, and solve the problem. We now know that to be wrong. Genes don't act independently of one another. What's most important is the complex set of relationships that determine how they get turned on or off.

And it's not just the medical sciences that have moved on. Businesses, government agencies, and educators have moved away from erecting silos between departments, encouraging more creativity under the logic that innovation prospers when ideas can connect and recombine. It's all very logical, really, except that most of agriculture is still mired in seventeenth-century ideology. Diversity has been replaced by specialization; small, regional networks have given way to consolidation. Farming has been broken into component parts in pursuit of growing more food.

One hundred years ago, Steiner saw this thinking as folly. To break nature into its component parts to solve problems, as you would go about repairing an old watch, is to go about it in entirely the wrong way. That isn't how biological systems work. It's how computer programs work.

What's become clearer to me, after spending time with farmers like Miguel, Klaas, and Eduardo, is that farming with nature's frustrating complexities—even, or especially, with supposed enemies of the system—is inherent to their success. True, their systems are "artificial" (Veta la Palma's pump-moderated estuary, Klaas's intricate crop rotations, Eduardo's man-made *dehesa*), but human intervention, in each case, is in service to the ecology rather than in opposition to it. They embrace the diversity of the natural world; they work within the constraints of nature—and, in the end, benefit from them by producing food with great flavor.

I often think back to what Miguel admitted to me in the first hours of our meeting. It was so true, and so humble—and, when you think about it, so Steineresque, with its hint of spirituality. He said that most of what happens between the species at Veta la Palma he couldn't see. "But," he added, "I am absolutely sure they are allies of the system."

CHAPTER 20

THE DAY after Carl and I visited Veta la Palma, Lisa arranged for Ángel León to join us for lunch at Sant Pau, a seafood restaurant in Sant Pol de Mar, near Barcelona. I was thrilled to see him again, especially in the company of Carl Safina. I'd imagined Carl was likely something of a hero to Ángel. And I figured Carl would enjoy meeting a chef with a passion for the health of the oceans that matched his own. Lisa also happened to be a great admirer of Carl's writing. It had the makings of a memorable gathering.

By the time we received menus, I was convinced I wouldn't be able to forget it soon enough. Ángel, it turned out, had never heard of Carl. Sitting there at the table across from me, he looked distracted. He had heard of Veta la Palma—maybe once, he said—but though his restaurant is only one hour from the farm, he had never visited. I asked Lisa how this was possible. How could Spain's apostle of the oceans not support such an enlightened fish farm, especially one that existed a few miles from his restaurant? And, more to the point, one that produced such delicious fish? She told me that Ángel was adamant in his opposition to aquaculture, which was when I heard Ángel's Churchillian decree from across the table.

"Never, never, never," he said, his dark eyes squinting in my direction. Lisa described Veta la Palma—the water purification, the natural feed, the birdlife—and I spoke about the flavor, but Ángel only shook his head. "I've spoken to fish farms," he said. "They all talk about their uniformity. 'I can get you a perfectly uniform fillet'—they tell me this all the time, as if that's something to celebrate." Beads of sweat formed above Ángel's upper lip. Lisa argued that Veta la Palma was the exception to the rule, explaining how, with its limited feed and cleansing waters, it could be a model for other fish farms around the world. Ángel shrugged. "There's already plenty of fish—if we cooked with the bycatch."

As Ángel excused himself to make a phone call, I told Carl about my meal at Ángel's restaurant, Aponiente—about the unnamed fish and the innovative cooking methods, and about the consortium of local fishermen Ángel had cultivated to help create a market for bycatch. Carl nodded without looking up from the menu.

"The problem," he said, suddenly peering up at me over his glasses, "is that your friend over here is going to create a whole new level of demand. It's just a matter of time before these smaller fish become fashionable."

Safina saw the flip side of Ángel's logic: by bringing awareness to new species, you initiate their decline. "It's the classic story of fishing down the food chain," Carl said. "We exhaust the large predators and move to the smaller fish for substitution. Of course, the technological advances of the last fifty years have meant that we've been moving up the chain as well, as technology and access have improved. So basically, it's been a round-trip."

Except that, according to Carl, we've got nowhere else to go. He predicts that our children will be living in a world of radically impoverished marine communities, dominated by simple forms of life, like jellyfish.

Just then the future came to us in the form of our second course: It was fideua, a traditional Spanish noodle dish, in which the chef, Carme Ruscalleda, had replaced the usual seafood with jellyfish. It was as delicious as it was imaginative, but would it be as delicious or imaginative when jellyfish was all that was left for us to catch?

Ángel returned to the table just in time to overhear Carl's predictions of what would be left of the ocean. "This is why chefs are so important," he interjected. "With less resources, we're the ones who are going to have to make what's left taste good." He looked at me meaningfully from across the table again. Carl appeared befuddled, and perhaps annoyed.

"Unless more farms are built like Veta la Palma," Lisa said quickly, breaking the growing tension at the table.

Ángel didn't play along. He mentioned Kindai tuna, a method of bluefin tuna "ranching" marketed as a sustainable solution to dwindling numbers of wild tuna. "It tastes terrible," he said. "Way too much fat. Let it sit in your hand even for a minute and you're covered in oil. Eating it is worse. Just thinking about it is giving me a stomachache. It's an insult to tuna," he said, sounding so much like Eduardo that I looked at him in bafflement. "When you go against nature, you always get it wrong. I only cook with the real thing."

Carl adjusted his glasses and looked at Ángel. "You serve bluefin?"

Ángel looked confused by the question. "Yes, yes," he said, turning to Lisa to make sure he hadn't misunderstood. "It's the greatest fish in the world. Of course I put it on my menu when it's in season. We celebrate it."

It was an odd moment. Both Carl and Ángel turned to look at me, their expressions each saying the same thing: *Who is this guy?* It must have struck Carl as especially strange. I had described Ángel León as a visionary chef in tune with the health of the oceans and the future of sustainable seafood. And here he was declaring bluefin, the most seriously depleted of fish, the species Carl had spent his life protecting, fit for celebrating on his menu. It was as if you were arguing for the slaughter of chimpanzees with Jane Goodall. I clutched my glass of water.

No one spoke until Lisa broke the silence. "Your tuna is coming from the *almadraba*, isn't it?" she asked, referring to an ancient system of tuna fishing practiced in Spain.

"I only cook with *almadraba* tuna. It's the best tuna in the ocean," Ángel said.

The *almadraba* is a web of large nets along the southern coast of Spain, around the corner from Veta la Palma and along the Strait of Gibraltar. The nets are hung every year, from May until mid-June, just as the tuna are leaving the Atlantic to spawn in the Mediterranean. On their way through the strait, some tuna lose themselves in the maze of nets and are corralled into an area small enough for the waiting fishermen to haul them to the ocean's surface. It's a passive pursuit compared with the sonar-seeking trawlers that have been so destructive to bluefin populations around the world. The *almadraba* is also part of a long, deeply held Spanish tradition, a tradition that's especially revered in Ángel's coastal area of Cádiz.

Carl wasn't impressed by the distinction. Though he had heard of the *almadraba*, he argued that we were at a point where the decline of bluefin was so severe that greater pressure, however passive and traditional, wasn't possible to support. Ángel leaned forward in his chair. His darting eyes made him seem both wary and mischievous.

"I have something I want to show you," he said to Carl, reaching into his pocket and removing a coin. "Do you know what this is? This is an original Phoenician coin. Look here," he said, leaning over. "Phoenician," he repeated, pointing to the coin's faded depiction of tuna enmeshed in the netting of the *almadraba*. "It's very emotional. I carry it with me everywhere I go. The *almadraba* is 2 percent of the tuna taken in the world. I want to be an apostle about protecting this 2 percent. The *almadraba* is not the problem. The sonar-seeking trawlers are the problem," he added, in a tone that suggested anyone who thought otherwise might also be a problem.

When Lisa finished translating, she nodded and added, "There's something inherently unfair about telling *almadraberos* that they have to stop doing what they and their ancestors going back millennia have done because other tuna fishers have overexploited and endangered the stock. Ángel is right; the *almadraberos* are not the problem. It's like banning sex because some men can't be trusted not to rape."

Carl said the debate was moot: the world was running out of tuna. "You can keep saying, 'We didn't create this problem.' I've heard the Japanese say that, too. This is the tragedy of the commons—everyone is responsible for tuna, which means no one is responsible. The idea that we can continue to take what is going extinct, whatever the reason, is crazy," he said.

There are moments in Carl's presence when you wonder if he's agitated about an issue or just agitated because he's talking to you instead of doing something more productive. I got the feeling that both were true at this moment. As Lisa translated Carl's response, Ángel sat so far forward in his chair that I thought he might fall off. Both men looked pissed.

"I want to tell you something," Ángel said forcefully. "I want to tell you about the *almadraba* fishermen. I know of many, the older ones especially, who can rub the skin of a tuna"—Ángel rubbed the tablecloth with the tips of his fingers—"and when they smell it they smell the fat," he went on, his fingers at his nose, his breath drawn in dramatically. "They can tell you just from that how old it is, and how it will grade out at market." He lowered his hand from his nose. "Can you imagine how many years of culture that takes to understand?" Ángel didn't wait for a response. He excused himself to smoke a cigarette outside.

Carl calmly finished eating while Lisa leaned back in her chair and shook her head, looking at me. "To tell the people of Cádiz that they have to stop eating tuna means they have to stop being themselves," she said. "The *almadraba* is more than an annual ritual. It has influenced class structure, leisure activities, religious beliefs, cooking techniques, and probably even mating rituals as well, if you looked into it. It's so fundamental to the mores and traditions of this area that it isn't just a part of the culture, it *is* the culture." Lisa turned to speak to the waiter.

Carl leaned over to me. "Sure," he said, "but without biology there is no culture."

CHAPTER 21

B ACK WHEN chefs like Alice Waters and David Bouley began naming the sources of their ingredients in the early 1990s, it was a refreshing novelty. But that kind of proclaimed sourcing has reached a fever pitch: we get menus now written in a way that all but begs for ridicule. They declare the provenance of their vegetables with holy pronouncements ("Farmer Dave's biodynamic turnips"). Or they announce their do-good nature with a Girl Scout–like earnestness that's hard to stomach ("We source only ingredients that are good for the planet"). Menus like these strain to be virtuous.

I got to thinking about menu descriptions because, after months of back-and-forth, Miguel and several others from Veta la Palma decided to come to New York City to introduce themselves to a few seafood distributors, with the hope of exporting their fish to the United States.

For Blue Hill New York, which still features a traditional à la carte menu, I struggled for words when it came to Miguel's fish. Do you say "Sea bass" and leave it at that? Bass sells itself, thanks to Le Coze's popularizing the fish at Le Bernardin many years ago. But I wanted people to know that what they were eating was not just stunningly delicious but also a hopeful sign for the future of aquaculture. "Sustainable sea bass" is what I think I settled on, because it sounded the *least* annoying, and it had nice alliteration.

But it didn't matter, because the meetings did not go as planned. When I called Miguel after their appointment with the first purveyor, he said it had gone well (though Miguel seems constitutionally incapable of saying something negative about anyone). After a little prodding, he told me that the samples of bass they had brought were never talked about or tasted. When I asked if headway had been made in importing the bass, he said he couldn't be sure when it would happen, if it were to happen at all.

I had planned to meet Miguel and the others for lunch at Esca, chef David Pasternack's seafood-centric restaurant in midtown Manhattan, but a problem at Blue Hill developed in the late morning, and I knew there was no way I'd make it out. I called Pasternack to explain who was coming in.

Before I could get very far, he interrupted me. "You're sending me a couple of goombah fish-farming guys?" Like Ángel León, he found the mere idea offensive. Pasternack, a lifelong fisherman from the North Shore of Long Island, is almost pathologically obsessed with great seafood. He regularly fishes from Montauk to the Far Rockaways and has been known to ride the Long Island Rail Road into Manhattan with the previous day's catch packed on ice.

He proceeded to stake out Miguel as soon as he arrived. Miguel answered questions about the farm, then handed over a sample of the bass to Pasternack, who called me from his kitchen, having just filleted it and tasted a piece. "Fucking *very* good," he said, his voice low and controlled (likely not wanting to admit in front of his staff that he was impressed by a farm-raised fish).

Pasternack didn't waste time. One taste of Veta la Palma's bass and he had what he described to me as "a small epiphany" about the potential for farm-raised fish. While Miguel was still eating, he called Rod Mitchell, the owner of Browne Trading Company, in Portland, Maine, to convince him that this was a fish he should import without delay. Mitchell is among the most important seafood purveyors in the United States, and though it sounds falsely modest when he insists he's merely "a fish picker," it's also true. Mitchell picks fish for the country's best chefs, from Maine to California.

It started in 1980 when he met Jean-Louis Palladin. Through a mutual friend, Palladin discovered that Mitchell was a diver, so he went to visit the wineshop in Camden, Maine, where Mitchell was working at the time.

Palladin examined the store, but according to Mitchell he kept looking out the window.

"It was one of those foggy, rainy days in Camden," Mitchell told me. "The first thing he says to me, pointing out to the shore, is that the area reminds him of back home. He says he bets there are some great scallops out there. I told him he was right about that."

Palladin knew that at that time of year the scallops would be at their peak flavor. When the Atlantic waters turn cold in October and November, phytoplankton drop from the water's surface and fall to the ocean's floor. Scallops, like geese, gorge on the excess nutrients to fatten themselves for the winter months.

The next day, Mitchell dove for scallops in his favorite area of the bay. "Jean-Louis took one look at them and nearly cried. In that thick, deep, gravelly-voiced French accent of his, he said, 'What else can you get me?' That started the business." And an industry. Just as Palladin helped establish John and Sukey Jamison for their grass-finished lamb—influencing a generation of small farmers—and just as he converted mushroom hobbyists into full-time foragers and small milk producers into cheese artisans, he transformed Mitchell from a recreational diver into one of the most important seafood distributors in the country.

"All of a sudden we got run over with demand," Mitchell told me. "It was like Jean-Louis had a front-page ad in every chef's morning paper. I kept hiring divers. These guys couldn't believe it. We'd dived for scallops all our life, just for fun, and now chefs in New York and Boston wanted to pay us a lot of money to do it."

Now on the menus of countless restaurants across the country, "day boat" or diver scallops are well-known to be the sweetest and most flavorful of their kind. They're among the most sustainable, too. Until Mitchell began paying divers, scallops were almost always harvested by dragging nets along the floor, up to one hour at a time. The divers may yield smaller harvests, but they are more discriminating (and less destructive) in what they catch.

When the scallop season ended, Mitchell sought out other fish to sell to chefs. "I just purchased the best of everything I could find," he said. "I had chefs, the best chefs in America, wanting to know how many scales were on the damn fillet. You couldn't fool these guys. So we always bid first and last, always, because I couldn't afford not to have the best stuff."

It was the beginning, Mitchell told me, of the specialty seafood business. An outlet for fishermen developed in the same way farmers' markets created an outlet for small farmers. Before long, Mitchell was delivering seafood to all fifty states. Even Gilbert Le Coze, who hadn't been interested in meeting Mitchell until the early 1990s, finally invited him to Le Bernardin. ("He brings me to the kitchen, stares at me, his right eye twitching in the way he did when he was nervous. 'If it's not the best fish you have,' he said, 'if it doesn't look like it just came out of the water, my name should not enter your mind—I do not exist.'") The restaurant would soon become Browne Trading Company's largest single purchaser, and remains so today.

The specialty seafood business gave a coterie of American chefs impeccable fish. It also gave small fishermen a viable business. And yet their thoughtful practices did not reflect the general ethos of the fishing industry.

With advances in technology and the scorched-earth tactics of drag-netting, seafood catches looked like those towers of shrimp at an all-you-can-eat buffet: bounteous and inexhaustible. The fishing spree of the 1980s and '90s decimated fisheries, kicking off the kind of lavish party that never ends well. Among those who understood that this was too much of a good thing were fishermen themselves. They warned of future problems. Many clamored for regulation. (And when fishermen want more government involvement, something is terribly wrong.)

Mitchell is adamant that if Palladin and Le Coze—and all the other chefs who followed in their footsteps—hadn't created the specialty seafood market, the big boats would have cleaned out the seas long ago. "They still might," Mitchell told me. "Chefs helped good fishermen distinguish themselves from the ones who weren't. These small fishermen are not the problem. They're part of the solution. They generally catch the fish at the right time, when they're fat and full of flavor, not young, tasteless, and unable to reproduce. You can't do that with large boats."

But chefs are also to blame. Palladin and Le Coze helped create a demand—and a chain of supply—that would eventually cripple the industry. It's not hard to see the irony. The two men who perhaps did the most to help chefs access quality seafood and introduce Americans to unknown flavors of the sea catalyzed the decline of many of the fish they promoted. (They never lived to see it. Le Coze suffered a heart attack in 1994, at the age of forty-nine, and Palladin died of lung cancer in 2002, at the age of fifty-five.)

No one could have predicted the declines—except perhaps Carl, who warned Ángel of his efforts to popularize the unpopular. But even Carl is amazed at the rapid destruction of once plentiful fish like monkfish and skate, which Le Coze single-handedly popularized twenty years ago. The chefs' influence was profound—and it still is.

Mitchell called me recently after visiting the Portland Fish Exchange. Daily catches now average about seven thousand pounds. "When there's a twenty-thousand-pound day, fishermen say, 'Look, plenty of fish!' Which is what we used to say in 1988 after a *two-hundred*-thousand-pound day."

It's brought chefs into a kind of meta moment: prices have risen so drastically in the face of steep declines that a new generation of chefs is struggling to include fish on its menus. They're asking questions about a future that appears grim. Is the answer to identify and support better fisheries management? Or are we going to have to cook with less fish? Or less local fish? Or more of the less desired fish? Or more farmed fish?

Over the past twenty years, we've used our influence to identify and boycott particular species of fish—a thumbs-up for fish A, a thumbs-down for fish B. The "Give Swordfish a Break" campaign of the late 1990s was one such example. It worked spectacularly well. More than seven hundred chefs around the country pledged to drop one of their best sellers from the menu. Thousands more soon followed.

Calling attention to certain well-managed fisheries ("Day Boat Chatham Cod," "Alaskan King Salmon," "Maine Diver Sea Scallops," "Hook and Line Haddock," "Sustainable Sea Bass")—the widely imitated Bouley form of menu writing—may feel contrived, but it's been another significant step toward raising the public's consciousness and preserving the right kind of fisheries.

And yet the question remains whether, in advertising these delicious and sustainable alternatives, we may be unwittingly endangering them for the future.

CHAPTER 22

When Ángel described the ancient tuna-fishing ritual of the *al-madraba* over lunch with Lisa and Carl, it had sounded interesting and rewarding in the way that important museum exhibits often sound interesting and rewarding—but don't necessarily compel you to go. It didn't help that since the lunchtime debate between Ángel and Carl, I had gotten the suspicion that Ángel, notwithstanding his good intentions, had it wrong about the *almadraba*. Without seeing it, I had sided with Carl, who has an uncomplicated view of fishing bluefin (don't do it) and sustainability (biology trumps culture). How can you disagree with outlawing the killing of a species that's going extinct?

But six months later, Ángel called Lisa to announce that he'd partnered with Miguel and Veta la Palma for an exciting new project. Partnered? The last I'd heard from Ángel, he hadn't even deemed Veta la Palma worthy of a visit. I was a little jealous, then impatient to learn more. A partnership between Ángel and Miguel seemed to me inevitable, a meeting of complementary minds.

Ángel called me himself a week or so after that to say, first, that he wanted me to see his "revolutionary new project before the world learned of its existence" and, second, to inform me that the captain of one of the *almadraba* boats, an admirer of his, was willing to allow me on board during the catch.

Lisa (who would also be allowed on board) called me as soon as I hung up the phone with Ángel. "It's incredibly rare for an outsider to get to witness the *almadraba*," she said. "Especially a foreigner!"

I reached out to Miguel, wondering how he felt about the opportunity. There was silence on the phone, and then Miguel said, "I don't want to impose, of course, but is there any way you think I might be able to go with you? I mean, this is my dream."

Later that week, Ángel called one last time. "Come. Do as the Romans did," he said. So I went.

—◦—

Lisa, Miguel, and I planned to meet the evening before our *almadraba* expedition at El Campero restaurant, in the town of Barbate, in Cádiz, the southernmost province of Spain. Shoehorned between the Atlantic Ocean and the Mediterranean Sea, the area evokes a forgotten time of small towns and picturesque coastline, before high-rises and cheap tourist attractions blighted the country's seaboard. There's a warmth to Cádiz, too, and it's not just from the blazing Iberian sun. The people seem friendlier, more relaxed. Maybe too relaxed: Cádiz's staggering unemployment rate has for decades been Spain's highest.

Barbate is the most famous of the *almadraba* towns along the coast, and El Campero, as restaurants go, is bluefin ground zero. Nose-to-tail dining is how a place like El Campero would be described in Brooklyn or Berkeley. (And that may be underplaying it; you can order tuna face, heart, ear, and semen.) But the restaurant is bright, nondescript, and totally without pretension. There is no big to-do, no fuss over a cuisine that utilizes the whole animal.

As we waited for Miguel, Lisa explained how bluefin tuna is a way of life there. "Like Eskimos and their fifty names for snow, the people of Barbate have twenty-five words to describe parts of the tuna," she told me. "This isn't a trend or a fad. Tuna is to the people of Barbate what *jamón ibérico* is to most Spaniards—culture by way of gastronomy, with deep ties to the region's identity."

While she spoke, I started to realize something that had eluded me until this moment. Miguel and Veta la Palma (and Eduardo and the *dehesa*) had made such strong impressions on me at least in part because of Lisa herself. Her translation made my visits possible in the first place, but more than that, I was able to better understand them thanks to the framework of information— gastronomic, historical, religious, and cultural—that she provided.

Farm-to-table cooking, or any of the gastronomic variations associated with sustainability, usually involves something larger than delicious food. The relationship between a farmer and a chef, or the connection between a community and an ecology—the story behind the food—can be as impor- tant as the food itself. (*Jamón* is delicious, but *jamón* as an expression of a two-thousand-year-old landscape is something to savor.) The waiter is often that conduit. Food writers can be, too. At Stone Barns, the education center plays that role, bridging the gap between farm and restaurant. But here in Spain, it took someone like Lisa, a deeply knowledgeable interpreter of the language and culture, to illuminate the meaning of the experiences.

Miguel arrived looking exuberant. He and his wife were adopting a baby from China. He said he had been busy lately, taking a course on Chinese his- tory. He had also spent the past year learning Mandarin, which he found tir- ing and difficult but worth the effort. "She will be Spanish, of course, but she will know who she is as well."

It was a beautiful early evening, the sun's brilliant light settling into the town. I asked the waitress if we could move to one of the outdoor tables, which were all unoccupied. She looked at me, puzzled. When we arrived at a table, it seemed as though she still couldn't believe we actually wanted to sit outside. She spoke quickly in Spanish, then cried out, "*Levante!*" I must have seemed confused, because she looked at Lisa and Miguel with an expression that said, *Tourist?*

Miguel said he'd heard warnings of the *levante* on the radio while driving to meet us. He explained that the *levante* is one of the strong winds that his- torically plagued the region's fishermen. One belief has it that during the worst of the winds, dead people's souls are being blown from their graves;

another tradition, like that of the French mistral, suggests that the winds make you a little loopy. Either way, their existence, defined by the direction of the wind, determines the fate of the fishermen's catch, and therefore their survival.

The waitress returned with our beers, inquiring about our comfort. I didn't even feel a breeze, or the hint that one might appear. But she was very concerned. When we assured her we were fine, she shrugged, washing her hands of responsibility.

By the second beer Miguel had relaxed in his chair and looked out over the town, the fading light painting the sides of Barbate's ramshackle homes. I asked about Veta la Palma. I knew that changes were imminent, especially with the farm's sea bass on the verge of being available in New York City. Miguel didn't look very excited by the prospect.

"Any day now they tell me the papers will clear," he said. "We've entered into a kind of spiral that we can't get out of." And then, as if catching himself, he added, "In a good way."

He got up and excused himself for the bathroom. I asked Lisa how a spiral could be good—maybe the stress of the job was getting to him. The waitress appeared with more drinks. Lisa shrugged and whispered, "Maybe it's the *levante*."

When Miguel returned, I asked him if he had second thoughts about selling his fish abroad. "No, no, this is very exciting. But I worry about how much bass we will have to sell." Since I had been under the impression that Veta la Palma could sell a lot more fish if only they had greater demand, I found the admission surprising and asked him to clarify.

"True, absolutely, we do have a lot more fish to sell, except chefs want the sea bass. Right now the farm produces twelve hundred tons of fish per year—nine hundred of this is sea bass. We don't have trouble selling bass."

I told him I didn't see the problem. "I think the carrying capacity of the system is about two thousand tons," he said. "I don't believe we can go beyond this without compromising the quality or hurting the farm." Two thousand tons did not amount to a lot of fish. Should New York City chefs taste what I tasted, the farm's supply would be exhausted in a few days. Factor in chefs from Las Vegas, San Francisco, and Los Angeles—all popular spots for Rod Mitchell's deliveries—and it would last a few hours. Miguel nodded.

We sat in silence until Lisa asked, "Could Veta la Palma produce more if some of the other fish you raised were popular?"

"Yes," he said, looking at her with an expression of relief and bewilderment—relief, perhaps, that such a bewilderingly obvious question hadn't been asked before. "Yes, absolutely. The sea bass, as I have explained, is raised in a semiextensive regime. That is to say, from at least March until October, and sometimes longer, the natural productivity within the system—the phytoplankton, zooplankton, crustaceans such as shrimps, small wild fishes, et cetera—feeds the bass. For the other times, the feed is complemented with a supply of dry food, or fishmeal."

"Isn't this true of all your fish?" I asked him.

"Oh, no, not at all." Miguel pushed his beer aside and sat forward in his chair. "The mullets, for example, their regime is absolutely extensive." He paused to make sure I understood. "No feed. Nothing. It is important to remember that the bass are active predators. They are carnivorous and rank high on the ecological network—the amount of energy needed to raise them is greater than the mostly herbivorous mullet. Mullets need less energy to live and reproduce. This is the second law of thermodynamics, and it's applicable to the laws of ecology—well, actually, it's a *principle* of ecology, as ecology doesn't have a lot of laws."

Just then, a strong wind arrived so suddenly, it was as if a switch had been flipped. A brutal current of air threatened, for a moment, to levitate our table and the three of us along with it. And then, just as suddenly, it died. The calm returned, only now it felt a little eerie.

Miguel hardly noticed. "Mullets are filter-feeding fish. They are all the time removing excess nutrients—nitrogen and phosphorus, to name the two most important. If they don't get taken up by the mullets or the other filter feeders in the system, these nutrients become concentrated."

"Algae blooms," I said, wanting to show him that I followed his logic.

"Yes," he said, sitting back a little in his chair. "They do the work for you. This is the ecological network. They are the keystone to the network. My belief is that within this network we could double the rate of mullet production, if we had a market for mullets."

"That's the problem," I said. "Mullet. Tough to sell mullet." Miguel nodded sympathetically.

"It sounds like I fell for the wrong fish," I said.

"You fell in love with a fish that has the highest commercial value but the lowest ecological value. So, maybe," he said. "Yes."

❧

We moved inside for our dinner reservation. Ángel had arranged for the mayor of Barbate to welcome us at some point during dinner. Lisa was on the phone with him when we sat.

"By now I'm sure you've heard," the mayor said into the phone, without introducing himself. "The *levante* is fucking us."

"Does this mean the *almadraba* is going to be impossible?" Lisa pressed. "Because we've come all the way—"

The mayor interrupted her. "Please, Miss Lisa, we're in Barbate. Nothing is impossible," he said. He promised to join us soon.

As we waited for menus, Lisa diagrammed the *almadraba*. She made a rough outline of southern Spain on the back of her napkin. "We're here," she said, marking an x near the southernmost tip of Spain. "Africa is right down here." She sketched Africa. Morocco's shores, at the northwest tip of Africa, were not far away. The Strait of Gibraltar, dividing two countries—and two continents—measures only eight miles wide, making the waterway more riverlike than the geographic distinction seems to warrant.

"The Strait of Gibraltar, of course, connects the Atlantic over here"—Lisa drew an x for the Atlantic, to the left of Barbate—"and the Mediterranean over here," where she drew another x, across the Strait and far to the right of Barbate. "So the tuna enter from the Atlantic, looking to spawn in the Mediterranean, which is what they're programmed to do. The labyrinth of *almadraba* nets is set up along the coast here." She made a line that passed several towns along the coast, including Barbate.

"As the tuna flood into the Strait—well, they don't so much flood as sort of trickle these days—some of them swim closer to the shore. That's when they enter the nets, which are placed in the water from the middle of May until sometime in June or July, depending on the quotas and the weather and the number of tunas. They move into successively smaller nets, until they reach the last chamber, roughly the size of a football field. And that's when the *levantá* [literally the raising or lifting up—and not to be confused with *levante*] begins."

The mesh holes of the nets are large enough to allow smaller tuna, the ones that haven't reached maturity, to escape and forge on to the Mediterranean, where they continue their life cycle.

A waiter arrived with raw, thinly sliced bluefin tuna belly, perplexingly served with chopsticks, soy sauce, and wasabi. For a restaurant specializing in classic southern Spanish cuisine, it seemed incongruous. I hadn't cooked bluefin (or eaten it) since that fateful day I prepared it for Caroline Bates. The memory, and all that I'd learned over the past few years, didn't make me especially hungry. I allowed myself the thought of not touching it, but then what was I doing at a restaurant devoted to bluefin tuna?

Miguel, apparently free of guilt, had nearly finished by the time I picked up my chopsticks. "I like this very much," he said between bites. "The wasabi reminds me of the Indian food I had in Tanzania. I like it!"

If there was anything disconcerting about watching a champion of sustainability tear into the *toro*, I didn't get far forming the thought. The chef of El Campero, Pepe Melero, suddenly appeared at our table. Pepe was short and stocky, with a round, weather-beaten face. His outsize mustache was exaggerated by small, recessed eyes that darted around as we introduced ourselves. "Ángel asked that I prepare a little tasting menu of tuna," he said shyly. Miguel smiled, nodding approval.

I asked about the Japanese influence, pointing to the pool of soy sauce on my plate. Pepe explained that the Japanese arrived in Barbate and the surrounding towns about thirty years ago as bluefin in the Sea of Japan became harder to find. Impressed by the quality of the *almadraba* tuna, which they began buying in large numbers, Japanese boats arrived to survey the scene. At the time, tuna in Spain was still mostly preserved or canned. There was no tradition of raw tuna. Sensing an opportunity, Pepe invited the boat captains' personal chefs into his kitchen.

"They were incredible to watch," he said, admitting that he couldn't do much else—they spoke no Spanish, and he didn't speak Japanese. The chefs showed him a completely new way to butcher tuna, starting with the belly. "They were so fanatical about cutting according to the fat. It changed the taste of everything."

So did the method of killing, which for about three thousand years had been a barbaric affair: the tuna were hooked by the neck, dragged onto the boat, and violently clubbed to death. "A bloody, horrific thing," Pepe said mournfully. The Japanese showed the Spaniards a different method that involved raising the nets, tying ropes around the tails, lifting the tuna, and plunging them into ice before slitting their throats.

"This is precisely what we do with all of our fish," Miguel said urgently. "This is most important." The humanity of the minimal-stress slaughter yielded significantly better-flavored tuna, which only raised the price.

"Now the death is cold and sweet. Before it was much more bloody," Pepe said, pausing to consider the history of tuna's last moments. "It was more spectacular, too."

Lisa asked Pepe if the Japanese ever explained the reason for preferring the *almadraba* tuna. "Because of the fat!" he said. Then he became very serious. "The legend is that the tuna hear a siren call from the Mediterranean at a certain point in their lives. It's at that point that their meat is at its best—the most optimal point to eat tuna. So they go to spawn. But what brings them into the nets?" he asked, his eyes scanning each of us. He appeared delighted that no one ventured even a guess. "There's a legend for this, too. It's that the tuna's fatty belly gets an itch, like a pregnant woman. They are drawn to the shallow waters to satisfy the itch. This is when they stumble into the nets."

While Pepe shared his legend of the tuna, I impulsively dunked a slice of *toro* in the soy sauce and dropped it into my mouth. I couldn't believe the flavor. It was richer and more intense than any tuna I had ever tasted, a fact that I noted to Pepe.

"Hmm, yes," he said knowingly, "the *almadraba* tuna are at the peak of flavor. All of the energy goes into great intramuscular fat, producing a tuna that is at the moment of perfect flavor." I thought of the Copper River salmon that David Bouley had exalted in his kitchen. I had always assumed the Japanese purchased from the *almadraba* because of dwindling stocks—a desperate move to satiate their world-leading appetite for tuna. But now I realized that the superior flavor must have driven the interest as well.

Pepe nodded vigorously in agreement. "Yet over the years I've noticed a severe reduction in the amount of fat. It's hard to believe, but it's true. Just the other day, I was standing next to one of my cooks as he prepared to sauté the neck of the tuna. He got his pan nice and hot and added a good amount of olive oil. Nothing wrong with that, right? Except I got so angry I almost kicked him out of the kitchen. It wasn't his fault. I became a little overwhelmed because when I was his age, the *morilla* was so full of fat, you didn't even put oil in the pan. The natural fat just poured out, more than enough to fry it. The tuna cooked itself."

The mayor arrived. He moved quickly across the room, shaking hands, grasping an old man behind the head and kissing him on both cheeks, pointing his index finger like a gun to another. As he exchanged cheek kisses with Lisa, his hand jutted out from the side for me to shake.

"Thank you," he said, turning to me by way of an introduction. "I know of your work. This is a big day for Barbate. Tomorrow you will help us preserve tradition. Three thousand years of tradition. I think I must fight for this, and so I do." He motioned for us to sit back down, a pastor in command of his congregation. "Of course, if the wind doesn't stand in your way," he said to Lisa as he removed his coat. "How do you say in English? It's totally fucked."

Another waiter cleared our empty plates of *toro*. "Did you like?" the mayor asked. "Incredible, incredible, I know," he said, without waiting for a reply. "When I was a child, there were so many tuna running at this time of year that I remember seeing the tuna washed up on shore, lined up on the beach one after the other, with huge bites taken out of their bellies. Even the sharks knew!"

The second course arrived. *"Mojama,"* the waiter said, gesturing to the plate of cured tuna loin. A well-known Spanish delicacy, the loin is salted for several days, washed, and then laid out to dry on rooftops against the blazing sun and strong breezes. Miguel reached for the thin slices. "The ham of the sea," he said. The mayor shot Miguel an approving glance.

Examining a piece between my fingertips, I was struck by tiny, capillary-like streaks of white fat. I'd never seen them running through the loin, a traditionally lean cut. It reminded me of the *jamón ibérico* Eduardo had held up to the light, its weblike striations of fat a "perfect expression of the land." I was devouring what had to be the perfect expression of the sea.

The mayor continued talking. "Back then, up through the 1960s, everybody lived off the tuna. There were a million workshops, places where the fish would be butchered and processed—preserved, canned, that kind of thing. The people of Barbate, we really only ate the scraps. Everything that could be preserved was sold throughout Spain. No one wanted to eat the profits! We were left with incredible cuts, though, like the *morilla*—that was always very popular. Did Pepe tell you about the *morilla* having so much fat it cooked itself?" We nodded. "He loves telling that one. I even think it's true."

The mayor popped out of his seat to greet two locals who had just entered the restaurant. As the plates were cleared, he sat back down, resuming where he had left off. "We were drowning in tuna, but then in the late '6os the industry suddenly collapsed."

"Because of depletion," I said knowingly.

"No, because of anchovies!" he said. Again the mayor stood to greet guests, shaking the hand of an elderly man.

He sat back down, motioning for the waiter to bring more beer. "No, the people of Barbate started working with anchovies because anchovies paid better. That was then. They paid a lot better, and plus, tuna was just three months of work. It's three months of very, very hard work. Only now, since the government is further restricting the number of days to catch, and the world is calling for an all-out ban of tuna fishing, the people here say, 'Hey wait a minute, this is our tradition!' It became important the day they tried to take it away."

Tuna heart was the next course. "Eat this one," the mayor said as his cell phone rang. "You're going to like it very much." The heart was glazed and sliced thinly. It tasted like a chewy fillet of beef, with a wash of sea flavor to finish.

"It's the captain of your boat," the mayor said, covering the phone's receiver. "The wind is terrible." (Lisa threw up her hands and looked at me. "What wind?" she whispered. "It's a goddamn *breeze*.")

The waiter returned, this time with a flight of *hueva* "lollipops" served on sticks. Lisa and Miguel couldn't agree on the definition of *hueva*, the debate turning on whether or not they were the same thing as gonads (the sac that holds the semen). "No, no, it's not semen. It is certainly more refined than that," Miguel said.

The mayor covered the phone again. "Balls," he said authoritatively.

"Please," Miguel said with great seriousness, "we cannot make an analogy with a mammal."

The mayor shrugged when he finally got off the phone. "It's out of my hands, really. They will make a decision in the morning, by ten o'clock. Pray for no wind."

Yet another tuna course appeared: the *morilla*, which, sautéed as Chef Pepe had described, looked to be oozing with plenty of fat. "This one I'm going to eat," the mayor said. "Because I approve of this." The waiter brought over a plate for the mayor. "I'm in love with tuna," the mayor said. "But I only eat tuna three months of the year. From Easter until San Juan—the longest day of the year—all I think about is tuna. After that I don't eat it. I don't even think about it."

I asked if he felt bluefin would be around for his children to enjoy. He answered carefully. "Basically, this whole tuna thing isn't a problem. The quotas are working—there have been big improvements in the stock."

Lisa, perhaps still smarting over the possible cancellation of the *almadraba* in the morning, recited a long list of statistics about the decline of bluefin. She pressed the mayor, who fidgeted and looked around the room to see if anyone was within earshot.

"Look, we don't want to take away anyone's business. We've been doing this for three thousand years. The Japanese have been doing it for thirty years. And that's when—so they tell us, anyway—the stocks became depleted by 90 percent." He looked at Lisa, with an expression that begged for some common sense. "The *almadraba* is seven hundred tons of tuna per year, which is about the same number caught by a big trawler in one day. Tell me something, are we the problem?"

I asked if there were any strained feelings toward the Japanese, who now purchase almost all the *almadraba* tuna.

"The Japanese have done nothing wrong," he said loudly. "We have a terrific partnership with the Japanese, absolutely terrific. There is no problem. Zero. I would like to set up a direct flight to Cádiz from Tokyo. They can bring their cameras and their little hats, visit the beach, and catch tunas." He was silent for a moment as the *morilla* was cleared. "The Japanese ambassador came to Barbate a few years ago. I kissed his wife two times. She was visibly upset. You know what? I don't care. When they come to Spain, we do it our way. When I get invited to Japan, okay, fine, no kissing."

As the last course was served, a small local fish roasted in its entirety, the mayor volleyed back and forth between gratitude and regret about the Japanese influence on the future of his town. His ambivalence makes sense. He's reliant on the Japanese to buy the bluefin at the highest prices. At the same time, it's the demand for tuna that's decimating the stocks and killing the *almadraba*—and the mayor's town.

"We should probably just shut down *almadraba* for a few years to recover the stocks. But if we shut it down, everyone has to shut down," he said quietly, with a look that acknowledged the impossibility of such a thing happening. "What can we do? We just fish our fish for three months of the year." The mayor paused and looked at me. "For three thousand years."

The next morning, Lisa, Miguel, and I were to meet the mayor at a bar along the docks of Barbate. The wind blew swirls of debris around the empty, brick-lined streets. Boats rattled against the shore. The decision about fishing for the day would be made at 10 a.m., but the mayor had arranged a breakfast meeting with Diego Crespo Sevilla, an owner of four *almadraba* boats who controls one of the major sections of the nets.

The bar was nearly empty, with just a few old-time fishermen clutching espressos and dangling cigarettes, and crumpled tapas receipts littering the floor from the night before. Black-and-white photos of the 1940s and '50s lined the walls, a shrine to past glories. The left side honored the bullfight—matadors brandishing swords, wounded bulls succumbing tragically. A set of pictures depicting the *almadraba* was on the right: enormous white waves framing heroic fishermen as they lorded over bloody and bludgeoned tuna. The montages told the same story—man's victory over nature—but as I faced the old tuna fishermen standing in the morning's empty bar, there was a feeling, as strong and undeniable as that stench of old fish, that neither side had won.

The mayor arrived, followed shortly thereafter by Diego, both men shaking hands with the old-timers as they made their way over.

The mayor introduced us. "Diego, say hello to my friends. They have come to experience the great *almadraba*. They await your—"

Diego ignored him and shook our hands. "We'll know in a few minutes about the decision, but I can tell you it doesn't look good," he said authoritatively. "I believe the winds are getting worse every year."

"Not good for the *almadraba*, but quite beneficial for the windsurfers," the mayor offered, careful to weigh the feelings of another one of his constituencies.

Diego's physique suggested a taste for sherry and idle afternoons by the dock, but he carried himself like a seasoned businessman, remote and elusive. In lieu of a suit, he was dressed in khakis, a sky-blue Polo sweater, and loafers without socks. I asked him if the closing of the *almadraba* during the height of the season meant economic hardship for the fishermen.

"As of today, we're at 78 percent of the quota for the season," he said. "The *almadraba* could run for at least another month, maybe more, but all we have is one or two more days of fishing and we'll be done." The bartender served Diego's espresso, refusing to take his money. At this, the mayor looked peeved. "When I was a kid, there were seventeen *almadrabas* along the Spanish coast," Diego explained. "We're down to just four."

"And we are stopping this madness at four!" the mayor broke in, dropping his cup heavily on the bar and getting the attention of the men at the tables. "We must defend what's left with everything we have."

Diego again ignored him. "It's a very passive art, the *almadraba*, so it's tough to predict, and to synthesize. But the trend has been way, way down."

Forty years ago, the International Commission for the Conservation of Atlantic Tunas (ICCAT, pronounced *eye-cat*) was formed to oversee and manage the dwindling tuna stocks. Perhaps a committee representing forty-eight member nations, with unequal influence and conflicting demands, was doomed to fail. Either way, it has failed spectacularly. (Carl Safina proposed several years ago that ICCAT be renamed the International Conspiracy to Catch All Tuna.) To this day, ICCAT's own scientists suggest quotas that the committee openly ignores, often doubling the recommendation. It's an organization that can't rule, exactly, but it can, and does, use rules to ruin.

I asked what Diego thought of ICCAT's work.

"ICCAT is a fine organization," the mayor volunteered. "Very nice people. I will say to you that, while ICCAT's been accused of lots of things, what's never been debated is that they are trying. But my God, they have helped make a mess of things."

Diego waved his hand. "There are two things to keep in mind here. First, ICCAT's numbers, inflated as they can be, are not obeyed. If the countries had obeyed even those ridiculous quotas, I think we would have been fine. But Spain overfishes. France overfishes. Libya—"

"Libya!" The mayor rolled his eyes and threw up his hands.

"Libya overfishes," Diego continued. "You see, the *almadraba* is the last of the honest fishing. Because it has to be. Inspectors meet us on the docks. They count the fish we're unloading. We have nothing to hide, because we cannot hide anything." The mayor grabbed the pockets of his trousers and pulled them inside out, showing what it meant to hide nothing. "Out at sea, the Japanese helicopters meet them on the boat. Or the boats dock in Vietnam—in Vietnam, they couldn't care less about counting the number of tuna caught."

Miguel, who had been listening attentively, neatly summarized the situation. "The *almadraba* is the best information source about the state of the stocks. It's a living lobotomy of what's happening in the ocean."

"Perfecto," the mayor declared. "It's like taking the temperature of the ocean."

I asked if it wasn't true that some biologists, like Carl Safina, had more or less predicted fifteen years ago what that temperature would be in fifteen years.

Diego was quick to reply. "Doctor Safina wrote *Song for the Blue Ocean* before tuna farming. It had just been invented. He couldn't possibly have predicted how awful the situation would become. This is the second thing to keep in mind," he said, wiping his mouth neatly with a paper napkin and searching my eyes for any trace that this was sinking in. "The farms are sealing the fate of tuna."

Diego explained that the bulk of farmed tuna is not coming from eggs. It is from tuna caught in the sea at an average of thirty-five pounds and fattened on these farms until they double in weight. The practice not only does nothing to save nature's dwindling stock, but it hastens the decline. Bluefin are removed from the wild before they have the chance to spawn.*

He told us that tuna farming is worse than industrial tuna trawlers, an argument that at first I thought might be far-fetched. It was not. "The way you fish tuna, even on the large boats, is that after the fish are caught, the boat needs to unload. Catch, unload, go back out—there's a natural resting period. For the farms, they bring out enormous nets, drag schools of tuna into the farms, and then go out and drag in another school. There's no recovery time."

The mayor: "Again, madness."

"We should perhaps not call these places tuna farms," Miguel said. "They aren't farms. Farming is a closed system. If you're talking about fish, you

---

* Kindai tuna, a method of bluefin tuna "ranching" developed by Japanese scientists in 2002 (and criticized by Ángel during my meal with Carl), accomplished what was once thought impossible: producing bluefin from hatched eggs rather than captured juveniles. But their success doesn't do much to solve the problem inherent in aquaculture. The amount of wild fish required to fatten bluefin in captivity makes the practice seem exorbitant. Farm-raised shrimp require two pounds of fish feed for every pound of harvested shrimp; by comparison, tuna, as a top predator on the order of a tiger or lion, has a ratio that's closer to twenty to one. And they're pickier about what they eat, preferring delicacies like sardines, anchovies, and herring. So while Kindai might have resolved certain issues, it continues to gnaw at conservationists, who see the ocean's limited resources dwindling in the face of increased demand.

hatch them, you grow them, you feed them. The tuna farms are just for finishing."

I had never thought of farmed tuna as similar to American-style grain-fed beef, and yet Miguel was right. We remove cattle from ranches and fatten them quickly on exorbitant grain diets before their slaughter. We don't call the confinement beef operations "farms," because they aren't farms; they're feedlots.

Diego answered his phone, excusing himself for a conference call on the final decision about the *almadraba*.

I asked Miguel if he had ever tasted farmed bluefin. Once, he said, and he didn't care for it. My experience was the same. The difference in flavor is not unlike the difference between confinement-raised Iberian pigs and the real thing. The fat, abundant because of acorns, never fully incorporates in the muscle without the exercise required to forage. Tuna raised in pens have a similar kind of fat—abundant but not well dispersed. Muscle activation is key to distributing the fat and carrying the flavor, whether it's a pork chop or a porterhouse—or a loin of tuna.

"Muscle activity is of course very important," Miguel said. "Stress is just as important. Maybe more. How the fat is distributed is one thing, but stress affects the *kind* of fat. The flavor of the fat from a stressed fish is very different.

"I see it with our own fish—especially the mullet I spoke to you about yesterday. At Veta la Palma, they must be the least stressed fish in world." Lisa laughed. "No, it's true," Miguel continued. "They enter Veta la Palma of their own will, because they are hungry and they know the estuary is a healthy environment. This is very important. We don't take them from the ocean and put them into our system. They come because they want to be there. And they are of course safe from predators—except some species of birds—but otherwise we're like a nursery. They feast on the abundant health of the system and in the process help maintain the health of the system. Sometimes I get the feeling they know all this and reward us with flavor." I could have been speaking to Placido about his hams, or Eduardo about his livers.

Diego returned and shook his head apologetically. "I'm very sorry," he said.

"*Levante!*" the mayor cried, throwing up his hands.

TWO MONUMENTS TO TUNA

The consolation prize: Diego and the mayor took us to a museum. They led us into a sleek, modern building, out of step with old Barbate and aggressively contemporary in a way that screamed tourist attraction. It is the first, and probably the last, museum in the world dedicated to the *almadraba*.

The wisdom of such a major investment was lost on me—and, to be honest, it made me kind of sad, too. We seemed to be the only visitors, for one thing. And judging by the generous greeting of the docents, it felt like we were the first in a long time. I somehow doubted that an *almadraba* museum would draw even one more visitor to Barbate. We were standing in what may soon become a mausoleum to bluefin.

We passed a display of the tuna netting, a thick, woven web of rope. Diego stretched one of the holes to demonstrate the wisdom of the system. "You see? Smaller tuna are free." Diego swam his other hand slowly through the hole, and then did it again, with the neat efficiency of a stewardess demonstrating how to fasten your seat belt. "Only the older fish get trapped. An eighty-year-old woman doesn't have the potential for so many babies, does she? It's the same with tuna."

A film played in a small room, where I saw the only other museum visitors—a Japanese couple with their young son—sit silently and stare up as a giant bluefin filled the screen. We learned about bluefin in the tone of a "Did You Know?" kindergarten lesson. I mostly did not know.

I didn't know, for example, that tuna can grow up to twelve feet and weigh on average 550 pounds. Or that they can live for thirty years. I didn't know that, like sharks, if tuna stop swimming, they suffocate, making them machines of motion. I didn't know that their dark red flesh, coveted by chefs for its meaty flavor, is the product of blood supply to the muscles. Unlike most fish, bluefin are warm-blooded and can also thermoregulate, adjusting their body temperature so they're always warmer than the surrounding waters. Warm blood is key, because it means their bodies expend less energy as they hunt for food, which they do at speeds of up to thirty miles per hour.

As we left the museum, I asked Diego about the legend Pepe had recounted about the larger bluefin entering the nets to satisfy their belly itch.

"Pepe?" he said, smiling. "No, that's crazy, of course. It's a fable." He put his hand on my arm and stopped me in the parking lot. "Do you want to know why the tuna come to the nets?" I assured him I did. "They come because they've had a great life. They've reproduced. They've eaten well. They've traveled. Now they want to die a dignified death. They know they will be treated with respect. Dignity in death is important to bluefin, because they are so dignified in life."

There's a black-and-white photo I remember seeing as a young boy, of a group of men standing over a slaughtered buffalo, rifles slung over their shoulders, expressions smug. You can't help but look at it and wonder: *What were they thinking?*

In *Song for the Blue Ocean*, Safina compares the massacre of sixty million buffalo that once roamed America's prairies (and supported Native American cultures for centuries) to the present-day destructive hunt for bluefin tuna. The comparison struck home a few years ago when I saw a photo of a record-breaking bluefin tuna just off a Japanese auction block. Around it stood a half-dozen fishmongers, all of them wearing that same self-satisfied look. Our greedy appetites made for the buffalo's quick decimation; we've effectively done the same to tuna.

Carl's prediction came to mind as Lisa and I stood in front of Baelo Claudia, a Roman ruin just to the west of Barbate in Tarifa, another important *almadraba* town along the southern coast. Dinner that evening would be at Aponiente, where we would learn of the "revolutionary new idea" Ángel had been working on with Veta la Palma. Distracted by the promise of Ángel's food, and uninterested in another lesson in history, I arrived at Baelo Claudia in the wrong frame of mind. It didn't take long for it to change. Having parked the car and walked just a few steps past a beach walkway, we came to the spectacular—and the spectacularly discordant—sight of the ruins, the remains of a 2,200-year-old salting factory for preserving the *almadraba* tuna. The industrial-size factory—enormous columns, salt basins, large, coffin-like drying areas—was surrounded by more relics of a once prosperous port town for the Roman Empire. It all looked like the backdrop for the production of a Classical drama.

There was something surreal about the scene, and it wasn't just that Baelo was ancient and yet so well preserved. (Americans are always gawking at such sights in disbelief, while Europeans seem to view them with cool nonchalance. True to form, we were the only ones actually looking around, while a group of bronzed sunbathers lined the beach nearby.) The surreal setting had to do with the striking prominence of the tuna-preserving factory, a prominence that Lisa pointed out as we approached. The entire town seemed to have been built around it, suggesting that the Roman Empire needed the *almadraba* tuna to feed its growing population. It was clear how abundant bluefin must have been. Was I looking at a Roman ruin, or the future of a ruined Barbate?

Past the line of sunbathers, past the approaching tide and the indigo water of the Strait of Gibraltar, I could see the vast silhouette of Africa, right there just a few miles from where we were standing. It was a vision, arresting in its simplicity, of a very connected world. The rushing waters of the strait are, of course, a major artery connecting the Mediterranean to the Atlantic Ocean, which eventually feeds into still more oceans: Pacific, Indian, Southern, and Arctic. You get the picture. It's one ocean.

At the foot of the strait—bluefin undoubtedly just below, jetting off to spawn, and the migrating birds from Veta la Palma just above—it seemed impossible to separate anything in nature, just as Rudolf Steiner had warned against doing. The bluefin below and the birds above were living metaphors for the ecological network Miguel believes in so strongly. And suddenly, though not by accident, I saw the network as larger—not just ecological, but cultural as well. It was everywhere around me, memorialized in the fading marble of Baelo, the menu of El Campero, and the Phoenician coin Ángel carried around in his pocket. Just as Lisa described, tuna are inseparable from these people and this place.

Look no further than the depleted *almadraba* fleets. These are the last vestiges of cultures that evolved in accordance with nature, of diets that listen to the seasons and the ecology and as a result are rewarded with the best possible flavor. Their demise, like our abuse of soil, is evidence of decisions made in the service of immediate goals (cheaper, more abundant food) rather than the future. It's Eduardo's "insult to history" all over again—an insult to culture itself.

By the time we left Baelo, I had changed my mind about the Barbate museum. If the *almadraba* is, as Miguel claimed, "a living lobotomy of the ocean," the museum was a peek inside the operating room. The passions of the people practicing a dying art were all along the walls—in writings, three-thousand-year-old etchings, and black-and-white photographs from the turn of the last century. The museum won't save bluefin from extinction, but for the children of Barbate yet to be born, it will help them know the culture from which they came.

CHAPTER 23

As we walked through the door to Aponiente, Ángel grasped my head in both hands. "Dan!" he said, shaking his head and apologizing for the missed *almadraba*. "Fucking wind!"

The dining room had been renovated since my last visit. Ángel said he had hopes of finally achieving a Michelin star, insinuating that he had been overlooked because of the restaurant's design, not the food. But once again we were the only diners.

Ángel sat down across from me. Slouching forward to light a cigarette, he brought his face under the light. He looked exhausted. His small, deep-set eyes were ringed with black.

"It's been so crazy," he said. I looked around the empty dining room and wondered what he was referring to. "I get these feelings. Thoughts in my head that I can't get rid of. I know something is going to happen, some idea is going to be born. It's going to come out of me. I say to myself, *You must relax*. But who can relax when you're about to give birth?"

I asked if he was getting enough sleep. "Sleep? Yeah, sure. Three and a half hours. This is optimal. I believe I'm getting that much. Anything less is no good." He extinguished his cigarette. "But anything more and I get very fucking crazy."

Lisa asked how he had become involved with Veta la Palma. It was, after all, the reason we'd come. He shrugged his shoulders and said they had simply approached him. It was "organic," he said. I reminded him of his rejection of fish farms at the lunch with Carl. ("Never, never, never.") How had he gone from refusing to even meet with Veta la Palma to becoming a champion of their fish?

"Actually, I signed on as a partner now," he told us, without a hint of embarrassment.

A small plate of *almadraba* tuna skin with tomato marmalade appeared. "If you can't get to the *almadraba*, I'm going to bring it to you," Ángel said. He had scraped the skin of its fat and impurities, boiled it, and then braised it. It was soft and gelatinous. Ángel told us that ancient Romans would dry the skin and use it for shields.

The tomato marmalade was very dark and intense. "Tuna blood," Ángel said after I took a few bites. "I added it to the tomatoes while they reduced. My parents are hematologists, what can I say?" He continued, impressed by his own train of thought. "Chefs know death. We know dead products. But in order to understand death, the chef also has to know the life." Ángel paused as I finished the marmalade. "So we're forensic scientists."

We are also naturalists. Not to get too high-handed about it, but chefs can do a pretty good job of translating the natural world. A delicious carrot communicates the soil it was grown in, a grass-fed lamb the kind of grasses it was pastured on, and so forth. The experience of a well-prepared meal can make these connections clear in powerful ways.

It really wasn't until my experience eating at Ángel's restaurant that I started to see just how powerful. Cooking with tuna skin and blood is daring, provocative, and in the vein of what's popular right now. If Ángel's food were defined only by this kind of nose-to-tail cooking of the sea, it would be just that—daring and provocative, and in vogue. But the totality of a meal in Ángel's hands transcends the individual plates of food, and even the craft of cooking. The meal becomes, like *jamón*, more than the sum of its parts. And you emerge from it with a sense of awareness you didn't have when you sat down—about fish, absolutely, but also about the fragile state of the oceans and our responsibility to keep them healthy.

In that sense, meals in the hands of chefs like Ángel are not only works of art; they can also be vehicles for igniting change in our food system. I know that's an unlikely idea—to sit in a restaurant halfway around the world (a seafood restaurant, no less) and come to the realization that a meal has the power to change the American food system. And yet that's exactly what happened.

Ángel, like Eduardo, is offering an alternative by way of consciousness. It's often said of restaurants that they are places of escape. Through Ángel I came to view them as places of connection, too. You eat, say, an unknown fish and want more of it. Wanting more of it invariably means becoming interested in how to ensure it survives. Knowing how a species survives requires an understanding of marine biology (like phytoplankton) and the politics around fisheries (like bycatch). Pretty soon you're interested in the cultural significance of preserving other species (bluefin tuna). Consciousness breeds what Aldo Leopold called a land ethic (and Safina later called a sea ethic), and he called it an ethic because he understood that ethics can inform and direct action.

Which isn't to say that the Third Plate, as I have come to envision it, exists solely within the world of haute cuisine. But it is to say that chefs like Ángel have an opportunity—and perhaps the responsibility—to use their cooking to shape culture, to manifest what's possible, and, in doing so, to inspire a new ethic of eating.

JAMÓN OF THE SEA

Across the small dining room, a waiter wheeled a cart. Ángel sprinted out of his seat and took control. Smiling broadly, he lowered his head to just above the cart, his eyes squinted eagerly, and drove over to our table.

The cart held three cutting boards, each with a different cured fish sausage. There was a *butifarra* (traditionally made with Catalan pork), a classically inspired chorizo, and a riff on a *caña de lomo* (cured pork loin). In each case, Ángel had substituted fish.

"*Jamón ibérico* of the sea," Ángel said, slicing a piece of each and handing me a plate of glistening meat. It was nearly impossible to distinguish them from their pork archetypes. Pimentón, the distinctive dried red pepper that's a classic seasoning for Spanish chorizo, perfumed the meat.

Here I have to credit Ángel with something I neglected to acknowledge at the table. No modern chef, as far as I know, had ever thought to fashion fish into charcuterie. There is of course cured fish, and there are fish "sausages"—fish mousse mixed with heavy cream and egg whites, stuffed into a casing, and poached (an invention of the nouvelle cuisine chefs in the early 1970s). But no chef had replaced cream or pork fat with fish fat and then actually hung it to cure. This alone was a pretty revolutionary idea. (Lisa saw it differently: "The Spanish mentality is to make everything resemble pork, since pork is the paragon of food. So in that sense, I don't know, Ángel actually did something rather predictable.")

Then Ángel said something truly revolutionary. "Dan!" he said, looking like a small child behind his sausage cart. "It's mullet!" Ángel explained that he chose mullet for the flavor, and the fat. "In fact, I no longer cook with the bass," he said. "It doesn't compare. The mullet is the most misunderstood fish in the history of fish."

To put this declaration in perspective, imagine a Texas rancher declaring his preference for boneless, skinless chicken breast over sirloin. This would be very much like choosing mullet over bass and boasting about it to another chef.

I first learned of mullet (this is grey mullet, not red mullet, the delicate and revered—and unrelated—fish of the Mediterranean) when I worked in Paris, where I saw their narrow, silvery blue bodies packed tightly together in fish stalls and markets. Alan Davidson, in his ocean guide and cookbook *North Atlantic Seafood*, writes that the mullet "swims along the bottom, head down, now and then taking in a mouthful of mud, which is partially culled over in the mouth, the microscopic particles of animal or vegetable matter retained, and the refuse expelled. When one fish finds a spot rich in the desired food, its companions immediately flock around in a manner reminding one of barn-yard fowls feeding from a dish."

The unflattering description is justified. The mullet's herbivorous diet translates into a flavor that's notoriously oily and off-tasting, often like the muddy water they inhabit. For bottom-feeders, that's to be expected. You are what you eat. Which is why, as Ángel explained, the mullet from Veta la Palma are so demonstrably superior to other mullet. Their habitat is clean, and full of the kind of varied diet that mullet—or any other fish, for that matter—rarely find.

"I could never get a cured sausage to work," Ángel said. "It's a delicate process, and it absolutely depends on the fat. The mullet have the right kind of fat."

He informed me (as Lisa and I had guessed by now) that he was collaborating with Veta la Palma on the project. Apparently convinced of the genius of the idea, the owners of Veta la Palma had committed to building a three-thousand-square-foot curing room and supplying the mullet for the operation. Ángel was donating the intellectual capital. He predicted that within a year they would be producing four thousand kilos (8,800 pounds) per week.

The next course was a fillet of roasted mullet with a puree of sea lettuce and phytoplankton. The dish was an ode to the mullet's diet. The garnishes were meant to instruct and amuse, and they were very tasty. But the mullet itself, the first fillet I'd tasted, was stunning.

I had reason to feel conflicted, having frequently declared Veta la Palma's sea bass to be the best piece of fish I had ever eaten, a fish that could change the world. I'd recently learned from Miguel that, as an herbivorous fish, the mullet is infinitely more sustainable to raise than bass, and now, thanks to Ángel, it appeared to be better-tasting, too—sweeter, richer, and flavorful in ways you don't associate with fish, especially not mullet.

Ángel took two large gulps of beer. "Mullet is not like other new fish.

You know how it is: you get all excited working with them for the first time, it feels so right, and then after a little while it gets stale and you start thinking about other fish. Every time that starts to happen with mullet, that's when they surprise me. They do something crazy. And I fall in love all over again.

"This reminds me of someone you should meet," he said. "His name is Santiago. He showed up at my door one day with shrimp from the Bay of Cádiz. He looked like he slept on the beach, he smelled a little, so I took pity. I bought the shrimp. And that night, just out of curiosity, I cooked them. These weren't just any shrimp. They were the most unbelievably sweet shrimp you've ever tasted. I served them that night. Customers thought we fried them in sugar batter. A few days later, he came again, with more of the shrimp—he knew I'd want to buy more. So I bought them all and I said, 'These aren't from the bay, are they?' My whole life I've gotten shrimp from every corner of the bay. I was sure these were not bay shrimp. He tells me that they are definitely from the bay. I said, 'No, they are not.' He said, 'Yes, they are.' I cooked up the new batch, and they were better even than the first. Unreal shrimp. Anyway, this goes on for a long time, until about a year ago, right after I started working with Veta la Palma. Then I got an idea." Ángel tapped his forehead.

"One day, Santiago drops off the shrimp and I say to him, really casual, I say, 'Hey, Santiago, let me buy you a drink.' And we go for a drink. Then I buy him another. Then another. Then, right in the middle of talking, I interrupt him: 'The shrimp are not from the bay, are they?' The poor guy starts stuttering. I have a hunch, so I just say it: 'You're stealing shrimp from Veta la Palma.' Without even pausing, Santiago says, 'I've been doing it for twenty years.'"

Armed with nothing but a small rowboat, some fishing gear, and a bottle of wine, Santiago works his way through Veta la Palma's canals, siphoning enough fish to make a small living. His deep familiarity with the terrain comes from years spent as a rice laborer in the surrounding fields. Apparently unable to make enough money to support his family, he ventured into illegal catches—something he had already indulged in from time to time on special occasions. He sells only to chefs, the most discerning ones in the area, but until Ángel no one had ever challenged him about his source.

"He goes to different ponds in Veta la Palma at different times of the year. Always at the full moon," Ángel said.

Thinking of Steiner and his lunar planting schedule, I guessed, "Because the fish have better flavor when the moon is full."

"No," he said, looking puzzled. "So he can see what he's catching. But he knows exactly when and where to get the fish at the height of flavor. Really, at the pinnacle—every time he brings me something, it's just a little bit better than anything Veta la Palma has ever sent me themselves."

"How has he not been caught?" Lisa asked.

"Caught? Caught how? It's thousands of hectares. Where are they going to look? Anyway, even if Veta la Palma were a swimming pool, they wouldn't be able to catch this guy. Miguel and the others, they know about him, they know this bandit exists, but they haven't even tried to stop him. Part of that is respect. They owe him, actually. That ice water method to kill the fish—the one that Miguel always talks about? 'Calms the fish, slows down the metabolism, and the quality is so much better.' That's not Veta la Palma's humane invention, or something they borrowed from the Japanese. That's because of what Santiago did! One night, I think it was ten years ago, Santiago caught some bass and left it for them in an ice bucket, to show these guys the right way to do it."

"It's right out of the picaresque," Lisa said to me later, explaining that "there's a tradition in Spanish literature of rascally characters—they do things like steal, cheat, and drink, but as long as they're smart, they're respected and given space to operate."

"I have never met anyone like him," Ángel continued. "He knows more about Veta la Palma than Miguel. One hour with Santiago and one bottle of wine is worth three months with Miguel." He smiled and shook his head. "Santiago doesn't know anything about biology, he knows nothing about ecology, nothing about Veta la Palma the company, nothing of what they've done. To him it's canals and ponds with lots of fish. Ángel lit yet another cigarette and seemed lost in thought for a moment. He recovered with a slap to the table.

"Anyway, anyway, why am I telling you all this about Santiago? Because he's really the one who helped me get so interested in mullet. He got me to get over my prejudice, the chef's prejudice, against a terribly misunderstood fish. One day, just before Christmas, he brings me the most incredible sac of roe I've ever seen." Ángel cups his hands and shows me how large. "I mean, I looked at this thing and it blew my mind. He told me to guess where it was from. I was sure it was from bluefin—it was that big. He said, 'No, not tuna.' I guessed every fish, and then he looks at me and says: 'Mullet.' I didn't believe him. 'Impossible,' I said. I called him a drunk. But he insisted. He said that the mullet get pregnant when the water turns warmer, during the summer. After that, they gorge themselves all fall, to support the baby and to anticipate the long winter."

"You're not going to tell us that the roe sacs of the mullets become like foie gras," Lisa said.

"Exactly! Yes, it's like foie gras. The mullet eat all the time, they eat everything in sight. The roe sac swells, and the fat around the sac becomes enormous—to protect it, I suppose—and then, sometime in October or November, ten days before birthing, they stop eating and they rise to the surface. That's when you know they're ready. They flap around and just hang out. If you want mullet roe that tastes better than foie gras, you have those ten days to get it."

"Freedom foie gras," Lisa said, shaking her head. "Of the sea." She and I sat there, dumbfounded, the connection to Eduardo and his geese almost too unmistakeble to be believed.

"He sells this to you?" I asked.

"No, he gave it to me. I tried to pay him. He wouldn't accept money. He was angry, actually. Maybe he was offended by my wanting to pay for his gift. We sat silently at the bar for what felt like a long while. And then he suddenly turns to me, his finger pointed. 'Ángel León, you have no fucking idea.'

I said, 'Santiago, what do you mean?' 'Until the day you can pick up a mullet, feel the slime, know how clean the waters are, know the age of the mother and the father of the mullet, know the temperature of the water it came from, know the number of moons that mullet has seen in his lifetime—until you can do all that, you have *no fucking idea.*'"

Ángel leaned back and wiped his forehead. "And do you know what? He's right. I know nothing."

———

The next morning, after Ángel got in the car for our trip to Veta la Palma, I had the thought that he might be suffering from bipolar disorder. The exuberance of the man driving the cart of mullet sausages to our table—so proud to be Ángel León at that moment—had dissolved into a valley of despair. He now wore the disturbed look of a marathoner bleeding in his shoes.

"There are days when I don't know if I'll be able to get out of bed and confront all the things I have to deal with," he said, lighting a cigarette. The cynical side of me wanted to point out that he had a thirty-seat restaurant, closed for three months of the year, that served no customers the last two times I visited. But my heart went out to him. I told him he was a genius, and reminded him of his fish sausages.

"Yeah, I've been told I'm a genius many times," Ángel said, in a tone only he could make modest. "All of my friends say I'm going to be a millionaire, but for some reason it's eluded me up till now." He stared out the window. "For ten years, I've been so disconnected from the people I care about. I only think of my projects and my work. My problem is that I don't remember why I did all of this in the first place. When I figure that out, I will be free.

"Dan," he said, turning to me, "have you ever cooked naked in your kitchen?"

"I have *felt* naked," I said, earnestly.

"One of my very dear friends, Moreno Cedroni—he has a small

restaurant on the beach in Italy—told me to do this a few years ago. I was in a bad state. I had very nearly given up on cooking. So I saw him at a conference, and he told me he had a theory about this being the most primitive thing to do—*Homo sapiens*, fire, naked man alone in his kitchen—and that if I did, I might emerge with a different mind. Moreno, he's a pretty weird guy. He's totally nuts, actually. But one night, after everyone had gone home, I was packing up my things and I thought, *What does Ángel León have to lose?* So I closed the curtains and turned off all the lights except for the kitchen. I stripped. Everything. Socks and shoes, too. I picked up my knife and I started cooking."

I asked how he felt. "I felt like an asshole," he said honestly. "After an hour, not so much. And then something clicked." He was silent for a few moments. I got the sense that the experience, rather than actually changing him, as he seemed to believe, had merely reinforced his need for a kind of freedom he had always longed for. "What I can say is you should try it. Because all kinds of emotions are going to come out of you."

❧

Passing through the large metal gates that separated the old town from Veta la Palma, we were suddenly moving along the familiar dirt road, with the marshes and canals and the dense vegetation on either side of us. The vastness of the open space, with views that stretched on interminably, struck me yet again, even on my third visit, as very nearly unreal. I have never seen anything like it, and I doubt I ever will again. Ángel stretched back to grip his headrest, suddenly more relaxed.

Miguel had called earlier and asked that we meet him at one of the mullet ponds. We made a sharp left turn on the way there and came to a small collection of flamingos gathered in shallow water. Ángel sat up in his seat and quickly rolled down the window as we drove past. "*Hola!*" he said, waving. When the flamingos ignored him, he reached over to the steering wheel and pushed on the horn. "*Hola!*" he said, authoritatively.

At the mullet pond, Miguel greeted us with three men who proceeded to wade into the pond with a large net. With water up to their chests, they dragged the net, slowly and very gently, across the placid water. There was silence all around us, except for the sound of parting water as the men tiptoed forward. It had the aura of a baptism. I had missed out on the *almadraba*, but here I was witnessing a fishing technique steeped in a similar kind of tradition.

Miguel confirmed the correlation. "The mullets come to us. They enter from the Atlantic and fight their way upstream."

"It really is like the *almadraba*," I said.

"Exactly like the *almadraba*, yes. Veta la Palma is a big farm in the middle of a delta. The mullets pass through in their migration from the sea to the lower coast of the river. It's a self-selecting art," Miguel said. I asked if any of the mullets were born on the farm.

"Some, but most of the mullets—that is, 99 percent—that are born in the ponds are taken by bird predation. On the other hand, we don't lose mullets when there's a sudden change in temperature. During the summer's worst heat, for example, we lost 40 percent of our sea bass. They can't adjust to shocks in the system. The mullets adjust very well." I told Miguel I hadn't known that fish could adapt to large swings in water temperature.

"Very few fish can, really," he said.

The fishermen walked toward each other in the pond, hands over their heads, dragging the nets and orchestrating their movements so seamlessly, it brought to mind synchronized swimmers. They emerged with three impossibly large mullets.

Ángel hurried over to inspect them. "Dan!" he said, laughing. "Look at this fish. Climate-change-resistant, low ecological impact—no, sorry—ecologically *beneficial*, unbelievable fat distribution, and the ability to produce foie gras of the sea. The greatest tragedy would be to not study this fish." He turned toward a bass pond in the distance. "Bye, bye, bass! I no longer have interest in you," Ángel said, waving to the pond. "See you never again."

I stared at the gigantic mullet. I admitted to Miguel (making sure Ángel couldn't hear me) that I hadn't known mullet could grow as large as sea bass, especially in half the time.

"Primary consumers that feed on the basis of our system are able to assimilate more energy than bass or other carnivores," he said. "Which is why we can raise mullets in half the time it takes to raise the bass. That's true of any ecological network. Lions assimilate less energy from their food than zebras, who feed on grass." I saw how the mullet, with their gentle ecological impact, might have reminded him of his formative years studying in the Mikumi National Park.

"And also," he added, "well, this is a theory I have, but I am quite sure it is true: when stressed—and for mullet, the greatest stress is crowding—the first thing mullet do is stop eating natural food." I asked what they ate instead. "It's funny. They'll eat artificial feed—chicken pellets, for instance. This is what I've heard most conventional aquaculture operations feed their mullet."

"So they refuse natural vegetation and choose . . . McNuggets?" I said.

"Yes, I know, it's so crazy. But when you cultivate mullet in a low-density, complex ecosystem, such as what we do here at Veta la Palma, they eat all natural food. They really do grow larger and stronger. Disease outbreaks are nonexistent."

I asked Miguel why he allowed me to get so excited by the bass on my first visit, when it was so clear that the mullet were the real stars. "We knew a lot of things about these mullet, especially how they work to promote good ecological conditions. What we needed," Miguel said, looking over at Ángel, "was someone to tell us they tasted good, too."

Ángel held the mullet at arm's length and smiled broadly, a magician with his rabbit. You couldn't help but be dazzled by the sight of the two of them—the proud, hyperkinetic Ángel León and the hulking yet humble and agreeable mullet. It was hard to say who was stealing the show.

CHAPTER 24

As promised, Rod Mitchell imported the first batch of Veta la Palma's fish only a few months after that lunch with David Pasternack at Esca. By mid-July we were receiving regular deliveries of mullet, which, for the record, tasted every bit as delicious as I remembered. Other restaurants ordered Veta la Palma's fish through Mitchell as well—almost exclusively bass. Within weeks, Mitchell was selling thousands of pounds to chefs in New York, Las Vegas, and San Francisco.

It was not cheap. The bass came in at $18 per pound, which, compared with the wild stuff—$7 per pound—was exorbitant. But chefs were willing to pay for the flavor. One of the most renowned, Eric Ripert of Le Bernardin, who assumed the role of chef after Gilbert Le Coze's death, declared the bass extraordinary.

Ángel, for his part, devoted himself to Veta la Palma and the sausage project for more than a year. He also dug deeper into his experiments with plankton, managing to successfully cultivate six new types. One was yellow and apparently loaded with 50 percent more carotene than a carrot, and another Ángel said almost eerily encapsulated the essence of pure shellfish.

"That something so small could express so much gastronomically and ecologically is really powerful," he told me over the phone, as enthusiastic as ever. "And when it's that delicious, telling the story of the sea is a lot easier."

Ángel made it even easier by introducing a new menu format at Aponiente, which was why Lisa—her own talent for identifying a great story still very much intact—made another visit to the restaurant. It was the off-season, but she was not the lone diner this time. Ángel had earned a coveted Michelin star in 2010. "Everything's changed," he told her. "People put themselves in my hands, they trust me."

The menu depicted a cross section of the sea. Different fish were shown at varying levels, with the names of the dishes next to the fish. *Almadraba* tuna, when it was available, appeared near the top of the chart. (Ángel serves it whenever possible, a practice that still offends Carl Safina. Despite recent reports of larger tuna schools, especially along the Atlantic coast, Carl believes bluefin consumption should be banned completely in order to allow stocks to recover. "I certainly feel that they are at, or near, the bottom of abundance in their history—a history that's hundreds of thousands of years old," he told me by phone, exasperation in his voice. "I don't think that any bluefin tuna from any source is currently an ethical choice.")

The mullet from Veta la Palma anchored the bottom of the menu. Ángel continued to find new ways to glorify the lowly fish. He featured it in several dishes, including one that treated the mullet as if it were a pheasant, hanging it for nine days to deepen the flavor and impart a meatier texture.

And then one day, Ángel stopped cooking mullet. There had been a falling-out with Veta la Palma, I learned. What exactly happened is unclear. Veta la Palma shorted him fish for a delivery. Ángel called and exploded, words were exchanged, and he swore to never order from the company again. Ángel said the relationship had been fraught for some time. He was continually frustrated—disrespected, or perhaps misunderstood; either way, he said his ideas were never fully implemented at the farm. (When I asked for examples, he didn't give any, except for a cryptic critique of the culture of Veta la Palma. "It's all hippie," he said. "No woman, no cry.") And the fish sausage business turned out to be less successful than he and Veta la Palma had hoped, further straining the relationship. Several months later, blaming the abysmal economic situation in Spain, Veta la Palma closed the factory.

Ángel told me all of this during a visit to Stone Barns, where he toured the farm and prepared a special tasting menu in the restaurant. He looked even more exhausted than usual—his eyes were like black holes, devouring everything in sight. He smoked furiously. It seemed as if his demons, which so fueled his creativity, had begun to consume him.

❧

Less than a year later, an e-mail arrived, with a picture attached. It showed the ruins of an ancient building, and in the foreground was Ángel León, wading in the shallow water of a canal, his arms stretched out in a wide embrace.

I called him. His voice still possessed hints of his itchy impatience, but he sounded calmer and deeply focused. He even sounded happy.

He told me that the picture had been taken in front of his new project. He was converting a deserted piece of marshland very near to Veta la Palma into a fish farm. There will be direct access to the sea—where the Atlantic meets the Mediterranean—and a series of small canals will be built to utilize the tides and bring fresh seawater into the farm. The site will also be home to Ángel's new restaurant, located in a two-thousand-year-old Phoenician salt mill (the ruins I saw in the photo). The floors will be glass so the diners will see the fish swimming below, he said. And he's going to have his own salt flats and a large vegetable garden to supply the restaurant.

I also learned that, while Ángel might have calmed his demons, his feverish creativity hadn't left him. He described a new dish on the menu at Aponiente: *"A Squid That Wished to Be a Carrot, an Homage to Dan Barber."* The squid is soaked in carrot juice for several days, turning it bright orange, then rolled and stuffed with minced cooked carrots. A piece of dill mimics a fresh carrot top. Ángel told me the inspiration came from his visit to Stone Barns. Jack had pulled one of his impossibly sweet mokums from the greenhouse soil, wiped it on his shirt, and handed it to Ángel.

"I never told you this," Ángel said, "but up until that moment I did not really like vegetables all that much. I stayed away from them, for the most part. But that carrot changed me."

As for the mullet, our beloved, mercurial fish—it hasn't fared well, especially without Ángel to take up its cause. Sales of Veta la Palma bass outpace mullet fifty to one, according to Mitchell, even though the price is half that of the bass. The perception of mullet as a trash fish, with insufficient fat and a muddied flavor, is so ingrained in the minds of chefs, it's been difficult to overcome. It hasn't been easy to convince diners, either.

Miguel isn't disappointed. Ever the optimist, he is quite sure the mullet will eventually sell. In the meantime, he has begun to travel and talk to other aquaculture biologists. I recently received an e-mail from Costa Rica, where a philanthropist was hosting him for a week on his preserve.

"Dan, I am so happy right now," Miguel wrote. "Just to tell you a fact about me, I had never been in the rainforest before. The feeling, when you are walking through the muddy and difficult trails of the deep rainforest, listening to dozens of different noises and having glimpses of rainbow-coloured frogs, snakes, hand-sized butterflies, birds and many other unknown creatures, is simply superb. . . . One can understand that thought of the Yanomami people: God is great . . . but the forest is greater."

One evening in the beginning of December, I watched from my office as Steve, our fish cook, prepared a newly arrived Veta la Palma mullet for service. He scaled the fish, carefully removed the first fillet, and began to roll the fish over to get to the second fillet—a procedure he'd performed countless times before. But this time he stopped in his tracks and peered down at the half-dissected mullet. His eyes widened and he scratched his head, much like a man who had forgotten where he had parked his car. Dropping the knife, he carefully removed what looked like—I swear—a large lobe of foie gras. He held it up for me at arm's length.

It was an extraordinary sight. Though we'd enjoyed Veta la Palma's mullet for nearly six months—working with it every day, savoring its flavor, advertising it to the staff and to our diners—I have to admit I'd become, in the words of Ángel, just the slightest bit bored by it. As with a date who wears the same outfit and tells the same stories, you might find your eye wandering a little—until one day she shows up carrying a purse bursting with roe and a cashmere coat of blanched white lard. And you fall in love all over again.

I asked Steve to sauté a section of the fillet so we could taste it. (We cured the roe for several months and then shaved it over savory carrot cookies—an ode back to Ángel.) He placed the mullet in the pan with a squirt of grape-seed oil, as he always does, and turned away to attend to the cooks gathered around. Thrilled, he demonstrated how the enormous sac was able to fit inside the relatively small cavity of the mullet. He claimed it was twenty-five times larger than anything he had ever worked with in the past, and for the first time in my career as a chef, I didn't think a line cook was exaggerating.

As Steve repeated the demonstration, placing the roe sac inside the mullet and removing it, over and over again, each time to the amazement of the cooks, I noticed the fillet of mullet in the sauté pan had become nearly submerged in its own fat. Steve turned around and raised his hands to his head. "Jesus," he said. "It's frying in its own fat." I wish that Pepe could have been there at that moment, to see the tradition of the self-sautéing *morilla* of tuna reborn in the lowly mullet.

Next to the pan I noticed a block of white flesh and examined it, standing off to the side. It was a solid chunk of mullet lard, glistening under the kitchen lights. I'm sure that anyone, even a salumi master, would have had trouble differentiating it from the highest-quality pork fat. I held it in my hand as it melted slightly. I smelled the sea.

A chef's life is filled with small pleasures—a successful new dish, a happy diner—but the truth is, there are few real satisfactions. I think back to this moment as having been one of the most satisfying, in part because I imagined how those mullet, free of stress, must have furiously gorged on the lush ponds of Veta la Palma—preparing for the long winter, and improving the system in the meantime. Santiago was right. So was Eduardo. It was a gift from nature.

Blueprint for the Future

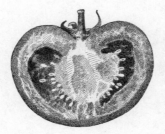

IT'S A STRAIGHT SHOT down Route 54 from Klaas and Mary-Howell's farm into the town of Penn Yan. The highway runs through a patchwork of fields and pastures, the horizon rippling with glacier-cut ridgelines. Seneca Lake sits in the middle of all this, the largest of the Finger Lakes.

Klaas credits the lake with helping to moderate the climate. The lake effect, as it's called, increases precipitation, warming the air during the coldest months of the year. It also cools the air on the warmest days; even so, the temperature on this particular morning had already reached ninety-five degrees.

As I leaned forward to turn up the air-conditioning, a police car appeared, coming toward me from the other direction. I quickly slowed, but the policeman made a U-turn, flashed his lights, and pulled me over. I watched him in the rearview mirror as he approached. He was well over six feet tall—trooper hat, black-tinted sunglasses, the works. I rolled down the window and the warm air flooded in.

"Eighty-five in a fifty-five," he said. I feigned shock (*Really, officer?*), then bafflement (*Wow, I've never driven eighty-five*). Finally I apologized, sounding a little desperate. He stayed silent and motionless as I fumbled for my driver's license and registration. "Sorry, officer, I'm harvesting wheat with Klaas Martens today. Rushing a bit." The trooper lowered his face to the window.

"You know Klaas?" he asked. I nodded.

The officer smiled. "All right, then, have a good day."

Klaas and Mary-Howell's influence is inescapable here.

They were the first in the county to give up farming with chemicals on a large scale, and, while neighbors initially doubted they would survive, the farm's success slowly convinced naysayers. A year or so after Klaas and Mary-Howell went organic, a dairy farmer named Guy Christiansen—an elementary school classmate of Klaas's, with land just west of his—started noticing the success of Klaas's crops. It was hard *not* to notice: Guy's conventional corn, which he grew to feed his cows, abutted Klaas's organic corn.

"The fact that Guy could *see* my corn—that he couldn't help but see my corn—made a difference, I think." Klaas's crop was thriving.

Guy, whose own profits were dangerously low, decided to switch his entire dairy to organic. Not long after that, Floyd Hoover, whose farm abuts Guy's, switched his corn, soy, and beef cattle from conventional to organic. Aaron Martin, a neighboring dairy farmer, took note of the prices Guy was getting for his organic milk and decided to convert as well. So did Eddie Horst, a Mennonite dairy farmer bordering Guy and Klaas, and Ron Schiek, just north from Klaas. One after another, in a growing circle, the Penn Yan farmers transitioned away from chemical agriculture. Each one would see a neighbor succeed and follow suit.

Mary-Howell began holding meetings in her kitchen for the newly converted. "You know, at the time, everything was so new," she said. "We were just trying to get information. But it turned into a small community of support."

In the mid-1990s, these pioneer organic growers got a lucky break. Monsanto, the agricultural chemical and biotechnology company, developed a genetically engineered growth hormone called BST, which increased the production of milk in dairy cattle. It offered farmers an opportunity to increase their profits in a notoriously slim-margined industry. But many consumers were wary of an engineered additive in their milk. Demand for organic dairy—the only kind assuredly free of artificial hormones—suddenly skyrocketed.

"It was so fast. All of a sudden there was strong demand for organic milk, and the demand for organic grain to feed these dairy cows went through the roof," Klaas remembers. "I've always wanted to say, 'Thank you, Monsanto.'"

Other farmers considering the switch to organic began attending Mary-Howell's kitchen meetings to listen and learn. As the group kept growing, they rotated to different homes, and then a few years ago Mary-Howell used her connections at Cornell to secure a large hall at the New York State Agricultural Experiment Station in Geneva. Now nearly one hundred farmers regularly attend, and the meeting is teleconferenced throughout upstate New York, spreading information and inspiring even more farmers to make the switch.

The last stretch of Route 54 before the town of Penn Yan attests to this flourishing community. There are five thousand acres of nearly contiguous organic farmland—all converted within the past two decades.

Penn Yan itself has benefited from the farmers' success. The town earned its name from the Pennsylvanian (Penn) and New England (Yankee) settlers who came to the area in the late 1700s in search of farmland. Rather than fight over the settlement's name, the two groups split the difference. Today it is the image of an American ideal, a postcard from a gentler, simpler time. Pleasant storefronts line peaceful streets, with names like Liberty, Elm, and Main. Traffic lights are few, crosswalks are wide, and stores are clean and inviting. Penn Yan's largest business, Birkett Mills, in operation since 1797, displays a twenty-seven-foot black griddle on one side of its building. It's the same one the company used in 1987 to cook a pancake big enough to set a world record.

Small towns have always held an iconic place in American culture. They embody what we consider our country's best qualities: community spirit, work ethic, and solid moral values. Since they can no longer be said to represent America as a whole—more than 80 percent of Americans live in metropolitan areas, after all—the sentiment contains a good measure of nostalgia. Most small towns today are not as picturesque as Sinclair Lewis's *Main Street* or Norman Rockwell's midcentury magazine illustrations portrayed them. The view is marred by run-down stores and abandoned movie theaters, grubby diners, and seedy bars. Many have become ghost towns, without even a school, post office, or grocery store. The rise of industrialized agriculture and the rapid consolidation of family farms in the 1950s and '60s drove the decline of these small towns. Penn Yan was no exception. Older farmers retired, and the next generation moved away or wanted out of farming altogether, to the point that Mary-Howell once described Penn Yan as having been "a town with a bombed out center."

Mary-Howell attributes the turnaround to a series of events. In the 1970s and '80s, depressed land prices attracted large purchases from Pennsylvania Mennonite farmers. Then, in 1976, the Farm Winery Act allowed New York winemakers to process their own grapes and build wineries to sell to the public directly. By the time the backlash against BST revived the local dairies in the '90s, new businesses had developed in Penn Yan to support the emerging farm and wine industries—supply stores, repair shops, and welding services, to name a few.

It was Klaas and Mary-Howell who made one of the most vital contributions to the town's economy. In 2001, they bought a run-down Agway mill just off Main Street, renaming it Lakeview Organic Grain. Klaas remembers talking to neighbors who wanted to go organic but were locked out by infrastructure, especially the lack of available milling and proper storage facilities. Mill operators have generally been reluctant to serve an organic market, in part because thoroughly cleaning the equipment—a requirement if one is dealing with both organic and conventional grains—is onerous and expensive. Through Lakeview, Klaas and Mary-Howell could fill yet another niche in their community, providing milling and storage for organic grain and selling the grain to a growing market of organic dairies.

"You know that expression 'If you build it, they will come'?" Klaas said.

"It was like that. We grew by 20 percent every month for more than two years. Pretty soon we had a half-dozen full-time workers. We literally couldn't keep up with demand."

Klaas and Mary-Howell looked at operating the mill as one part business opportunity (they insisted on good margins to keep the mill profitable) and three parts responsible land stewardship.

"We encouraged farmers to improve their soil by creating a market for those grains that added fertility," Klaas said. "We paid—really paid—for so-called 'other' grains—like triticale, oats, and barley—because we knew these played a critical role in maintaining the health of the soil. Without a buyer, farmers can't justify planting them into the rotations. Without planting them into the rotations, sooner or later soil fertility declines."

Klaas acknowledged that if soil fertility in the region declined, the mill's profits would decline, too. "So in many ways we were acting out of self-interest," he said.

Creating a market for the less desired grains also helped the local cows. The standard feed mix Lakeview sells contains nine different grains. By dairy industry standards, most of those are considered superfluous, but the diversity does the cows good in the long run.

"Cows eating our diversified grain diet are getting minerals and vitamins that are not available to them through just feeding corn," Mary-Howell explained. "They may produce a little less milk—and that's debatable, especially over the long term—but they are healthier. Can I prove it? No, but a diversified diet means more amino acids, more minerals, less acidosis."

I can prove it tastes better. At Stone Barns, Craig switched the feed for his pigs to Klaas and Mary-Howell's mix a few years ago. The pork is more delicious than ever.

The mill, managed by Mary-Howell, now has eight full-time employees and has expanded into the seed business. "It was a natural progression," Klaas said. After all, the new crop of organic farmers needed a supply of organic seeds.

Why would there be a shortage of seed in the middle of the recent boom in organic farming? Monsanto again. Throughout the 1990s, Monsanto bought up small and midsize seed companies, eliminating many sources of organic seed.

"What we forget is that not so long ago, every farming community had a seedsman; some had several," Klaas said. "This was an exclusive club, made up of the most cerebral, honest farmers. In fact, you had to be voted into the seed improvement co-op to become approved." These farmers paid close attention to things like germination percentages, and they were especially vigilant about disease, weeds, and any contamination possibilities.

From an early age, Klaas was drawn to the wisdom and honesty of seeds-men. "I'd spend as much time as I could learning from them when I didn't have to do farm chores," he said.

One morning in 1983 when Klaas was harvesting soybeans on his own farm, he spotted a plant that stood out from the surrounding field. As he got closer, he realized it was soy, just not any kind he recognized. "It was an off type," he said, "a mutation of sorts. It was an incredible plant. I stopped the combine just in time." He ripped the plant from the ground by its roots, saved the seeds, and the next spring planted them in his garden. He wanted to see what would come up. But he also wanted a reason to consult with a young woman, a plant breeder he had met at Cornell University, less than an hour away.

"I pretended I needed help and didn't know anything about seed propaga-tion," Klaas said. "Compared to her, I actually really didn't know much at all. She was a terrific breeder." *She* was Mary-Howell.

Their shared interest in breeding comes in handy in their burgeoning seed business. The mill encouraged more farmers to convert to organic; the seed business allows them to grow organically with the kind of crop diversity that will help soil fertility. The network sustains itself, and continues to grow.

THE AGRICULTURE OF
THE MIDDLE

The more I visited Klaas, the more I realized how difficult it was to fit him into a recognizable model of farming, especially the kind we're most drawn to: the small family farmer, who tends the harvest and sells what he grows at the farmers' market. Supporting these farmers is a good idea—they produce tastier food, for one thing, and since less than 1 percent of the population currently makes a living from farming, rewarding their efforts through direct transactions has made a difference. But they aren't the whole story.

In 2007, the U.S. Department of Agriculture (USDA) announced that the number of farmers' markets nationwide had doubled in four years, a sign that the sustainable food movement, long a fringe idea, was gaining traction. I was interviewed about the news on PBS, along with Dennis Avery, an agricultural analyst, strong promoter of agribusiness, and author of *Saving the Planet with Pesticides and Plastic: The Environmental Triumph of High-Yield Farming*.

The host asked if, in a perfect world, we would do all of our food shopping at farmers' markets. The question was a little silly, I thought, and after fumbling over my answer a bit, I ended up just saying yes. The host turned to Mr. Avery, who, sitting in his office in Swoope, Virginia, looked very stately in his leather chair, a library of scholarly books lining the wall behind him. He smiled.

"Well," he said, "I like farmers' markets, too." He said he shopped at his local farmers' market occasionally—for things like cinnamon rolls, sausages, and peaches. On the other monitor, I saw the host smile. Avery continued. "But this country needs hundreds of millions of tons of food every day. If New York tried to supply itself from farmers' pickup trucks, the traffic jam to end all traffic jams would lock up the city." He deftly painted my perfect world as a disaster. I had lost the exchange at "cinnamon rolls."

Large-scale agribusiness should be difficult to defend. It has helped bring about ecological problems of unparalleled scope and significance. The costs to soil fertility alone are too great to sustain for the long term. Yet it's hopeless and naive to try to argue that farmers' markets (and boutique farmers like Eduardo) can feed all of us. Asking every farmer to plant, harvest, drive his pickup truck to a public market, and man the cash register would be like asking chefs to cook, serve, and then wash all the dishes every night.

The more I hung around Klaas and the community of farmers in Penn Yan, the more I learned that the answer to my interviewer's question lies somewhere between my idealized vision and Mr. Avery's reality. There is actually a *middle* of agriculture, and it's worth examining. These are farmers with midsize operations—farmers like Klaas, who are too big for farmers' markets but a bit too small to compete with the mega-scale food system. Unfortunately, most of them are, unlike Klaas, caught up in the commodity game. They're planting single crops because that's where the money is. (Behind on a bank loan for a combine, with a mortgage to pay on top of that, how many of us wouldn't play it safe to keep up?)

Fred Kirschenmann, the president of the board of Stone Barns Center and himself an organic wheat grower, is the country's greatest advocate for this disappearing contingent. He told me that, though midsize farms cultivate more than 40 percent of our farmland, in another decade most of them will be gone. As their land is consolidated into ever larger farms, the prospects for diversifying what's grown continue to narrow.

In our search for solutions to modern agriculture's seemingly intractable problems, we would do well to consider the ingenuity and nimbleness of midsize farmers like Klaas and Mary-Howell as examples of what's possible for the future of agriculture. A sustained food system is more than a set of farming practices, as the *dehesa* and Veta la Palma proved, and more than an attitude toward food production and consumption, although both of these are central to it. It's a culture, too, and while we may idealize small farmers for their rugged individualism, as agents of change their power is limited.

The culture became clear to me during my first visit to Klaas and

Mary-Howell's mill. It was an early spring morning, the parking lot filled with pickups. A large flatbed truck was pulling in with a delivery, and another, larger one was leaving. Inside, Mary-Howell worked the phone, taking orders, scheduling seed inspections, and accepting calls from farmers seeking her advice. Forklifts carrying bags of grain crisscrossed the crowded floor. Klaas held court with several grain farmers, who were amazed by his conviction that *Fusarium*, a debilitating and costly fungal infection, is avoidable through properly managed soil. Another group of farmers, double-fisting coffee and doughnuts, stood with the mill manager and debated market opportunities for organic soy. One man searched through the clutter of a large message board for a deal on used machinery.

It struck me that the mill served a purpose beyond enabling organic agriculture to thrive in Penn Yan; it created a kind of social fabric, too. Much like an old grange hall, it was a place to vent frustrations, exchange ideas, and escape the isolation of farm life. There was a feeling of shared purpose, if not the thrum of mutual regard. From the truckers to the mill workers to the farmers to the revitalized town of Penn Yan itself, it was all around me, thriving.

On the drive back to Stone Barns from Penn Yan, I started to see Klaas's story as really several stories nesting inside one another like Russian dolls. On the outside, the framing story—the one I first planned to write about— was Klaas's decision to break with his brothers and embrace organic agriculture. It is a large enough story to stand on its own. But the quality of his farming and the masterful rotations that feed his soil suddenly gave birth to another story: a whole community of like-minded farmers. Which led to more stories. The farmer as seedsman. And then as miller and distributor. Put them together and they start to take the shape of a sustainable food system.

But I was reminded of my conversation with Wes Jackson, who wouldn't call it that at all. "Because it won't last," he had told me, pointing out that sooner or later, as the history of agriculture shows, someone comes along and makes a shortsighted decision, degrading the land and compromising its health for the future. Even with all of Klaas and Mary-Howell's good intentions and hard work, eventually, Wes predicted, the system would unravel.

He had a point. This wasn't the *dehesa*, after all, where generations of Spaniards protect the land because it is a fundamental part of their history and culture. As Lisa had explained to me, the *dehesa* operates as a mythical landscape in the Spanish imagination largely because of its relationship to *jamón ibérico* and the ways of farming and eating that have evolved around it. A place like Penn Yan, and a whole category of farming—the midsize kind—doesn't resonate as strongly with Americans because there is no food culture attached to it, no *jamón* equivalent.

Generally speaking, the organic farmers of Penn Yan aren't really feeding people, at least not directly. They're producing the grain that feeds animals that feed people. Klaas admitted this to me several times over the course of my visit, but it didn't resonate until I drove south along Route 54 on my way out of Penn Yan. The evidence was everywhere. I saw milking cows (organic milk is still the economic driver for many of these farmers) and endless fields of feed grain and cover crops. Compared with my rooftop view of the *dehesa*, I would argue that the views were no less beautiful. But what I didn't see were crops that I could cook with. Which meant a vital part of the story was still missing.

If the true sustainability of a food system is about the strength of its disparate parts, and the way to measure that strength is to examine how deeply they penetrate the culture, then Wes was probably right. As explosive as the sales of organic milk have been, what will happen when another trend comes along to replace it? Or when more farmers in the Midwest, where land is cheaper, convert to organic grain and the dairies in upstate New York can no longer compete?

By doing what the conventional grain market refused to do—pay for crop rotations that produced healthier soil—Klaas and Mary-Howell addressed an agronomic problem, and by setting up the mill and seed businesses, they solved an enormous economic problem for these farmers. (They also, quite by accident, cooked up a recipe for feed that ended up producing better-tasting milk and pork.) But as Wes might argue, one hundred years from now, Klaas's farm could look a lot different, and not for the better.

What was missing from his story—crops ingrained into the mores and traditions of our culture—was no fault of Klaas's. At least some blame falls to chefs and restaurateurs like me. Because those rotations of wheat, millet, flax, soy, buckwheat, rye, and dozens of other grains and legumes—the things that give soil its fertility, and give us the best-tasting food—are largely grown for animal feed. They could be grown for our menus instead, and in the process provide much greater profit for the farmer.

"If there was the demand and the infrastructure to make it happen," Klaas told me once, "the farmers of Penn Yan alone could bury New York with local grains." I'd seen enough to know that he wasn't exaggerating.

Ángel León saw mountains of discarded fish while working on commercial fishing boats and created a market for them by broadcasting their culinary value on his menu. Many of Penn Yan's soil-building rotational grains and legumes are the agronomic equivalent of bycatch. Throwing them into animal feed, while serving a better purpose than tossing unmarketable fish back into the sea, is not a way to build a sustainable food system. If I was truly in pursuit of a Third Plate, I needed to take a page out of Ángel's playbook and learn to incorporate these less desired grains into Blue Hill's cuisine. More than incorporate, I needed to make them as essential to my menu as they were to Klaas's rotations.

I would start, I decided, where Klaas himself had started—with wheat. In working with his heirloom emmer and spelt, could I finally recover what he called the lost taste of grain?

CHAPTER 26

THE DAY KLAAS'S wheat first arrived at the restaurant, no one knew what to do with it.

"Can I tell you something?" Alex, the Austrian-born pastry chef at Stone Barns, admitted to me later. "That day I was handed the bags of whole wheat—the spelt, the emmer—I did not know what the hell these are. I said, 'Sure, I make something with this here,' but the only thing that I think to make is *knödel*." Alex pretended to lift something very heavy, to show that whole wheat dumplings wouldn't have been right for our menu.

"So the first thing I do is I call my grandmother," he continued. "I tell her to help me. She must have worked with these grains here. 'No!' she says to me. She says she would have to ask *her Grossmutter* about such grains. For whole grains, this is how far back we need to go."

That night, Alex couldn't sleep. "Because I am the pastry chef," he said. "So of course, I should know flour. A carpenter knows a hammer, doesn't he? If you hand him another kind of hammer, an older hammer, maybe it's longer and heavier and really weird, and maybe it takes him a moment to adjust, but he can still use the hammer, no? He doesn't put the hammer down and say, 'I am sorry, but I can no longer build this table here.'"

The next day, he rebounded with an idea. He made a classic brioche loaf but substituted the whole emmer wheat for white flour. Had he asked, I would have warned him not to do it. Brioche, like coq au vin or tarte tatin, is the kind of classic that does not need reimagining. Why bastardize a perfect bread?

Alex milled the emmer wheat in the same tabletop grinder we had used for the Eight Row Flint corn. "The more we ground," he said, "the more the kitchen started to smell like dirt. No, not dirt. Nature. It smelled like nature, like going on vacation with my parents in the summertime when I was a kid, in the field when the wind blows through the wheat."

For several days, Alex had nothing for me to taste. "The first loaves, this was a disaster. It was too—how do you say this? *Schwer wie ein Stein*—like a rock, this is," he said, remembering the weight of it. "This was a learning process. My brioche was always very basic, straightforward—eggs, flour, yeast, butter, mix everything together, proof it, and bake it. This is it. Perfect every time. But this, now, this was more sophisticated." He mixed and proofed the dough for longer periods of time. He adjusted the quantities of butter (not quite as much) and eggs (a little more).

A week later, just before dinner service began, Alex finally showed me the results of his efforts, cutting thick slices of warm whole wheat brioche. A swirl of steam rose from the bread and wafted toward the ceiling like an image from a children's cartoon, its nutty apricot scent lingering around us. The bread looked airy and light like classic brioche, only the loaf was a rich russet brown.

Alex handed slices to several of the cooks gathered around. It is not difficult to get hungry cooks to eat anything, especially warm, buttery bread. He could have offered them a tray of melba toasts. But the usually impartial cooks were encouraging ("You rock my world, Alex") and genuinely impressed ("Who knew a German could bake bread?").

The brioche *was* delicious, comforting in the way bread should be, but also a little exciting, with a flavor of toasted nuts and wet grass. Just as the Eight Row Flint polenta tasted of corn—reminding me (because I needed to be reminded) that dried corn should actually taste of corn—the whole wheat brioche tasted distinctly of wheat.

The experience reminded me of my first taste of raw milk. I was with my brother, David, in Mr. Mitchell's kitchen, a few miles from Blue Hill Farm. We had just finished morning chores, and Mr. Mitchell's son Dale reached into the refrigerator and opened a box of chocolate chip cookies for breakfast. I was ten years old; it was a dream. Just as we dug in, Janet, Dale's sister, appeared with a steel jug of the morning's milk—unpasteurized, butter yellow, and still warm. Dale took a long swig and passed it to me, like moonshine. "Udderly delicious," he said. I was fixated on the cookies, but I took a gulp. I couldn't believe what I was tasting. It was creamy and sweet, but also tangy, with a scent that reminded me of morning pasture. It was milk *unplugged*, and it made the thin, pasteurized kind seem like a poor facsimile, in the same way that frozen orange juice doesn't compare with freshly squeezed. Alex's brioche was like that. It hardly resembled the version I knew before.

By the third or fourth slice, I'd gained an entirely new appreciation for the term "whole wheat." The technical definition is that the entire wheat kernel is preserved in the flour. What's missing from this definition is the feeling of satiation that whole grain provides. Classic brioche is rich, even decadent, but it doesn't do much more than fulfill a craving. It's a pleasure, not a satisfaction. It struck me then that Alex's whole wheat brioche was so delicious for a reason that went beyond Klaas growing the wheat in the right way, and beyond fresh milling. It was delicious because, as the name implies, it was wholly satisfying to eat. It was a complete food.

———

We began serving the new brioche on the tasting menu at Blue Hill at Stone Barns. I didn't want it to get lost in the breadbasket or treated as an afterthought between courses. So I borrowed that lone perfect peach concept Alice Waters had used at Chez Panisse. I sliced the brioche, toasted it lightly, sprinkled it with a bit of salt, and served it alone on a small white plate. The bread showed off Klaas's wheat perfectly—why not let it speak for itself?

But the waiters spoke first. When I did a test run at the expediter's table, that narrow landing pad between the kitchen and the front of the house, the waiters stared skeptically at the naked slices of bread. They knew Klaas, and they were not ignorant about the importance of whole wheat, but they shook their heads in disapproval. *It won't work*, they said. *Diners won't get it.* Having expected some pushback, I told them about the lone Mas Masumoto peach, and the revelation of serving something so perfect and so unexpected. One of the lead waiters took me aside and said, definitively, "That only works in California."

There is a misconception about the power of the chef, even in his own restaurant. We are often thought of as generals, leading an army through the blitzkrieg of dinner service. But a chef is really more like a chairman of the board; the waiters are the board of directors. With no menus at Blue Hill at Stone Barns, the power tilts toward the board, and the waiters know it. They develop an intimate connection with each table—befriending, comforting, informing, even lightly provoking. They are ambassadors for both sides— they represent the restaurant, but in our system they become emissaries for the diners' interests as well. In many cases they will follow the ticket into the kitchen and write the menu themselves. So when our waiters are not excited about a dish I've conceptualized earlier in the day—a whole roasted beet with homemade molasses yogurt, for example—the message on the ticket arriving in the kitchen might be: "*Table has an aversion to beets.*"

I pushed ahead anyway with the naked brioche, but that night our diners mysteriously suffered from a "whole wheat aversion."

So I modified my approach. A few days later, I met with several of the waiters before their shift. I had them taste the whole wheat brioche just out of the oven. I added a spoonful of homemade ricotta, still warm from the stove, and a marmalade made from salad greens grown in the greenhouse. The waiters spread it over their brioche. "This works," one of them finally offered, anointing it worthy for service.

That's when the trouble began. "Table requests brioche" started appearing on the tickets, because the waiters were enthusiastically preselling the bread, showing off Klaas's grains at the table and warming guests to the idea that one of their courses would be, as one waiter put it, "a celebration of freshly ground whole wheat."

It did not take long for the supply of Klaas's wheat to run low, so Alex purchased some conventional whole wheat flour, already ground, while we waited for the next delivery. But he dismissed the replacement with the first batch of bread.

"The brioche came out *looking* nice," he told me. "Very puffy. But when you open the oven, it does not smell aromatic." He complained that it smelled "dusty—dusty like an old closet."

We stopped serving brioche entirely until more of Klaas's emmer finally arrived. Only it, too, proved to be problematic. The new batch, from a different harvest, wasn't compatible with the recipe Alex had perfected. The dough was sticky and difficult to work. At first, it didn't rise. Then it would rise but quickly collapse.

"The wheat, it seemed to want my attention," Alex said. He modified the recipe again, raising the water temperature, mixing the dough even longer, and changing the size of the mold. After several days of testing, he reported that he finally understood the new harvest.

"It's like a good wine," he said. "Every year is different. So you have to adjust. Every time you have to relearn it. But that's the nice thing about these flours. You just have to follow, basically, nature's instructions." Within a few days, his patience had paid off. He brought the brioche for me to taste.

The bread had the same nuttiness we had tasted in the first batch, but with a new honeyed spice flavor that was uniquely its own. And it seemed just a little lighter, with more cavities in the crumb, or interior, of the bread. Holding a piece up close, I thought back to the subterranean reaches of Jack's vegetable field, with its miles of routes for microorganisms to maneuver. Back then, Jack had explained how the organisms' ceaseless activity provided the plant with everything it needed to produce delicious flavors. I bit into another slice of the brioche and tasted exactly what he meant.

—

We suffer from a prejudice against whole grains.

You could say we were born that way. We covet white flour because it is sweeter than whole wheat. Humans evolved with powerful preferences for sugar because of our need for energy-rich foods. In his book *Breaking the Spell*, cognitive scientist Daniel Dennett argues that our sweet tooth is really evolutionary biology at play. There is nothing "intrinsically sweet" about sugar molecules—rather, we developed an instinctual liking for sweets because they provided more energy. It's the brain's way of rewarding us for seeking out calories. Refined white flour satisfies those same urges by delivering a hearty shot of glucose more efficiently than whole grains. With no fibrous bran, there is nothing to slow the conversion of starches to sugars. A nutritionist I know once compared refined white flour to crack cocaine— a pure, immediate hit that only makes you crave more.

Another theory sees our preference for refined wheat in sociocultural terms—a food trend that has endured for thousands of years. Ancient Romans ground and filtered flour through fine linen. The whitest, softest loaves of bread were reserved for the privileged classes; poor peasants ate coarse loaves made of mixed grains. Over time, this mark of social distinction became entrenched. Dishonest millers added mashed potatoes, chalk, sawdust, or even dried bones or poisonous white lead to get whiter flour at any cost.

But there's a culinary logic to our preference, too, one that is reinforced today by bread bakers who believe that great bread has more to do with craftsmanship than with wheat. Jim Lahey, the head baker and proprietor of Sullivan Street Bakery, in New York City, once told a group of bakers and farmers, "You could give me dog-shit wheat and I could make it taste great." I believe him. In the hands of a skilled baker, denuded wheat, stripped of its bran and germ (and therefore its flavor), is no hindrance. In fact, when the baker is after the open, airy "French crumb," it helps to work with refined flour.

Flour contains proteins that produce a magical substance called gluten, which, under the right conditions, stretches like a rubber band. The magic here is how it keeps its shape without snapping back to its original form. (The gluten in pasta dough is what allows it to be rolled out thinly without it shrinking back or falling apart.) As a general rule, the higher the protein content, the stronger the gluten, and the more space the dough has to capture the carbon dioxide released from the yeast as it grows and divides. This is what makes light-textured bread.

The bran in whole wheat flour acts like little shards of glass slicing through gluten, resulting in heavier, denser breads. (In part, Alex's brioche is so light because of the butter, eggs, and milk.) Even the most experienced bakers find it hard to make an airy whole wheat loaf.

The packaged-bread industry, capitalizing on our preference for light-textured breads, has pushed wheat breeders to select for higher protein percentages and stronger gluten, ensuring maximum dough strength. It doesn't hurt that dough strength is also essential for speedy baking. Industrial bakeries demand that dough act like steel scaffolding—holding its shape under the assault of large-volume mixing and lightning-fast proofing on its way to a fluffed, full loaf. The faster the flour moves from dough to loaf, the more loaves can be baked per hour.

Perhaps they took our preference a bit *too* far. In his 1969 book *The Making of a Counter Culture*, Theodore Roszak charged the food industry with destroying our staple food. "The bread is as soft as floss," he wrote. "It takes no effort to chew, and yet it is vitamin enriched."

Ask any chef today about the effect that packaged breads have had on dumbing down the American palate and Roszak's claim doesn't seem far off.

The countercuisine movement of the 1960s and '70s tried to redeem whole wheat. Hoping to reclaim the flavors of a pre-industrialized food system, it demonized white foods, which represented not only heavily processed, sanitized, and denuded ingredients but also the blandness of modern American culture. ("Don't eat white; eat right; and fight.") The new philosophy was codified in the best-selling *Tassajara Bread Book*. Part gentle manifesto (the book is illustrated with drawings of laughing Buddhas and cats), part cookbook, it perfectly captured the longing for a wholesome alternative to industrial food and farming.

Countercuisine began to embody ethical eating in broader terms—not just by embracing particular foods and habits (vegetarianism, communal dining) but by scrutinizing the entire food chain. Who was growing your food? How was it getting to you? Everything mattered. But not everything tasted good. Especially the bread. Whole wheat loaves from the countercuisine era often seemed more like bricks, better suited to building forts. They were virtuous, but they weren't always delicious.

"Anything that had 'whole wheat' or 'whole grains' on the label, you knew right away had all the implications of bad natural foods," Nancy Silverton, the founder of La Brea Bakery, in Los Angeles, told me." And a lot of them *were* bad.

Jeffrey Steingarten, the food critic for *Vogue*, once wrote about his experience baking whole grain bread. "The worst loaf of bread I ever baked," he begins, "was the Tibetan Barley Bread in *The Tassajara Bread Book* . . . so heavy and dense that it was an accomplishment to slice and unpleasant to eat without first slathering it with butter." Steingarten did say the book included other rewarding breads. But he tapped into the Achilles' heel of countercuisine—bad cuisine.

—

And yet the cuisine doesn't deserve all the blame. For me, the tip-off came from the dusty-smelling brioche that Alex made with conventional whole wheat flour. It didn't taste good, either. The truth is, most of the whole wheat grown in this country does not taste good. And we're not going to be drawn to whole wheat, or persuaded to change our bread preferences, unless it does.

So the question is why—why did that batch of pre-ground conventional whole wheat flour taste so different from Klaas's? If it was the same ingredient (sort of), why couldn't it approximate the version we were grinding ourselves?

One answer has to do with fresh milling. The natural oils in the wheat germ are what imbue it with flavor, but they have a short shelf life. *Seriously* short—they begin to spoil as soon as they are released. Which means that in order to capture the grain's aroma, the flour has to be fresh. That's true for nutrients as well—flour has been shown to lose almost half its nutrients within just twenty-four hours of milling. Truly *whole* whole wheat flour, you could argue, means milling it yourself. If that sounds like the worst sort of food snobbery, think of coffee: no self-respecting barista uses pre-ground beans. Increasingly, neither do serious home coffee drinkers.

Another answer is soil. Whereas Klaas employed thoughtful crop rotations and careful soil management to ensure that his wheat had great flavor, the conventional batch undoubtedly came from chemically doused fields, starved of nutrients.

But I've come to understand that even soil doesn't dial back far enough. Its microorganisms can be thriving, and you can still end up with flavorless wheat—because modern wheat (unlike Klaas's heirloom varieties) is not bred for flavor. It's bred for monocultures and high yield, and for industrial milling and baking.

We've lost the taste of wheat, in part, because we've stopped breeding it for flavor.

CHAPTER 27

To WHAT EXTENT do genetics decide the taste of wheat?

I knew from experience that certain varieties—an heirloom to-mato, for instance, or the Eight Row Flint corn I had tasted many years earlier—could have drastic implications for flavor. But it took another reve-lation for me to grasp what good breeding looked like in the wheat world.

It happened the next time Alex ran out of Klaas's emmer. Once again he ordered a batch of whole wheat flour to replace it, but this time the flour came from Anson Mills, an artisanal grain company run by Glenn Roberts, the same philanthropist and seedsman who had sent us the Eight Row Flint corn.

"Okay, we try this here," Alex said the day the Antebellum Style Rustic Coarse Graham Wheat Flour arrived. "'Graham.' This I do not know, but we try."

I didn't expect much from a graham flour brioche. Can you blame me? Named for nineteenth-century dietary reformer Sylvester Graham, who preached the virtues of coarsely ground whole wheat flour (especially when baked at home by a wife tending to the nutritive needs of the family), graham flour is about culinary asceticism, not excellence. It brings to mind good di-gestion, cardboard flavor, and a dry mouth. I figured we would end up pre-ferring the dusty-smelling whole wheat flour Alex had used a few months earlier.

But later in the day, Alex brought over a loaf. "This I love," he said. After a slice, then two, then three—just out of the oven, the bread was impossible not to eat—I thought about Graham. His wholesome prescription for a simpler life turned out to be very tasty. It was softer, sweeter, and in some ways even more flavorful than Klaas's emmer. The graham flour made its way into the small bakery we operate next to the restaurant: graham cookies, flatbreads, scones, and, in what I thought was a novel twist on an old southern staple, graham biscuits.

When I called Glenn to learn more about his process, he told me that he grinds all his flour by hand, at cold temperatures, to preserve the flavor and the nutrition—Graham's original intent. The industrial process does neither.

But that was only part of the equation. To Glenn, "graham" designates more than just a style of milling; it also refers to a particular kind of wheat. Glenn's graham flour comes from Red May, which, he explained, was one of the varieties originally used in the nineteenth century. Ill-suited to roller milling, it was mostly grown in small kitchen plots and ground by hand. It had nearly disappeared until Glenn himself managed to resurrect it.

Glenn told me his goal was not only to provide the best-tasting varieties but also to explore how and why these varieties were created in the first place. "No one gets back to the place-based idea of what cuisine means—why a bread is in fashion, why it's diverse, or even what a local loaf looks like," he said. "If you're not selecting for any of these important things—in other words, if you're not selecting for flavor—your wheat won't taste good." His interest in wheats like Red May was only one part of a larger project to resurrect a whole system of forgotten crops—wheat, but also corn, beans, and rice—that had once formed the foundation of southern food culture. "My life's work is the repatriation of a lost cuisine," he told me. Just like that, something clicked. If I was after not only great-tasting wheat, but also a way to integrate all of Klaas's rotations into my cuisine, Glenn might provide some much-needed insight.

THE RICE KITCHEN

Glenn once thought he knew all there was to know about southern food. In the spring of 1997, working as a consultant for a large hotel chain, he was hired to arrange a historic dinner for the Smithsonian's board of directors in Savannah, Georgia. It was an important gathering, and Glenn worked hard to make it successful. He spoke to locals, read the latest cookbooks on southern cuisine, and settled on an ambitious menu that included Savannah red rice, a legacy as much cultural as gastronomic, from what was once the world's most prosperous rice-growing region. He bought the best local tomatoes and pork for the dish and made sure the rice was dried slowly on top of the stove and finished in the oven.

Glenn knew rice. Though he was raised in La Jolla, California, his mother had grown up in South Carolina, at the epicenter of southern cuisine known as the Carolina Rice Kitchen. Culinary historian Karen Hess defines a rice kitchen as a place of rice worship, where the grain was on the table with every meal. For Glenn, it describes his childhood.

"Breakfast, lunch, and dinner," he told me. "No matter what we were eating, it was always, always served with rice. And there was always rice on the stove." Glenn was allowed to cook rice only for the dogs. And sometimes the cats. "If I even touched the rice pot on the stove, I didn't get dinner," he said. "It was hell on road trips, too, FYI. Everyone wanted hot dogs, but Mom prepared rice stews."

Glenn remembers his mother's complaints about the quality of the rice that was available in supermarkets. "It wasn't the garden-variety carping you get from older generations," he said. "It was deeper than that." Having grown up on hand-milled Carolina Gold rice from the Lowcountry region (named for the low-lying coastal areas of South Carolina and Georgia), she felt offended by "machine rice," which was all there was in California in the 1950s. "She hated the aroma right out of the box. She said it smelled like vitamin tablets."

The dinner for the Smithsonian board went well, or so Glenn believed, until a letter arrived a few days later. "I can still remember holding that letter in my hand," he told me. "It was a searing critique. 'Do you even know anything about Southern foodways?' Apparently I did not." The writer lambasted him for every course he had chosen to serve.

Humiliated, Glenn began reading about the history of southern cuisine—and was shocked by what he learned. "*Jesus Christ*, I said to myself. *This all happened here?* Some of the best wine in the world was once grown in Savannah? And the best food was produced and sent all over the world from right here? Like the red rice, which was *actually* red, and not made by adding ketchup to the white stuff."

Glenn kept reading, gathering source material from more than 140 texts. The more he learned, the more astonished he became. Not only had the South established the preeminent cuisine of this country, but the cuisine had evolved from a supremely advanced farming system unparalleled in nineteenth-century America.

"It was the best market system in existence," Glenn told me. "Everybody came here to study it. We were the seedsmen to the planet, the envy of the world for our cuisine. How the heck did that happen?"

It happened in part out of necessity. By the 1820s, the South's main cash crops, such as tobacco, cotton, and corn, had exhausted the soil. Large tracts of land throughout the East were worn out, and many farmers moved west, where they would repeat the same model of extraction agriculture (leading to the destruction of the prairie, among other tragedies).

Glenn claims that it was this soil crisis that forced the farmers who stayed behind to try something new. From 1820 to 1880, agriculture in the South became largely experimental. Farming journals were printed and widely dispersed, agricultural societies formed, and prizes for the best varieties were routinely awarded at fairs and exhibitions. Model farms and plantations demonstrated the virtues of crop rotations, intercropping, and green manuring. If you were a farmer during this time, you were likely a breeder, too, and something of a soil scientist. Integrated farming became a necessity for restoring fertility to the soil.

"Call it the age of scientific farming, or just a time when our backs were against the wall," Glenn told me, "but the antebellum South was a watershed period of intense, almost frenzied experimentalism. It led to the most important era of vegetable breeding in the history of this country."

The desperate attempts to plant crops that would succeed, and work in tandem to repair the soil, turned Charleston and the surrounding Lowcountry into a center of seed diversity. Crops were imaginative and varied: African rice, Italian olives, South American quinoa, and Spanish Seville oranges. Glenn found farm journals documenting fields filled with forty kinds of rutabaga or dozens of varieties of sesame seed. And the successes were broadcast quickly, ushering in a boom of new vegetable and grain cultivars.

"The really interesting part," Glenn said, "is that unlike any other time in American history, and I would argue any time since, taste was the determining factor. *Taste*. Even a good-yielding crop—if it didn't taste great, it didn't get replanted."

Lowcountry cuisine came together out of this cornucopia, as well as through the collision of Native American, European, and African cultures. Gastronomic societies formed. Cookbooks were quickly written to store the recipes.

Of course, there was an uncomfortable duality to this short historical period. Flavorful food was widely available; in many places, it was the *only* food available. But it happened at a time when the South was still in the grips of slavery.

"At least initially," Glenn said, "everyone experimenting with kitchen gardens had a slave. They did the work. They almost always did the cooking, too. We had a bunch of wealthy white people who could afford to write books about what some really smart black people were doing."

The Carolina Rice Kitchen—what Glenn refers to as the "belle epoque of good eating," with rice as the standard-bearer—evolved out of the soil crisis and slavery, which is to say out of both ecological and human desperation, providing the impetus for America's first complete and distinct regional cuisine. Glenn's mother merely knew it as the food she was raised on.

The beloved cuisine, and the farming system that supported it, didn't last. They disappeared—first, as larger-scale farms that supplied produce to the East Coast expanded and overtook the market in the 1800s, and then in a more complete sense during the Civil War, when nearly 112,000 acres of rice fields were abandoned. With the advent of chemical fertilizers, farmers abandoned time-consuming rotations and focused on money-producing staple crops. Pest problems developed, which gave way to the need for pesticides. Soil health declined.

There were other problems, too. California and New Jersey became the main produce-supplying states. Corn and wheat moved to the Midwest, further depressing prices for the once profitable southern crops. David Wesson's method of turning cottonseed into cooking oil solidified cotton as a staple crop, ending the era of experimental agriculture altogether. By the start of the Great Depression, Carolina Gold, the rice Glenn's mother cherished, had all but vanished.

When Glenn, a California surfer with a talent for the French horn and mathematics, accepted a full scholarship to the University of North Carolina, he was reunited with the cuisine of his youth. He began sending packages of grits, wheat, and—when he could find it—native rice to his mother in California. But it wasn't what she remembered.

"*It sucks*, is what she'd say," Glenn said. "So I started sending what looked like really good collards, and occasionally field peas. But she never liked anything. The flavors had disappeared."

It was then that Glenn began to grasp the ramifications of what had happened—the disappearance not just of certain flavors, but of a whole way of cooking. The Smithsonian dinner and the research he did afterward only crystallized his understanding.

"The idea that everything had disappeared was unacceptable," he told me, explaining his decision to leave the hotel business. "It was that simple. It just clicked. I knew what I needed to do."

——

Glenn started Anson Mills in 1998. "I began with the notion that Anson Mills would repatriate, grow, and sell rice. No feasibility studies, no forecasts, no budget. I was so proud. And so ignorant. After a few weeks, I discovered that no one had any rice seed. Because no one was growing rice! Um, *duh . . .*," he told me, smacking his forehead.

Glenn quickly adjusted, deciding to grow heirloom corn for grits. He predicted that eventually he would earn enough to begin the long and expensive process of growing rice. "It was a sub-moronic plan," he said. "But that was my plan."

He soon discovered that heirloom corn seeds were also in short supply.

So Glenn sought out moonshiners in the region, figuring that their illicit distilleries used corn they themselves had grown and passed down through generations. His mother reminded him that the best grits from her childhood came from the coast. "So of course I went to the coast," he said, nodding dutifully. "Who knew bootleggers worked the coast? Well, Mom did—and they do."

He found one moonshine operation completely "off the grid." The family had been farming the same land since the late 1600s. But this wasn't just a moonshine operation. The family also raised pigs, goats, and sheep. And they grew countless crops for food. Everything was intertwined, and everything grew together.

"So, you didn't have field peas as field peas; you'd have field peas and corn together in the same field. You didn't grow just wheat; you'd grow wheat that was, say, thirty inches tall, and then you maybe grow rye above it that was seven feet tall; harvest the rye first, cut it high, then cut low for the wheat, and then they'd have clover down at the bottom, or winter peas, or whatever. No one thing was growing in a field."

It was unlike anything Glenn had ever seen. "Like an idiot, I said to the father, 'You can't machine this.' By that I meant you couldn't run a tractor combine to harvest it. He looks at me real funny and says, 'Why would we want to machine this? This is eating food.' He wasn't growing for animals. He was saying, 'This is our kitchen food. This is what we eat.' He would just as soon run a combine on his field as he would grow GM [genetically modified] corn. These people were frozen in time."

Glenn ate lunch with them. "Holy crap! I mean, everything on the table—everything—had been grown and processed on the farm," he said. "It was unbelievable food. Breads, butters, jams, hams, even wine—you name it. Oh, the best corn grits I'd ever eaten. Unreal, impeccable flavor, simply and honestly prepared. I sat there with the bootlegging family in the middle of this food paradise and felt in reality what my mother had always said about southern kitchen gardens, and about the food she grew up on. A terrific feeling came over me. It was a wonderful epiphany, the realization that my mother had been right."

The family agreed to sell him some corn seed, and even allowed him to grow it on their land. Glenn calculated the size of the crop based on a profit that would allow him to begin growing rice.

The corn he harvested that first year was delicious, but the yields were low. And since Glenn was hand-milling it to make grits, the price was high. Supermarkets and even specialty retail stores balked at the price and were befuddled by Glenn's insistence that the fresh grits be refrigerated.

"The store managers looked at me like I was from outer space. Refrigerate grits? They had never heard of fresh-milled anything, so the idea that it could spoil, that it *would* spoil very quickly, was absolutely foreign. I wasn't just selling heirloom grits, which people had heard about from their grandparents. I had the flavorful grits *and* the fresh milling process to preserve that flavor. You couldn't have one without the other. No one knew what the hell that meant. Grits were grits."

Glenn knew he needed a language to describe the quality of his crop, and he needed a market that spoke that language. So he called on chefs, approaching them as if he were a boutique winery selling sommeliers a limited selection of the very best of a vintage. Some southern chefs bought the grits, but their reach was local, and back in the 1990s the high-end southern restaurant market was small. Then he called chef Thomas Keller at the French Laundry, in Napa Valley, widely considered one of the best restaurants in the United States.

When he began his pitch about his premium grits, Keller interrupted him. "I can't sell grits," he said. Glenn proposed artisanal polenta, and at that, Keller was interested. Glenn promised to send some fresh-milled organic polenta from corn-seed stock that originated with the Native Americans.

"He agreed to try it," Glenn said, "which is when I knew I had the sale. Because once chefs try this stuff, once they cook it and taste it, it's sold. A chef like Thomas Keller has a vetting process that relies on his tongue. It's the final word."

Glenn was right. A week later, it was on the French Laundry menu. And within a few months, other chefs around the country called about the polenta. Anson Mills began to appear on menus and in cookbook recipes. Glenn's business started to grow.

Then another problem. The bootlegging family, which by this time had virtually adopted Glenn, told him he couldn't just grow corn in a field by itself. In order to produce great-tasting corn—and one day produce great-tasting rice—they said he'd have to start growing other crops and selecting rotations to boost soil fertility.

"Even though—and this is just too ironic almost to mention— surrounding me was the family's farm, with unreal crop diversity, the incredible soil tilth, and the stunningly drop-dead-delicious food that came out of all that diversity, I just never put it together," Glenn said. "How thick can you be?" (I didn't mention that I'd never put it together, either, until I met Klaas.)

Glenn surveyed social club and church archives and collected oral histories from local cooks, farmers, and the community of commercial Sea Island fishermen, to develop a catalog of ingredients from the nineteenth-century rice kitchen.

Which is when he realized he had also missed the point of the Carolina Rice Kitchen.

———

The southern obsession for rice may seem single-minded, but in fact it was built on a whole system of interrelated crops. After the soil crisis of the early 1800s, farmers had learned which plants and animals worked well together. Buckwheat, peas, corn, barley, rye, sweet potatoes, sesame, collards, and livestock worked in tandem to improve the soil and produce superior harvests.

Farmers discovered, for example, that rotating sweet potatoes and sesame after a rice harvest helped boost rice yields and suppress disease and pests the following year. So new varieties of sweet potatoes were bred for better flavor and better yield. The rotations continued to improve and, along with them, the Carolina Gold rice. Favored for its delicate grains, Carolina Gold was exported to China, Indonesia, Spain, and even France, where legendary French chefs like Marie-Antoine Carême and Auguste Escoffier made it famous.

"Crème de riz? That was a dessert based on the sweetness of the rice. The sweetness was helped along by the soil fertility, which came from the proper rotations," Glenn told me.

Seeking to diversify his system, Glenn planted field peas to supply nitrogen for the soil. He settled on the Sea Island red pea, because he remembered his mother describing it as an ingredient in one of her favorite rice dishes— Hoppin' John, a staple southern dish of rice and beans. The recipe most likely emerged from successful plantings of beans followed by rice.

Soon Glenn added barley and rye to his rotations, and Anson Mills expanded into peas and other grains. As his company grew, Glenn was finally able to begin investing his profits in rice. Only, just as before, he first had to find the seeds.

—✺—

There were once more than a hundred varieties of Carolina Gold rice grown in the Lowcountry, but by the time Glenn came along, they were gone.

"So I had to start over," he said. He obtained seeds for Carolina Gold from a seed bank at the Texas Rice Improvement Association, but upon growing them out he realized that they didn't express the characteristics he was seeking. After decades of neglect, the rice more closely resembled Carolina White, a related—but less delicious—variety.

Glenn called the top rice geneticists in the country, one by one, and explained what he was after—to repatriate the rice culture of the Carolina Rice Kitchen and replicate the flavor of what once was the most sought-after rice in the world. He asked for their help in tracking down and identifying the right genes. The response was tepid and often dismissive.

"They were all preparing rice for market—shelf-stable, machine-milled rice—the kind of stuff my mother said tasted like vitamin pills. They weren't thinking about the potential for flavor if you kept the bran intact, and the dimensions of that in breeding."

Glenn quickly learned how little geneticists think about nuances of flavor. "They talk about mouthfeel, they talk about millability, they talk about cookability, but they never get back to flavor profiles, beyond the idea of 'aromatic' rice," he said. "They use the same note with various volumes, but the note never changes. It's one gene, and it's the only gene they talk about."

Carolina Gold is categorized as a "non-aromatic rice," meaning it doesn't exude the perfume of, say, jasmine rice. But Glenn found that a few of his best harvests, hand-milled to his exacting specifications, produced floral and nutty aromas. And through his research, he knew that soil variation and even water quality could cause those flavors to express themselves differently.

"All of our rices in the South had distinct flavors according to which river they were grown on and how the soil was managed," he told me. "There were people on record who could tell where rice came from just by tasting it. They could tell you which river it was grown on—even what part of the river it was grown on."

Glenn did what his mother had taught him to do—he cooked different varieties of rice and described their flavor to the geneticists. The scientists, some of whom had worked with rice all their lives, learned about characteristics they had never considered. Working with the geneticists to breed for his system, he pushed the flavor to become better expressed. And, in the meantime, he began doing more of his own seed work in the field, painstakingly selecting varieties of Carolina Gold from a large pool of possibilities, winnowing out what didn't meet his specifications for a flavor ingrained in his mother's memory.

"I became a breeder," he told me. "I mean, I had no choice."

It was almost by accident that Glenn added wheat to his rotations.

"I liked to ask older southern ladies what they thought of my grits. A few of them said, 'Yeah, great, but where's the graham flour for my graham biscuits?' I was like, *What?* Biscuits made with whole wheat flour?"

Glenn was surprised. Everyone knew southern biscuits were made with white flour. He told the ladies they could get whole wheat flour from the large mill just outside of town. But they said they didn't want whole wheat flour; they wanted *graham* flour.

"As far as I knew," Glenn said, "graham flour was just another name for whole wheat flour. But one day, out of the blue, while I was waiting to pick up my dry cleaning, I remembered Mom talking about graham biscuits. It just popped into my mind. I asked a few of the ladies working there—they were all like ninety years old—if they remember having graham flour as children. 'Oh, yes!' they said, 'graham biscuits. We lived on them.' I nearly fell on the floor."

Glenn researched the history of graham flour in the South, which was when he learned that wheat was part of nearly every kitchen plot rotation. The nineteenth-century southern kitchen gardens brought Sylvester Graham's ideas to reality. Wheat was initially grown to restore carbon lost from the rice harvest, but the farmers selected a variety called Red May for its flavor. Traditionally, after the wheat berries were harvested, they were milled in the yard with a hand grinder called a quern. Biscuits—and, later, things like Triscuits and graham crackers—were made from this home-milled wheat. White flour biscuits were in fact a rare indulgence, or they were part of wealthy white southerners' tables.

"I knew we had to try it," Glenn said. "We grew the Red May, milled it nice and coarse, and made graham flour for biscuits. It was friggin' extraordinary."

CHAPTER 28

I ARRIVED IN CHARLESTON, South Carolina, on a warm and muggy morning in early July. Glenn picked me up just outside the terminal in a small rental car. "It's all I do, rental cars," he told me, "I'll drive it for a few days and then run in and change cars."

Glenn is an oddity, even in a place like the Lowcountry, where eccentrics grow like beautiful weeds. He is tall and silver-haired, with an expression forever on the brink of enthusiasm. On this particular day, he wore khakis and a short-sleeved white polo shirt, looking very much like a man on the way to a Sunday outing at the yacht club. To be fair, we were on our way to Clemson University's Coastal Research and Education Center, which for Glenn is a little like a weekend getaway. Anson Mills donates money to the university, and, in return, the university provides land for Glenn's crop experiments.

As we drove, he excitedly outlined the day's itinerary, and I was quickly reminded of his habit of dispensing information in rapid-fire bursts of arcane facts and dizzying non sequiturs. Glenn drops names and historical events as though you should know them, but shrugging as he does it, as if he doesn't mean any harm. And he doesn't. He delights in surprising people. His thirst for knowledge is matched only by his desire to show it off. He can tell you about water-driven machinery (he worked for a time as a doffer in a twine factory), topology (his major in college), the diaspora (not the Jewish one, but the Abenaki Native American one), and his recent interest in John Letts (a British archaeobotanist who documents the history of cereals). A conversation with Glenn can feel like an airplane flight interrupted by fierce, inexplicable episodes of turbulence.

A simple question, like the one I asked soon after I got into the car—"*Since I'm in Charleston, would it be possible to visit that family who helped you get started growing your corn?*"—led to Glenn mentioning something called the ATF, followed by a short history of South Carolina market farming in the 1820s, the observations of Harder, and then suddenly to an extended arm and a *"By the way, if we were to follow this road here, the destination would bring us to Doc Pasaventos's olive trees on Folly Island."*

ATF? Harder? Doc Pasaventos? I picked one. "Harder?" I asked.

"Jules Harder," he said. "Which is what I meant by the reference to the Bradshaw collection."

"I don't know Jules Harder," I said (but wondered about the meaning of the Bradshaw collection, which he had never mentioned before).

"Harder was the chef of Delmonico's," Glenn said, in the way of a gentle reminder. "In the early 1870s."

Clarity remained a far-off cousin to whatever exchange we were having. I tried to climb my way back to where I started. "What about visiting the farm family of yours . . ."

"That's what I'm saying: not a good idea to discuss, the ATF notwithstanding."

We arrived in front of a large padlocked gate. I asked what ATF stood for. "Bureau of Alcohol, Tobacco and Firearms," he said, referring to the fact that the family had an illegal moonshine operation—not to the place we had just arrived.

I shook off the exchange as Glenn tapped the steering wheel and smiled broadly at the large, open fields on either side of us. "Welcome to the second-oldest farming research center established after the revolution," he said. "Way cool."

Glenn sped along one of the enormous fields and spoke excitedly, if not altogether clearly, about an ongoing experiment with cowpeas. "We've got fourteen varieties of peas doing allelopathic suppression and fertility," he said. "We mix up the varieties, which is mass population genetics. This is all prep for wheat. No clover, that's the goal. Because we're gonna do total legume wheat."

I unraveled this to mean Glenn was experimenting with intermixed varieties of cowpeas, an animal-feed crop, and evaluating the plants' ability to resist pests ("allelopathic suppression"). He was also measuring the benefits to the soil that would come from planting a leguminous crop before wheat. The tradition in organic farming, as Klaas illustrated with his rotations, is to precede crops like wheat and corn with clover, a legendary nitrogen fixer, but Glenn had a hunch that cowpeas would be just as beneficial. More important, he suspected that cowpeas would make the wheat taste better.

The research site was surrounded by large, manicured fields and high-tech greenhouses with commercial plant-breeding experiments. Thriving in between were Glenn's chaotic experimental plots—the cowpeas, various inter-plantings with rye, and patches here and there of ancient varieties of wheat. These wheat experiments, in particular, stood out in striking relief: varieties like emmer, so old they're referenced in the Bible, allowed— actually, encouraged—to express their unique traits next to university-controlled seed varieties so new they hadn't yet been named and grown under a regime of strict uniformity and control.

"They're doing their thing, and I'm doing mine," Glenn said. "Which is celebrating landrace farming and honoring a tradition of seed saving and seed improvement that reaches back into prehistory."

By "landrace," Glenn meant a kind of farming that encourages variation in the field, with less distinct and less uniform varieties. He wasn't overstating it. Though the first breeze of the morning made the hodgepodge of ancient varieties sway and rustle in unison, nothing about it suggested cohesion or uniformity. You'd be forgiven for thinking that it looked a little all over the place. Which is the point, really, of landrace farming. Glenn's not bothered by the diversity. In fact, he's banking on it.

Plants in a landrace system are different, but only slightly so. As opposed to a modern field of wheat (or corn, or really any cultivated variety)—all the plants identical in size, shape, and growth patterns—a landrace crop's in-built diversity allows it to thrive under a variety of circumstances. It's a natural insurance policy for the population, ensuring that, while some of the crop may succumb to a disease or a natural disaster, some of it will not. In periods of drought, for example, most of the wheat will fail, but the plants with greater drought tolerance will survive and pass their advantage on to future generations.

I once heard a lecture given by Abdullah Jaradat, a USDA agronomist, to a group of grain enthusiasts. "When you domesticate a plant like wheat, you spoil it," he said. "You have to provide it with all its needs; otherwise it will not produce what you expect."

Glenn's old-world, chaotic plots were exactly that—unspoiled. And they offered a glimpse into what had been, until recently, the only farming system possible.

SAVING SEEDS

From the beginning, which is to say from around 8000 B.C., when agriculture is thought to have begun, farmers knew to save at least a small portion of their seeds to plant for future harvests. By the time agriculture replaced hunting and gathering as humanity's primary source of food, seed saving had emerged as one of the community's most important responsibilities. With each community preserving and selecting its own seeds, thousands of locally adapted landrace varieties evolved across the globe. These varieties were not static. They adapted and changed depending on the environment and the preferences of the culture, producing the characteristics most likely to thrive under the circumstances. It was a rich reservoir of diversity that came to a very sudden end.

At the start of the twentieth century, plant breeders discovered a way to farm more efficiently. They learned that two distinct lines of corn could be crossed with each other to create a new genetically uniform generation imbued with "hybrid vigor"—making it faster-growing and more robust than plants left to pollinate naturally (the same idea that led to the foie gras industry's Moulard duck). The vigor would last only a year; subsequent plantings would not be as successful. So farmers bought new hybridized seeds every year to maximize their yields and turned away from the ancient practice of seed saving. Commercial seed companies came to dominate the market for corn and then, as the trend toward hybrid seeds continued, for most other grains, fruits, and vegetables as well.

In some ways, wheat was an exception. Bread wheat is a hexaploid, which means it has six sets of chromosomes (and therefore six copies of each gene), whereas corn and most vegetables—and even human beings—possess only two. So it doesn't open itself up to easy manipulation. It is also self-pollinating, each wheat plant containing both male and female parts. Since the plants fertilize themselves, crosses between different varieties are less likely to occur, either naturally or through deliberate intervention. Saving seeds is always possible, without any loss of genetic integrity.

Which isn't to say that farmers continued to do so. As breeders began to develop new and improved varieties (albeit without the staggering success of hybrid corn), more farmers began buying their wheat seed. Why engage in the laborious ritual of saving seeds when a better-yielding, more consistent crop was now readily available for purchase? Genetic uniformity became the status quo.

In Glenn's landrace system, by contrast, every grain of wheat contains a germ with a distinct destiny. It's impossible to know exactly what the wheat will be like until you cast the seed on the ground and see what grows. Most of it will look quite uniform; a good percentage might even remind you of a monoculture. But there will be the inevitable wild cards, the offshoots—called "sports"—and they provide not only an insurance policy for the crop but also the potential for new flavors.

"Going sportin'," which means venturing out into the fields to find these irregular plants, re-creates what farmers did throughout history—seek out the one plant in the crop that doesn't look like the others, the ugly duckling of the bunch, and celebrate it for its distinctiveness. Should that offshoot turn out to taste good and be encouraged by the farmer, the entire crop and cuisine might change, at least slightly, to include this distinctive first cousin.

Sporting is also, one could argue, the most democratic of farming practices, because it allows recessive traits (those qualities we all have hidden in us somewhere) to express themselves. No one knows why certain genes lie dormant for hundreds, or even thousands, of years. But landrace farming leaves open the possibility that an unexpected trait might reveal itself at any time. An environmental event, such as a heat wave or even a sprinkle of rain at the right moment, can trigger a genetic awakening.

Glenn's goal is not efficiency. What if, he asks, instead of forcing nature to go in a particular direction, we allowed nature to dictate how the seeds should evolve, and then adapted to those changes? It would mean more variation in each crop, which would, in turn, cause there to be slightly different ripening times and different kernel sizes. Without identical characteristics, our plants would have better disease and pest resistance, more vigor, and greater resilience in the long run. It might also mean the discovery of a superior flavor, a concept I had always found a little abstract, until Glenn told me that the Eight Row Flint corn Jack had managed to cultivate so successfully at Stone Barns had come from a landrace farming system—generations of farmers selecting ears by hand and tasting them.

"This is what farmers have always done. Throughout the ages they managed to broaden the genetic base and deliver a very rich source of variation,"

Glenn said. "You're not just looking for change," he added. "You're celebrating change."

But this rich source of variation, cultivated over thousands of years and supported by countless generations, changed irrevocably in the middle of the twentieth century. Wheat was transformed on a global scale, and practically overnight. It was a revolution that began, improbably, with a dwarf.

THE AGE OF DWARFS

No one was looking to grow short wheat—not at first. The beginning of the Green Revolution, that period of agricultural modernization and massive productivity gains across the globe, is often traced to dwarf wheat, but dwarf wheat actually came out of an impromptu visit by an American to the hillsides of Mexico.

In 1940, vice president–elect Henry Wallace attended the inauguration of the Mexican president, Manuel Ávila Camacho. The trip was a show of support, and Wallace, the former secretary of agriculture, seized on an invitation to visit the hillside fields of the local Mexican farmworkers. Before entering politics, Wallace had started the Hi-Bred Corn Company, which came to lead the industry, and soon the world, in hybrid corn-seed technology. Wallace was a wealthy man. He was also a progressive for his times—an early advocate for civil rights and government health insurance. His heart went out to the Mexican peasants who worked their small plots in miserable conditions. The soil was failing, their seeds were unproductive—they had no machines and no fertilizers.

After returning to the United States, Wallace persuaded the Rockefeller Foundation to support a special collaboration with Mexico to improve farmers' crop yields. (He had failed to convince Congress.) Until then, aid had come in the form of donations. Wallace's idea was to send the best American agricultural scientists to train their Mexican counterparts in the latest breeding science. A young scientist and developer of agricultural chemicals for DuPont liked the idea and agreed to join the effort. His name was Norman Borlaug.

Borlaug was born in Iowa and attended college in the Midwest at the height of the Dust Bowl. That great environmental disaster had many people reconsidering large-scale modern agriculture, but Borlaug saw it as proof that technology and a greater emphasis on high-yield farming were the only options for the future of food production. The International Maize and Wheat Improvement Center (CIMMYT), formed as a collaboration between the Rockefeller Foundation and the Mexican government, allowed him to put his ideas to work.

Borlaug was ferociously dedicated. He spent fifteen-hour days in the fields, examining different crops and soil conditions and heading a small team that managed to cross more than six thousand distinct varieties of wheat. His research showed that adding fertilizer to wheat production could triple growth, but the kick was so powerful that the wheat stalks shot up too quickly. Without full development and enough strength to support their heavy seed heads, they fell over and rotted on the ground. Harvesting was nearly impossible.

Then, in 1952, word arrived of a newly developed short-straw wheat from Japan called Norin 10. Using samples of Norin 10, Borlaug began growing new semidwarf crosses and found that fertilizer enabled this wheat to mature more quickly without falling over. Within a few short years, Borlaug had produced wheat that yielded three times more than its predecessors. By 1963, 95 percent of the wheat grown in Mexico was his semidwarf variety, and the country's wheat harvest was six times what it had been when he arrived. Encouraged by the results, Borlaug next sent his dwarf wheat to India, which was on the brink of mass famine. Farmers planted the new seeds and followed the fertilizer regimen, and within a few years the results were just as incredible: crop yields had more than tripled, and India became a net exporter of wheat.

The new varieties continued to spread throughout Asia, with the same effect—displacing local landrace varieties (and thousands of years of genetic refinement) and upending the traditional practices of millions of farmers. New strains of "miracle" rice soon followed, which matured fast enough to allow farmers to grow two crops in a year instead of just one.

Such was the power and, indeed, the aim of the Green Revolution: to increase food production without bringing more land under cultivation. From 1950 to 1992, harvests increased 170 percent on only 1 percent more cultivated land. Today, more than 70 percent of the wheat grown in the developing world carries genes that Borlaug developed in Mexico. And semidwarf varieties make up the majority of wheat in the United States as well.

It is estimated that a billion lives were saved by Norman Borlaug's work, which makes questioning the success of the Green Revolution complicated. How do you argue against a system of agriculture that saved a billion people?

One way has been to look at the global increase in diet-related diseases since the 1970s. Certain types of cancers, cardiovascular disease, diabetes, and obesity are, many argue, the enormous collateral damage the revolution inflicted. Lives were saved by providing calories, but the Green Revolution ultimately altered the way we eat, and, for the most part, not in a good way.

Without question, it altered the way we grow food on a large scale. The world is now awash in monocultures of genetically uniform varieties, fed by chemical fertilizers. And their legacy has been disastrous for soil health. The short wheat came with equally short roots, like those whispery filaments I saw on Wes Jackson's banner, diminishing important highways of bacterial and fungal activity. Soil became compacted, degraded.

"They took beautiful stuff like this," Glenn explained as he pointed to his test plot, "and dwarfed it, dwarfing the roots, too, limiting their ability to uptake micronutrients from the soil. Questionable nutrition and zero flavor."

(There was nothing dwarfed about Glenn's wheat. Every stalk reached to my chest, and a few them extended well above my head. "Tall straw, deep roots," he said.)

The dwarfed root systems also retained less water, a shortcoming that many countries have compensated for with enormous, government-backed irrigation projects. From 1950 to 2000, the amount of irrigated farmland tripled. One-fifth of the grain grown in the United States is irrigated; in India, it's more like three-fifths, resulting in the rapid depletion of the country's groundwater sources. According to author and activist Vandana Shiva, India's water crisis is clearly linked to the introduction of Borlaug's green-revolution varieties. "Although high-yielding varieties of wheat may yield over 40 percent more than traditional varieties," she writes, "they need about three times as much water."

Green Revolution varieties consume fossil fuels, in the form of synthetic fertilizers, just as greedily. As Cary Fowler and Patrick Mooney note in their book *Shattering: Food, Politics, and the Loss of Genetic Diversity,* the relationship between dwarf seeds and chemical fertilizers is "akin to the relationship of the chicken and the egg. The fertilizers made the new varieties possible. The new varieties made the fertilizers necessary."

By conservative estimates, more than a third of the Green Revolution's yield gains are owed to synthetic fertilizers, which, as many have pointed out, makes the revolution not exactly *green* in the environmental sense. Modern, commerically bred seed varieties depend on chemicals now more than ever. To get them to work, you need the chemicals; once the chemicals are in use, soil organic matter falls off, and the soil is less able to transport nutrients to the plants efficiently. The result is that more chemicals are needed to get the same kick.

All true. And yet . . . a *billion* lives.

Susan Dworkin, a former assistant to a breeder who worked alongside Borlaug for many years, once described how breeders working on hunger tend to see the problem purely in terms of yield. "How much food could you get out of an acre? How many people could you feed? That's where they are. That's what they think," she said. "They are not looking at the dinner table. They're looking at the swollen belly."

With a billion lives at stake, the single-minded pursuit of yield is both defensible and important. But what if, all along, our math has been wrong? What if, in our mad dash for greater productivity, we've miscalculated the *true* yields?

Consider the farmer who grows a variety of semidwarf wheat. He applies the requisite chemical fertilizers and sits back, with hopes of watching his yields (and his profit) soar. But shorter straw means less to plow back into the ground to become food for soil organisms. Or, if the wheat is being used as food and bedding for cattle, dwarfed straw means there's less feed for the cows. Either way, it amounts to less food for someone. And not just anyone. As Klaas liked to remind me, soil organisms and cows are partners in making a healthy system work. In the modern calculus of efficient farming, those things are left out of the equation because they're not feeding our bellies (at least not directly). And it's a critical omission.

There is another miscalculation, too. I once attended an agriculture conference where a scientist argued that organic agriculture could not feed our growing population. One of the studies he cited compared a small plot of conventionally fertilized corn with another small plot of organically grown corn. A photo showed the two plots planted right next to each other: same variety, same soil. The conventional corn was tall, vigorous, and thriving. The organic corn looked dry and stooped over, a sickly cousin. The photo, as convincing as Wes Jackson's side-by-side analysis of annual and perennial wheat, seemed to prove that yields for conventional corn far surpass those for organic corn.

It wasn't until Glenn pointed to his landraces, and then to the university trials, with their military uniformity, just beyond, that I could consider another side-by-side comparison. And this one looked different. Glenn's wheat wasn't a shriveled, sickly cousin (with its varying heights, it was more like a crazy uncle), because Glenn had prepped the soil. He had rotated in different crops—cowpeas, barley, and oats—to ensure fertility. He gave his wheat a fair shot at thriving.

Of course, even thriving landrace wheat might not yield as much as the conventional varieties, at least not consistently. Growing one variety with fertilizers usually wins. But that doesn't mean the corn study was right. It was wrong—and here's the heart of the bad math. Barley and oats make a good meal. They are delicious and full of nutrition. So, while an acre of conventional wheat may yield more than an acre of organic wheat, that does not mean it yields more *food*. It just yields more wheat. The equation is missing the sum total of its parts: barley plus oats plus wheat will yield more food than wheat on its own.

"It is often said," Vandana Shiva has written, "that the so-called miracle varieties of the Green Revolution in modern industrial agriculture prevented famine because they had higher yields. However, these higher yields disappear in the context of total yields of crops on farms."

But total yields of crops on farms matter only if we're eating all of the farm's crops—the math holds up only if we eat the barley and the oats. If the farmer can't sell the barley and oats because there isn't enough demand, the logic of growing wheat (or corn, or soy) in monocultures is difficult to compete with. It feeds on itself. As long as we don't eat the diversity, the pull to produce more of the primary crop is too strong.

———

Which brings us to cuisine.

The challenge of making delicious use of various ingredients is at the heart of all great cuisines, and it evolved from diversity. Cuisines did not develop from what the land offered, as is often said; they developed from what the land demanded. The Green Revolution turned this equation on its head by making diversity expensive. It empowered only a few crops. And in the process, it dumbed down cuisine.

Of all the arguments against the Green Revolution, dumbing down cuisine sounds like the most insignificant—an acceptable sacrifice on the road to bringing agriculture out of the Stone Age and feeding the hungry. But nature writer Colin Tudge reminds us that the world's population at the beginning of the agricultural age, ten thousand years ago, stood at about ten million. By the time industrial agriculture came into favor, in the 1930s, it was three billion. A three-hundred-fold increase, achieved with old-world farming techniques—organic farming before there was such a thing as organic. Not bad for an agriculture system now considered archaic. But more to the point: small-scale, old-world farmers produced not just a lot of food but a lot of really good food.

Legendary Soviet botanist Nikolai Ivanovich Vavilov, who traveled the globe in the early twentieth century mapping the world's greatest centers of crop diversity and collected specimens along the way, came to believe that the landrace crops he discovered were "the result of intelligent, innovative minds—and often the work of geniuses."

For most of history, these geniuses were peasant farmers working with nature to create thousands of new crop varieties. This diversity, in turn, launched thousands of highly distinct cuisines. Not merely Indian, Italian, and Chinese, but their more local and original incarnations: Punjabi, Sicilian, Szechuan . . . not to mention Glenn's Lowcountry.

Cuisine did not shift with fashion or preferences—that happens only with high-end cuisine, which has flourished over the past sixty years. Today's chefs have the freedom (and the imagination, if not the pretension) to mix ingredients and techniques from around the world into one meal, or onto one plate. But we're not reinventing anything. We may push new ideas forward, but we don't create new cuisines. We really just build on what other cultures figured out over thousands of years, when peasants farmed the land and what it could produce dictated what people ate. Location used to be everything; now it's just another ingredient.

True cuisine is more than just a style of cooking, or a unique combination of techniques and flavors. It is the foundation of culture. It determines a way of life. We are as complex—or as monotone—as the foods we grow and consume. As the great gastronome Jean Anthelme Brillat-Savarin said, "Tell me what you eat: I will tell you what you are."

The Green Revolution made that difficult to do. It forced farmers to reduce crop diversity—to specialize, monetize, and modernize. It pressured countries in Africa and Latin America to give up locally grown crops like chickpeas and beans, making people dependent on less nourishing (and less delicious) grains.

Losing indigenous crops didn't just change what people ate; it compromised people's cultural identities. Vice President Henry Wallace's alarm over the condition of Mexico's small farmers may have sparked the Green Revolution, but it was soon clear that there would be no revolution unless those very same farmers abandoned their efforts in favor of large-scale monocultures. Borlaug's modern varieties didn't work in small landrace plots, and the need for large amounts of expensive synthetic nitrogen drove farmers around the world from their land.* Since peasant farmers, from the beginning of agriculture, had also been breeders, a few hundred seed companies displaced the millions who had for centuries been saving seeds and breeding new varieties.

Glenn isn't turning back the clock as much as trying to even the genetic playing field.

As he turned to walk back to the car, he stopped and reached into his coat pocket for his cell phone. He showed me a picture of landrace oats he'd been experimenting with in another field. "I'm glad to have this picture, 'cause this

---

* "This was a classic case of unintended consequences," Peter Johnson, the Rockefeller family historian, told me. "The Rockefeller Foundation invested in Borlaug and agriculture because starvation, and the political instability that came with it, was the pressing issue of the time. No one understood how quickly this would accelerate urbanization—landless peasants with no skills, driven from the land and ending up in urban slums, living hand to mouth. The consequences of mechanization—the urban ghettos in Mexico City and Shanghai, and even the black migration from the South following the dwarfing of cotton—these were all the direct result of the Green Revolution, and we're still living with it today."

will save us thousands of words." In the photo, a single oat plant towered over the others, looking like an awkward teenager after a terrific growth spurt. "There's maybe six of those plants out there," Glenn said. "This one has three times the amount of seed at the head. *Three times*. Imagine that. It just appeared, which means it's been dormant, but for reasons no one understands—and trust me when I tell you no one ever will—it decided to express itself. We're going to save it in the middle of the field, put electric fencing around it, and go out and pray to it, pray that it makes it to harvest and we can save the seed and find out what the hell this is."

The excitement, tinged with urgency, in his voice suggested more than a discovery. He was showing me a frontier of possibilities, a *Who knows what's out there?* not only for oats but for all crops.

Glenn told me he sent the photo to Klaas's wife, Mary-Howell, since he values her expertise as a seedsman. "I asked her, 'What the hell is this?' And she says, 'Hell, I don't know. I've never seen that before.'" He shook his fist at the phone, not in anger but in glee, like a child who just scored a goal or opened a birthday gift. "That right there—*that's* landrace farming."

CHAPTER 29

O N O U R W A Y back to Charleston for lunch, Glenn made a detour just off Savannah Highway. We drove to the intersection of two large fields and got out of the car. The heavy air felt like it could suffocate the crops. South Carolina was in the middle of a long drought, which was why Glenn had brought me to the fields. They belonged to Tris Waystack, a lifetime cattle farmer who had once grown only corn and soy for feed. A few years earlier, he had signed on to grow other crops for Anson Mills.

"One day Tris comes up to me and he says, 'My dad's sick with cancer. I wanna go organic.' Just like that. Tris is a master Eagle Scout, straight as an arrow. I figured the kid could farm, but he couldn't even afford gas at that point."

Glenn purchased a combine and a grain bin for Tris and paid him in full before Tris had even started. He offered to consult on the planting as well, but Tris asked only for seed to start testing, and for assurance that Anson Mills would buy whatever he grew. Glenn gave him Hopi Blue corn, an heirloom seed famous for making blue tortillas.

"A year later I'm standing next to the corn, about to harvest," Glenn told me. "It was amazing. Eighteen acres of gorgeous blue corn. Instead of selling all the corn to me, he decided to donate some of this first crop to the Hopi nation, because he believed the people had good juju. He knew the Hopis don't trade maize for money. He was just looking for the karma."

We stood at the intersection of the two fields. On the right, Glenn pointed to a quadruple planting of cowpeas, sorghum, cane, and sesame seed. I didn't know you could even plant four crops at the same time in the same field. Wouldn't they compete for soil nutrients and water? Apparently not. Glenn explained that they grow at different rates and, as farmers discovered long ago, each crop has different needs. All four of them here looked vigorous in the withering sun.

On the left, just a hundred or so feet away, the other field was clearly suffering. The dirt was dry and cracked, the vegetation sparse, and many of the once green leaves had turned brown. I asked Glenn what the crop had been.

"Soybeans," he said. "Modern soy." He said Tris had gotten cold feet about the quadruple planting and, with this second field, went with soybeans alone.

Glenn pointed to the healthy field and then back to the failed field. "He's going to have to cut the soy under. He's going to get nothing. Not a thing. He just blew all that money on expensive new seed." The drought had killed the soy, he said, because modern soy has such small roots.

What I'd seen as old-fashioned that morning—a nostalgic attempt by Glenn to resurrect a forgotten system of farming—now seemed modern, complex, and even futuristic. In the face of weather that is less predictable and more unforgiving, a diversity of locally adapted crops is one way for farmers to hedge their bets. Glenn's landrace system isn't just repatriating a lost cuisine. It's gathering the seed stock for the future of eating.

THE GLASS ONION

Glenn took me to lunch at The Glass Onion, a mustard-yellow restaurant with generic-looking alliums painted on the sign. Located just off Savannah Highway, it looked like any of the casual chain restaurants that mark the exits of America's highways, but the chalkboard menu offerings were unexpected: Lamb and Oyster Mushroom Ragout, Fried Buttermilk Quail, Local

Shrimp with Farmer Benton's Bacon and Grits. There were local micro-brewed beers and biodynamic wines. Only a few dishes on the menu were more than $10.

"The owner trained in white-tablecloth restaurants," Glenn explained. "She was a real range rat for a while. But she decided she liked unpretentious southern food, no bullshit-bain cooking." ("Range rat": an experienced restaurant line cook. "Bain cooking": the use of premade sauces or garnishes held in simmering water in a bain-marie, or double boiler.) We ordered most of the menu, including a side of Anson Mills grits to share.

The dishes were as true to southern cuisine as I'd ever tasted, and as unpretentious as the surroundings. They were superb, too, every one of them. Glenn made quick, furious stabs at the food, but he chewed thoughtfully, almost reverently, as if his high expectations for deliciousness would be met, if not with this bowl of mushroom ragout, or that side of coleslaw, then surely with the steaming collards. His face looked worn from a lifelong battle with optimism.

"I'm not a foodie," he said. "I'm a food junkie."

Glenn is interested in the primal act of tasting—eating food directly from the plant, with little processing—but he's also searching for what he calls the "sub-taste threshold." I asked what that meant. "Below the palate sensitivity range. It's visceral, which good chefs can pick up on. It's a McGee thing." (Harold McGee: writer and food scientist.)

I asked for an example. "Well, look at the reaction you had to our graham flour. The flavor came from the healthy soil, no doubt about it. But let's build on that a second, okay? Because we harvest the wheat before it dries on the stalk. It's not green, it's not immature—that would be like 20 percent moisture—but we harvest it long before it fully dries out, around 14 percent. We catch it on the way down. It's like catching a ball in midair. Unlike 99.9 percent of the wheat harvested in this country. If your endgame is conventional wheat—dead wheat, whether it's white flour, or even whole wheat flour that's just been recombined after killing the germ—then you want wheat that's stable, that's safest from spoiling in the bin, which is as dry, or as near to death, as you can get a wheat seed."

"You're saying I can taste the difference between wheat left in the field for a few extra weeks?"

"Oh, hell, yeah. You can taste the difference even if it's overnight. If you're going for flavor—which is how wheat was bred, harvested, and milled since antiquity—then you have to watch it very, very closely. That's not sub-anything. That's above the line. A baby can tell the difference there," he said, extending the plate with the last of the shrimp over to me. I declined, so Glenn brought it back close to him and eyed the lonely crustacean.

"So now it's time to mill the wheat. We mill it to order. You call up your graham flour order, and the next day we mill and send it to you overnight. I'm looking at flour like you would a carton of milk or a bag of peaches. Because it's a fresh product. It's alive."

"Versus everyone else?" I asked.

"Everyone else pursues shelf life. Most flour preservation is done by toasting it slightly—it's called kilning—and what you're doing is drying out the grain further so there's absolutely no moisture. That's what we eat. Wheat picked long past ripeness, then broken apart, and then mummified. Mills are abattoirs for wheat."

I asked Glenn if people's attitudes were changing, if they were beginning to recognize the significance of fresh milling.

"Well, you tell me," he said. "Chefs like you are driving this thing. You're demanding it. You're making it happen. I can't do retail, because no one wants to refrigerate. They keep shelving the stuff, and I come into the store and see it and I say, 'Thank you very much and goodbye.'"

The waitress presented the steaming plate of grits and cleared the other plates. Glenn stopped her and grabbed the lone shrimp, popping it into his mouth.

"I'm not going to give up on the idea that we're in the epiphany business," he said. "No chef will pay the kind of money we charge unless we do everything we do. That's the way this works. If I flip a chef with these flavors, I'm set. Chefs are pit bulls. They knock down walls to get to something with great flavor. I hit a chef and the infrastructure disappears. Everybody else just comes out of the woodwork. I remind myself every day: *Glenn, you're in the epiphany business.*"

To what heights of deliciousness can grits rise? Tucked into a corner table in the epicenter of southern food culture, with the apostle of cultivating, harvesting, milling, and storing the best possible grains, I couldn't be faulted for expecting a small epiphany myself. The grits were very tasty, but then again, the best grits of my life were not much better than the worst grits of my life. I consulted the expert.

"They're okay," he said, looking away, with an expression of complete dissatisfaction. I got a glimpse of what Glenn's mother must have looked like when she rejected the rice he sent home from college. I insisted on knowing why.

Ever the southern gentleman, or perhaps not wanting to offend a customer of Anson Mills, Glenn lowered his voice before delivering his critique. The chef, he told me, had cooked the grits too quickly, at too intense a heat. "There's no art to high-heat cooking," he said. Packaged grits are like white flour—"dead, plain and simple, so go ahead and boil the shit out of that stuff." But for freshly milled grains like these, he said, "the roiling boil is a surefire way to blow out almost all the flavor."

So there, an epiphany. It's possible to go to the ends of the earth to recover a lost seed, research its history, and work it into a complex matrix of crops so that it's cultivated in the most fertile of soils and harvested at the perfect time—the perfect hour, even—but if a chef cranks up the heat to ready the grits for a busy lunch service: game over. Glenn sat back in his chair, allowing his sunny expression to show brief concern at the idea that his life's work could unravel so quickly.

Recovering, he pushed the grits to the side of the table. "The only reason the Italians knew this is because no Italian peasant was going to boil water for polenta," he said. "It requires way too much wood. *Basso, basso* polenta. That's slow cooking. Claudio says that all the time. *Basso, basso, basso.*"

I could not resist. "Who's Claudio?" I asked.

"He's Chris's chef."

"Who's Chris?"

"Claudio's boss." Glenn ate the last of his quail and declared it delicious.

BECOMING THE NEIGHBOR

After lunch, Glenn drove us to one of the oldest plantations near Charleston to see how eighteen acres of his Carolina Gold rice had fared during the recent drought. We passed several plantations, one after another, on a tree-lined road off the main highway. Walls thick with vegetation surrounded the estates. Large gates opened onto long driveways and, more often than not, perfectly preserved mansions. These were vestiges of the old South, heavy with the memory of slavery.

"We're a quarter-mile from where my mother grew up," Glenn said. "Isn't that beautiful?"

Glenn's mother was fourteen years old when the Depression hit and her father was wiped out. "She still talks about that time in moments of incredible clarity," he told me, explaining that she now suffered from dementia and lived in an assisted-living facility in La Jolla. "She went from being extremely well-to-do to having nothing, like that." Glenn snapped his fingers. "It just happened. Her father called her up and said, 'You gotta turn the car in, you gotta turn your clothes in. . . . That house that you're in, that's going away, too." He pointed to the lush surroundings. "She was actually raised on all this. I wasn't. I just had the stories and the cooking."

It struck me as we continued driving that even though he grew up in California, Glenn *was* raised on this. Indulging in a little armchair psychology, I came to see his mother's obsession with rice as more of an obsession with her lost childhood. Rice wasn't just rice; it was her resplendent youth in the Low-country of the ACE Basin. In that sense, Glenn's lifelong search has been about recovering both his mother's loss and the cultural experience that came with it—to fill a void that became his own.

"Oh, I don't know," he said, considering my theory momentarily. He looked like he wanted to move on. "Like I said, bringing back Carolina Gold on its own was impossible. Bringing back anything from the past takes more than just the seeds—if you want it to last. That's key. Do you want this to last for your children's children? If you do, then you have to complicate the picture more. That's what I've done. I've made it more complicated."

"By 'complicated,' you mean what exactly?" I asked. "The rotation experiments at Clemson? The seed work with the rice geneticists? The milling techniques?"

"Yeah, all of the above. Everything. And something more, too. What I discovered is tough to see when you're in the middle of a mad pursuit for rice. It's that modern agriculture has thoroughly separated the *agri* from the *culture*. They've killed the meaning of the word—bifurcated it, completely, in just the last thirty or so years. We're producing grain strictly as a commodity, without its cultural heritage. Uniformity has replaced excellence. It's replaced local distinction." He paused, weighing his next thought. "You can't just grow a bunch of heirloom vegetables. Not if you want it to last. That's key."

Since Glenn had never been in the business of "just growing heirloom vegetables," it wasn't hard to recognize Stone Barns as the reference. It was a fair point. We were, at the time, growing mostly heirloom vegetables, and no grain.

"So when you sent us the Eight Row Flint corn, it was with the hope that . . . what? We'd start growing grain?" I asked, reminding him that Jack has only eight acres to work with on the farm.

"I was hoping for exactly what happened. That it would trip something and the consciousness for landrace cuisine would click in." Landrace cuisine, he said, begins with fresh milling—"milling to order, milling à la minute, milling right into cookery." He was right. Until Glenn sent us the Eight Row, I had never considered milling anything. Milling the Eight Row got us to milling wheat. It's what inspired us to opt out of the conventional way of doing things.

"Did you actually think Jack would grow grain on just eight acres?" I asked.

"Well, yeah, of course I did. Hell, yeah. And you did it with the Three Sisters." Glenn smiled at the memory of Jack's riff on the Native American planting strategy. "I didn't expect all of a sudden that you were gonna end up in production and I could buy some polenta from you, or some grain from you. That wasn't the motive. The motive was to create community."

I must have looked confused. "Remember Tris?" he said. "Same thing. We were the neighbor—we gave him the seed for the Hopi corn, and now he's making a profit on it and saving seed every year. And the farms around him have gotten into the game, too. They saw what he did and they went for it. What's wrong with repatriation, to get started, is you don't have any neighbors. We were the neighbor. We were the community."

"But Stone Barns was already a community," I said.

Glenn gyrated his lower jaw, weighing how to respond. "Let me rephrase that: you already had the community—that's true. But I was after a larger community that recognized grain as instrumental to cuisine. That's what I meant by having to make it more complicated. It answers the question of why I didn't just plant the Eight Row myself, and why what drives me has changed. Because bringing back Carolina Gold rice by figuring out the right rotations and working with the right geneticists doesn't amount to a hill of beans if it isn't going to last. By 'last,' I mean a couple hundred years or more."

"So you paid us to plant Eight Row to . . ."

"To become the neighbor. The idea was for you to become a beacon. You're the little light I put down in the darkness."

We arrived at the plantation with the drought-threatened crop of Carolina Gold. Through the 1800s, the farm had grown rice, cotton, corn, and wheat. Glenn persuaded the plantation owners to farm just eighteen acres of rice in the same way he had convinced Tris to farm corn; he donated the seed and equipment, and guaranteed a market.

I asked what would happen to the rice if he succeeded in harvesting it. "Same as what happened to the Eight Row I sent you. I'll pay someone, somewhere, to grow this out." Glenn now distributes grain seeds in over thirty states, and as far away as Mexico and Canada.

"And the business model for all of this?" I asked. "I mean, for Anson Mills. It sounds like you've become a company that donates seeds and pays farmers to grow them out. Where's the profit in that?"

"There's no profit. We're up to three million in revenue, based on the sales to thirty-two hundred chefs. Chefs are driving this, literally. Every year we turn around and zero out the bank account. All the money goes to seed work at this point." It was humbling to see how, in Glenn's hands, the power of chefs could be harnessed for such good.

As we walked back to the rental car, Glenn tried to make sense of what Anson Mills had become. "Fifty years from now, that's when my work starts having some kind of meaning. And if I drop dead this instant, it carries on, because it's out there now. I don't even know who's growing my rice anymore. How cool is that? I send out tons of seed, I don't charge for it. I couldn't get anybody to take free seed five years ago. So I know it's working. It's happening at the speed of light, because people are putting it together and saying, 'Oh, these things made it through thousands of years of screwing up. And by golly, it tastes out of this world.'"

The seeds he distributes are blueprints for a certain kind agriculture, and a certain kind of culture, too. Which means, over time, a certain kind of cuisine.

Without cuisine, Glenn said, farming systems can't last. "They don't," he said. "Maybe for our lifetime, or for our children's lifetime, but eventually, forget it. Food and cuisine have to be an important part of our culture, and not just something that fuels the culture in one way. Food as fuel is a dangerous concept. That's where we are right now—food as fuel. It's why nothing tastes good, and why our farming systems are collapsing."

He stopped before we got to the car to underscore his point. "I've learned above all else that if what I'm doing is going to work, the culture of food is as important, if not more important, than the production of food."

Several months after my visit to Charleston, I called Glenn with a question about Wapsie Valley corn, a variety Klaas had recently delivered. "Wapsie put open-pollinated back on the map," he said. "Thank Steinbronn for that." ("Open-pollinated": varieties pollinated naturally, by the birds and the bees. Adolph Steinbronn: midcentury corn breeder, Fairbanks, Iowa.)

He said he was in La Jolla, and I joked about landrace field trials in Southern California. "Maybe someday," he said. "No, I'm here because Mom recently passed. We eulogized her this morning." He sounded nearly apologetic to be giving me the news.

I told him how sorry I was. I asked if his mother had ever had the chance to taste his Carolina Gold rice.

"She did, yes. I sent the grits to her first, which she liked a lot. Then she said, 'Now that you got the corn right, are you going to get the rice right?' That was just like Mom. I told her I would. It took a few years, much longer than I thought. But eventually I flew out for a visit and brought some newly harvested Carolina Gold. Of course, she wouldn't let me cook it. She did the cooking."

I asked Glenn what she said about the rice.

He was silent for a minute, something I couldn't recall having experienced with Glenn before. "She didn't say anything, actually," he said finally. "Not a word. We just sat and shared it. Quiet reflection over a bowl of rice is something to behold."

CHAPTER 30

IN THE SPRING OF 2009, several tractor-trailer loads of tomato plants left a wholesale nursery in Alabama, destined for large-scale retail centers in the Northeast—places like Walmart, Home Depot, and Kmart. There is nothing particularly strange about a phalanx of eighteen-wheelers trucking thousands of tomato plants. It happens every spring. Big-box stores routinely stock from large breeders rather than grow the plants from seed. And in 2009, the first full year of Michelle Obama's White House garden project, there was a significant spike in the number of people looking to create their own gardens, so an even greater number of tomato plants were on the road.

But there was a problem with the cargo. Many of these plants harbored a fungal disease called late blight, which attacks potatoes and tomatoes. Late blight is devastating. It looks innocent at first—a few brown spots here, some lesions there—but it quickly multiplies. Plants that appear relatively healthy one day, with abundant fruit and vibrant stems, can turn Chernobyl-like within a few days. Think Irish potato famine—it was caused by the same fungus.

Most farmers in the Northeast, accustomed to variable conditions, have come to expect late blight in some form or another. Like a sunburn or a mosquito bite, you'll experience it sooner or later. And while there are things farmers can do to minimize its damage and sometimes even avoid its effects entirely, the disease is almost always present, if not active.

But 2009 turned out to be different. For one thing, the disease appeared much earlier than usual. Late blight usually comes, well, *late* in the growing season as fungal spores spread from plant to plant. So the outbreak caught just about everyone off guard. And the perniciousness of the disease was unprecedented. The pace with which it spread (it covered the Northeast in just a few days, as opposed to over several weeks, which is the norm) and its ferocious strength (topical copper sprays, a convenient organic preventative, were less effective than they had ever been) were a shock to even hardened Hudson Valley farmers. Several described the damage as "biblical."

Organic farmers were forced to make a brutal choice: spray the tomato plants with fungicides or watch their crops disappear. (One local grower, who had never used a chemical application on his fields, told me he didn't hesitate to spray. "If I lose my tomato fields," he said, "my daughter doesn't go to college this year.") But fungicides only suppress the disease; they don't cure it. Even for farmers who routinely sprayed, or who reluctantly sprayed precautionary amounts, the blight drastically lowered yields.

Many blamed the weather for the outbreak, as record rain and high humidity that June made for the fungal equivalent of a four-star hotel. But even the weather didn't explain the severity of the disease. Early summers had been warm and humid before. Late blight is always destructive; it's rarely, if ever, utterly devastating.

One theory had it that those tractor-trailer loads of tomato plants were roving incubators of fungal spores. The infected plants dispersed millions of the spores in transit (spores travel up to forty miles), showering the Northeast. The infected starter plants arrived at the stores, and before anyone was aware of the danger, they were purchased and planted, transferring their pathogens like tiny Trojan horses into backyard and community gardens.

❧

By the second week in June, Jack had heard that there might be late-blight outbreak of historic proportions. He checked his plants carefully on a

Saturday. "They were clean," he told me. "Good fruit set. It actually looked like it was going to be a great year." Four days later, he called with a different story. "It's here, and it's spreading like locusts."

On a gray, rainy day a few weeks later, I stood in the field with him—the same field that Eliot Coleman had appraised several years earlier—in a spot once occupied by one thousand heirloom tomato plants: Brandywines, Cherokee Purples, and Black Krims. It now looked either tortured or bombed out. Jack had removed most of the infected plants to prevent the fungus from spreading to the potatoes, which happened anyway. Surveying the scene, with the full weight of a tomato-less summer menu settling in, I suddenly saw something at the other end of the field: a small row of what appeared to be perfectly shaped bright red tomatoes that had somehow escaped the carnage surrounding them. I walked over for a closer look.

"Weird, right?" Jack said, catching up to me. "I've been waiting to show you these. They're Mountain Magics, an experimental seed I got from Cornell, bred for resistance to late blight."

The new variety had been developed and trialed by land-grant university plant breeders to benefit farmers. And yet despite this, and against the logic that beggars can't be choosers, I eyed the tomatoes skeptically. They looked not only ripe and ready to burst, but almost *too* good. Too manicured, too uniform, too much fruit set per vine. In other words, too much like supermarket tomatoes. It was an ungrateful sentiment, considering the circumstances. But misshapen heirlooms, with their celebrated lumps and bumps, have become the standard for good flavor. We see the blemishes and flaws of heirloom fruits and vegetables—whether they're tomatoes or peaches or peas—as indicators of something more natural, and more delicious, than manicured, mass-produced varieties.

Jack tried to cure me of my prejudice as we collected a few of the ripe Mountain Magics. Heirlooms, he said, are like a set of sterling silver trays your grandmother might pass down. "They're great to have, and they were saved for a reason. But they capture a moment frozen in time. That can be a burden as well as a joy."

My expectations were still low. Assuming the too handsome tomatoes would carry all the flavor of a pencil eraser, I figured we would make them into sauce. But the Mountain Magics were sweet and fruity, with an aroma and an acidity that are sometimes missing even from heirlooms. They were drier than heirlooms, too, with a more concentrated flavor. And no bruising or wrinkling appeared for more than a week, sometimes two. By that standard they were, in fact, magic.

The tomatoes were so delicious, diners began asking about them. Many wanted to know how Jack could grow them organically in the face of late blight. We put together a small tray—Mountain Magics on one side and a few late blight infected heirloom tomatoes on the other—and brought it to the tables to explain the experimental trial. We might have trumped up the technological ingenuity of Mountain Magics too much, because often the reaction was startlingly negative. Instead of *Yum, can't wait for the next course*, we got *Yech, Frankenfood*. We somehow had diners thinking that Mountain Magic was a genetically modified tomato.

Given the misunderstanding, their reactions were not surprising. Genetically modified foods—bred by recombining genes in a laboratory rather than allowing varieties to cross naturally—have been controversial since the idea was first introduced in the 1980s. It was a tomato with the unfortunate name of Flavr Savr that gave American consumers their first real taste of the new technology, in 1994. By manipulating the gene that causes ripening, genetic engineers at the biotech company Calgene created a tomato plant in which natural rotting was delayed. So farmers could let them ripen in the field before harvesting them, rather than pick them while they're still green and push them to ripen by spraying them with ethylene gas, which is the norm. It was marketed as a coup, better for the farmer *and* the consumer. But while the political and bioethical debates raged about genetic engineering (Frankenfood, the popular term coined at the time, reflected the general horror felt by the public at the manipulative nature of the technology), the Flavr Savr failed for another simple but important reason: there was no flavor to savor. In 1995, Monsanto bought Calgene and discontinued the revolutionary tomato.

To avoid confusion, I realized I needed to better explain—and understand for myself—the difference between the Mountain Magics, bred the old-fashioned way at a land grant like Cornell, and GM tomatoes from a company like Monsanto.

LAND GRANTS

Land-grant breeding programs were, for many years, considered cutting-edge. You might argue that they were created, at least in part, to avoid the kind of disaster late blight brought to our tomato crop.

In the mid-1860s, agriculture was considered unworthy of study and rarely taught. Of the three hundred colleges in America in 1860, nearly all were private and rooted in the liberal arts. Congress recognized that in order to make farming more efficient, it would have to act.

Their decision culminated in one watershed year of legislation. The Homestead Act was passed in 1862, encouraging the swell of westward settlement. Also in 1862, Congress created the USDA. But in many ways, the most important legislation of that momentous year was the Morrill Land Grant Act, which provided public land for the formation of land-grant colleges around the country to educate Americans in agriculture and the "mechanical arts," such as engineering. Today, every state has at least one land-grant institution.

The Hatch Act of 1887 followed, mandating funds for agricultural "experiment stations" where researchers could investigate everything from new crop rotations to plant pathology. And in 1914, the Smith-Lever Act added a third component to the land-grant complex: the extension service, which brought the teachings of the college and the research of the experiment stations directly to farmers in the field. Agricultural extension agents visited farms to share the latest technological advances and, yes, check fields for early signs of disease. (Which in the case of our late-blight disaster could have made all the difference. To an audience with even the most rudimentary gardening experience, those purchased tomato plants weren't Trojan horses at all; they were sick plants flashing sirens of impending doom.)

Hoping to improve on farmers' efforts, plant breeders working for the USDA and state agricultural colleges used scientific selection methods to increase yields and improve pest and disease resistance. New varieties of grains, vegetables, and fruits were developed and trialed at regional experiment stations, and successes were shared with farmers, encouraging the exchange of information and better varieties.

The sweeping legislation of 1862 created a feedback loop of experimentation and learning for farmers in every part of the country. With improvements in seeds and the application of the most current technological advances, something profound happened: harvest yields increased; prices dropped. The quality of food improved. These advances did not come about without foreclosures and consolidation, and eventually—with the advent of chemical agriculture—great environmental destruction. But the land-grant system was conceived as an instrument for public good. And it was, for well over one hundred years, incredibly successful.

—◦—

A few weeks after our Mountain Magic revelation, Jack went to Cornell to meet the university's plant breeders in search of more opportunities to collaborate on new varieties of fruits and vegetables.

"They actually *want* to talk," Jack told me when he got back. "They're a bunch of foodies." The breeders needed to hear more from us, he said, and we—the farmers, the chefs, and the waiters—needed to hear more from them. When Jack suggested inviting the group for dinner, the answer was obvious. How had we not thought of that before?

A team of breeders arrived at Stone Barns two months later. After touring the farm, they joined the farmers, cooks, and waitstaff gathered before service to explain their work. Each breeder specialized in a particular crop. There was Bruce Reisch (grapes), Margaret Smith (corn), Courtney Weber (strawberries and raspberries), Walter De Jong (potatoes), and a young rising star named Michael Mazourek (squash, melons, and peppers).

Michael spoke about the challenges of breeding in a world disconnected from its food. "The breeder has to make decisions about which way to breed—it's like spinning a wheel and deciding which direction you're going to go," he said. "And the direction we're being told is: yield and uniformity." The other breeders nodded.

Mountain Magic, he pointed out, was an exception. Usually, breeders have to look for the largest market. "Actually, before we even begin with an idea for seed, we need to prove there is a market," he said. "One hundred years ago, we were breeding seeds for farmers in our region. Now we have to think about the seed growing not just in New York but in Texas and Oregon, too. With 1 percent of the population feeding the other 99 percent, I suppose it's inevitable." More nodding heads.

"We don't have a face to breed for," he continued, "so we breed for the two-sentence description on a seed package. That's what the whole universe of breeding possibilities gets whittled down to: two sentences, and one of them has to be about yield and the other has to be about uniformity."

Breeders are architects, and seeds, as I learned from Glenn, are the blueprints for the farming system. Even before farmers actually farm, before they employ their rotations, improve their soils, and choose their varieties, the seed sets the foundation. If yield and uniformity are the determining factors, then the system, from field to distributor to marketplace, pretty much falls into place. We say, "It begins with the seed," and really it does, but it also begins with the idea for the seed.

It's not a great reach to say that in the burgeoning local-food movement, breeders have been ignored. And while cooking the meal that night, I realized that chefs were partly to blame. We might have brought innovation and modernity into the kitchen, but where ingredients are concerned, we've tended to look to the past. We've idealized heirlooms and heritage breeds as paragons of flavor. And we've put farmers on pedestals as their custodians. But we've overlooked the breeders, the people writing the original recipes, and by virtue of that neglect we have helped prevent the development of new and delicious varieties that could thrive in particular regions.

At the end of the meal, I sought out Michael and asked him if a winter squash could be bred to have the concentrated flavor of Mountain Magics. "Would it be possible," I asked, "to shrink the squash but improve the flavor?"

He told me he was in fact working on such a variety and would send over a sample in the fall. And then he laughed and, fidgeting with his glasses, looked down at the floor. "It's a funny thing, or maybe a tragic/funny thing," he said, "but in all my years breeding new varieties—after maybe tens of thousands of trials—no one has ever asked me to breed for flavor. Not one person."

BREEDING FOR FLAVOR

Can wheat be bred for flavor?

The thought occurred to me not long after our visit from the Cornell breeders, when I received a phone call from Lisa. I was expecting an update on Eduardo or Miguel, but instead she was calling about Aragon 03, an old Spanish variety of wheat she had read about, which had all but disappeared in the past fifty years.

Farmers in Spain abandoned Aragon 03 in the 1980s because its yields were lower than those of modern varieties. But a family in one small northeastern Spanish town had continued to grow it for personal use, each year selecting the best seeds to plant the next year, thus making the wheat even better over time. Now it was being rediscovered, and demand from bakers had farmers wanting to grow more of it.

Aside from its relatively low yield, Aragon 03 exhibited all the characteristics modern breeders look for: high protein content (a whopping 17 percent), formidable disease resistance, and drought tolerance. (Lisa said Aragon 03 was described to her as wheat that "nourishes itself with the morning dew.") Even better, it was said to be delicious.

Lisa thought Aragon 03 would make a good idea for a story—a valuable heirloom that had almost disappeared in agriculture's single-minded pursuit of yield. And I thought it might make a good addition to our menu. Saving Aragon 03 meant flavors that would otherwise become extinct. The project seemed right up Glenn's alley, but my mind was still on breeders and the possibility that they could help not merely restore an endangered variety but reinvent it.

That's how I found myself e-mailing Steve Jones, a breeder for Washington State University whose name I'd heard tossed around by several people, including Glenn. Steve was renowned in the field for his interest in small-scale wheat breeding and his commitment to flavor. I introduced myself and told him I was interested in learning more about his work. His response came the same afternoon.

Steve described his efforts to empower farmers to breed their own varieties and recounted the story of a local farmer whose son had died in a tragic accident ten years ago. The farmer wanted Steve's advice on getting his twelve-year-old granddaughter, Lexi, involved with the farm.

"I told him: Why not have her breed a new wheat variety?" he wrote. Steve had Lexi assist in making some crosses in his greenhouse. And eventually she helped plant the wheat in her grandfather's fields, each summer selecting the plants that looked best to improve the variety. A few years later, Steve entered Lexi's wheat variety—Lexi 2—in a statewide program to test the top sixty varieties in the Pacific Northwest. The entries came from companies such as Monsanto and Syngenta, as well as the university programs. In Douglas County, Lexi 2 was the top-yielding variety, beating out the fifty-nine other lines.

"There are old, cranky wheat breeders who've spent their lives doing this work that have never, and will never, have a top yielding variety anywhere," Steve wrote.

Convinced that I'd found a kindred spirit, I sent the information about Aragon 03 to Steve the next day. And I offered to have Lisa act as an intermediary with the Spanish family that preserved it. But apparently he required no assistance. "Aragon 03 is in hand," he wrote back a few days later. "I'd like to make a cross with a wheat variety local to you. Please advise if you have something in mind." Then he added, "This could be fun," but I got the sense it was with the wan enthusiasm of someone agreeing to look after a neighbor's houseplant.

I phoned him the next afternoon and asked how he'd managed to find the seed. "After I got your e-mail, I called my friend at the seed bank. He mailed it to me," Steve said, his enthusiasm suddenly sounding more genuine. "I'd like to honor Aragon 03. I can do that by showing its worth as a parent." He said we could take the characteristics Lisa described and make them even better. By crossing Aragon 03 with a local wheat variety, we could ensure that its strengths would be well expressed in New York. "To get the qualities you're looking for, we may even ménage à trois it."

So I introduced Steve to Klaas—they had already heard of each other but had never met. Klaas spent several weeks researching varieties that Aragon 03 might pair well with. "Like a good marriage, you want the same worldview but distinctive traits," he said, a Jane Austen of good agronomy. In the end he recommended Jones Fife, a wheat highly adapted to the weather and soil conditions of the Northeast. "I believe Jones Fife is a worthy candidate for this cross," Klaas wrote to me.

I didn't object. Who was I to stand in the way of two interesting varieties creating a potentially valuable offspring? But the nagging worry that we— that is, Steve—had given up on perpetuating the pure Aragon 03 seed left me wondering if I had the stomach for modern breeding. Supporting Mountain Magics over heirloom tomatoes had felt ingenious and modern. But imposing on the actual breeding process suddenly felt clinical and, to be honest, a little creepy.

I thought of Glenn and his landrace breeding, the chaotic unpredictability and painstaking selection of admirable traits. The capability to bypass all that and select two varieties that had never felt the same breeze or touched the same soil, and *force* their offspring—I wasn't so sure this was the right thing to do. Glenn worked to honor what nature intended, just as Eduardo and Miguel did. Wasn't that the best recipe for the most delicious food? And wasn't I, as the sorcerer's apprentice to Steve, interfering with nature's will?

A few months later, Steve sent a photo of the Aragon 03 and the Jones Fife cohabitating under a blanket of thin plastic. "Making the cross," Steve wrote as the caption. "Come out and see for yourself."

It was late afternoon by the time I arrived at the Washington State Research and Extension Center, in Mount Vernon, an hour north of Seattle. Steve was standing outside the building's entrance with his hands in the pockets of his khakis, looking serious. He is six foot five, but with his baseball cap tipped back slightly, he appeared even taller. He welcomed me to the Skagit Valley, looking less like a brilliant breeder or a graduate professor than a high school basketball coach.

Steve walked me through the newly rebuilt research center, with its plush offices, carpeted hallways, and sleek lighting. It seemed like a tech company, flush from its IPO. In fact, that's a little like what had happened with this place. Lacking funds for research—a function normally filled by the state's land grant—the farmers of the surrounding Skagit Valley paid for much of the center themselves. They donate a certain percentage of their profits to a foundation and invest the money in the agricultural equivalent of start-ups: crop research, new seed varieties, and pest-resistance strategies. All the benefits flow back to the farmers, giving them an advantage in the marketplace.

You'd think ninety thousand acres of pristine farmland, with some of the best soil in the nation (Skagit Valley's soil is rated among the top 2 percent in the world), would ensure success, but the Skagit farmers lack the marketing muscle of the big commodity growers.

"Most of these farmers worked the land as grandchildren," Steve told me. "They're not getting any handouts or subsidies, so they have to work together and invest in their future. In a way, that's freed them from getting locked into some kind of funky system where it's one crop being planted again and again. It's forced them to be creative."

I asked how much of the valley is organic. Not much, he said. But he warned me against drawing conclusions about the quality of the farming. He said that all the farmers practice mixed agriculture and feed their soil through good rotations. They consider wheat a good rotation because it breaks up disease cycles and aerates the soil for their main cash crops, like fruits and vegetables and flowers.

"My job is to make wheat an integral part of a sustainable farming system, not just a good rotation crop," he said. "To do that, I need to breed wheat that tastes good and performs well in a bakery. In terms of how much is planted, wheat will always be a minor crop around here—it will never be the most profitable—but I want to make it a *major* minor crop, if you will. Wheat isn't a tomato, yet I'd argue that there's a lot of value in looking at it that way. That's my goal. That's why I'm here."

ON DR. JONES

Steve's interest in wheat started in college, when he entered a program that offered students five acres to farm whatever interested them and pocket the profits.

Steve planted wheat and dry beans and, on the side, marijuana. "All buds," he told me in a whisper. "It was good stuff." The marijuana netted

$400. He struggled with the beans: "I worked my ass off, waking up at 2 a.m. to irrigate those things. I never worked harder, and I made no money." But Steve's wheat, planted in November, was ready to harvest by June—with little effort on his part. He earned $450. "I was like, *I got to get into this.*"

It was Steve's grandmother, a Polish immigrant, who inspired his initial interest in farming. The week after they moved to Brooklyn, his grandparents ripped out the lawn and parking strip in front of their new home and planted potatoes and cabbages. His grandmother cooked often and baked Eastern European–style breads. When Steve was eight years old, she moved in with the family and taught him to make bagels, which he continued to do for many years. "Besides growing wheat in college, I ended up making a lot of bagels. People thought that was weird."

Other than his grandmother's influence, and a talent for mowing the grass of the local golf course, there was little in Steve's childhood that foreshadowed a life devoted to agriculture, and nothing that would have suggested his becoming one of the leading wheat breeders in the country. And yet that's exactly what happened.

Steve came to the Skagit Valley after heading the wheat-breeding program at Washington State University for nearly twenty years. In that job, he had worked in eastern Washington, where the wheat farms would look right at home among the tabletop monocultures of Kansas. At first he had felt energized and challenged to help the large, commodity wheat farmers improve their yields and increase their profits. By all accounts, he was very good at it. But he was also interested in helping small-scale wheat growers—farmers who were too small to compete in the commodity game—and organic growers, who have an even greater need for new seeds to keep ahead of pest and disease pressures. The university, however, didn't encourage these efforts.

"Maybe I was a little naive or whatever, but I considered my job to be a public servant, which means I serve the *entire* public," he said. "But the university—every land-grant university now—is locked in with the largest constituency. For us, that was the eastern Washington growers. Two and a half million acres of wheat destined for industrial flour—most of that leaving the country. Anonymous wheat is what I was expected to continually improve." Steve didn't fight to change the system, at least not at first. But quietly he began to work on alternative wheat breeds for organic and small-scale production.

Then one day, several years into the job, he was called into a meeting. "The department chair was sitting there, and the associate dean, and three guys from Monsanto. I walked in and I thought, *Oh, shit*. Right away, first thing they say? All the land grants are going to put Roundup in their wheat. It was like, *You're going to do this, too*—essentially change the way wheat has been bred for ten thousand years—*and isn't that great?*"

Roundup is Monsanto's best-selling herbicide, developed in the 1970s and used widely all over the world. At the time of the meeting, nearly all of Steve's wheat farmers used Roundup on their fields before planting (today, to farm wheat conventionally, there's really no choice), but the group was suggesting something momentous for wheat. They asked Steve to breed a genetic modification into his wheat seeds to make them resistant to Roundup. This would enable farmers to spray both their crops and surrounding cropland with the herbicide—killing the weeds without killing the crops. The logic is simple: eliminate weeds when they compete most with the wheat, and increase harvests.

Genetically modifying other grains to withstand a dousing of herbicide had already been done successfully with cotton, corn, soybeans, and alfalfa. Even though GM wheat was not approved for commercial release, Monsanto and others were lobbying for its development. Steve heard that they were reaching out to university wheat breeders around the country, confidently predicting an imminent change in policy. But he was still surprised by the tone of the Monsanto representatives, who assumed that Steve would happily comply.

"They said: *We want you to put these genes in your wheat, and then we'll commercialize them together*. They had the scientists; they didn't have the germplasm," he told me, referring to the genetic information stored in a collection of seeds. "That was their problem."

He added that private companies still rely on public genetics for their seeds. "So they needed university breeders with access to the gene pool. And they needed the trust we have with the growers. We've been breeding wheat at WSU since 1894. For a hundred years, we've developed that trust. So they needed me for that. And they figured I'd be a willing participant because they were offering royalties."

At the time, paying public breeders a royalty for their seeds was a relatively new idea. The 1980 Bayh-Dole Act made it legal. For the first time in the history of land grants, breeders entered into the business of commercializing their work rather than simply making their seeds available to farmers for free. As Steve described it, the Bayh-Dole Act amounted to a friendly takeover of public research, with the devastating, if unintended, consequence of tilting universities toward profit-making projects.*

By the 1990s, private industry had surpassed the USDA in the funding of agricultural research at land-grant institutions. And the spending gap continued to expand. In a little more than a century, the spirit of a regional food system encouraged by land-grant colleges was effectively turned on its head.

⸺

Steve had a more immediate—and, in many ways, more far-reaching— worry as he sat across from the Monsanto executives offering him the deal of his life.

---

* In 1983, U.S. Secretary of Agriculture John Block announced that from then on, federal research would be phased out of plant-breeding programs, so as not to compete with the private sector. This trend continues today. Seed company officials "prefer to talk in terms of a 'division of labor' in which public bodies develop new breeding material that is turned over to companies for 'final exploitation in the marketplace,'" explain Cary Fowler and Pat Mooney. "Translated, this means that government does the costly, basic and innovative research, while big companies pick up the profits in the marketplace."

Wheat was (and still is) the last of the big grains to be bred almost entirely through the land-grant institutions, without the influence of corporate money. Why? Because seed companies sell commercial hybrids, which provide both higher-yielding and more uniform plants, and force farmers to buy new seed every year. Since wheat is self-pollinating, saving seed for the next generation is cheap, easy, and free from private enterprise.

"Wheat is the last frontier," Steve told me. More than half of all wheat growers still save their seed, he said, "and if they're not saving seed, a land grant is saving or improving seed for them. That's an astronomical number from any modern agricultural context."

Monsanto's vision for a Roundup-resistant wheat would likely increase harvests and, at least initially, the farmers' income. But because the seeds would be patented, farmers would no longer be permitted to save them, and, having lost that tradition, they would likely never return to it.

"The only way to control wheat is to offer something farmers can't produce themselves," Steve told me. "You Roundup Ready it and all of a sudden you've cornered the market on the big enchilada. I mean, can you blame them for wanting to do this? If you're a shareholder at the annual board meeting and you hear the equivalent of 'Next year our company will begin to corner the world's wheat market,' you'd be saying, 'Right on, way to go, sounds great.'"

—

The longer Steve sat across from the Monsanto executives without showing enthusiasm for their proposal, the more confused they became. Were the Roundup-resistant wheat to succeed, Steve would profit enormously in the form of royalties for his work. Steve's department chair would win, too, since the university receives a share of the royalty payments, helping to offset the steep reductions in government funding. Which is why Monsanto felt so confident in arranging the meeting. Each breeding program they visited before Steve, and all the ones they visited after, entered into a partnership of some kind to profit from the research. Although at the time of writing this it is not yet commercially approved, genetically engineered wheat will likely flood the market—and dominate it—in the next decade.

The meeting descended into awkward silence. "They were basically waiting for me to jump up and down and thank them for the opportunity. And the head of my department, he just about wanted to kill me. But what I said is that for ten thousand years farmers have been improving wheat, saving these improvements and planting them again the following year. The right to replant what you harvest is among the oldest rights of humankind. Biotechnology removes that right. How can a land-grant university actively dismantle a ten-thousand-year-old tradition? The answer is: pretty easily. I didn't want to be complicit in that. So I said no, and I walked out. That started eleven years of unpleasantness."

After his meeting with the Monsanto executives, Steve became known as something of a renegade. He talked openly about his opposition to genetically modified crops, and large-scale farmers increasingly questioned his work. Convinced that genetically modified wheat would be accepted soon and that other university breeding programs were ready to release seed, they feared that Washington state wheat would be left behind.

Steve's colleagues, for their part, became more spirited in their opposition to his leadership. "Finally one day I went to my department head and I said, 'Look, I can fight this for another twenty years, or I can move on and make way for someone else to do this job.' He agreed. And that was that."

Steve decided to leave the department. Even though his relationship with WSU had become fraught, he landed another job with the university, as director of the Mount Vernon Research and Extension Center, a six-hour drive to the west. Given that he had chaired the famed wheat-breeding program on the main campus, the move was a little like leaving a vice president position at General Motors in Detroit to run one of the branch offices in Kalamazoo.

The position did not call for an expertise in wheat breeding. In fact, the research center had never in its history been involved with wheat breeding. "I was hired to direct the center. I was done with wheat. I was there to serve the needs of farmers with a very rich tradition of growing specialty fruits and vegetables."

But on the drive to his final interview for the position, Steve noticed a wheat field and pulled off the freeway. His wife, Hannelore, took photos. A few hundred yards down the road, they saw another wheat field. And then another. "I thought, *Holy shit. There's wheat all over this place. Why don't we know about any of this?*"

It turned out that the Skagit Valley wheat was all soft wheat bound for places like Korea and Singapore, to be made into noodles and soft breads. The farmers of Skagit were using wheat rotations in place of buying expensive fertilizers.

"What these farmers recognized," Steve said, "and I stress they recognized it not because they're modern-day hippie-dippies but because they're dead set on improving fertility, is that small-scale wheat harvests are both very good for the soil and surprisingly profitable."

Except that, before Steve came along, there wasn't that much profit. Exporting wheat on the cheap is more economical than purchasing fertilizer—but just barely. "When I grow wheat," one farmer confessed to Steve, "I just want to lose less money."

Steve started asking himself a lot of what-ifs. What if the wheat remained local? What if, instead of growing anonymous wheat in Skagit, wheat could be bred for good yield and exceptional flavor? What if there were a local market to support these distinctive varieties? Might it be possible to improve both the return for the farmer and the quality of the flour for the bakers? These questions had never been asked, because the operating assumption was simplistic: wheat was wheat, in the same way that that fresh fish had been considered indistinguishable before chefs like Gilbert Le Coze and Jean-Louis Palladin came along and created a market for small fishermen.

Steve learned that, in fact, there was a strong history of wheat in the region. From 1850 to 1950, he discovered, there were more than 143 varieties of wheat grown in Washington. And Whidbey Island, just west of the Skagit Valley, had set the world record with one wheat crop of nearly 120 bushels an acre.

"People say wheat is 'out of place' around here," he said. "I always get a kick out of that. A place like Kansas is considered real wheat country. But that's not correct. It's not that Kansas is the best place to grow wheat—it's that wheat is the *only* thing that will grow in Kansas."

Steve brought up his idea of breeding specialty Skagit wheat with the farmers and found them surprisingly open-minded. They had nothing to lose, after all, and there was a local tradition of innovation.

"We had a breakfast with all the farmers real soon after I was hired," Steve said. "A big, scary-looking farmer guy stands up and asks to say just one thing. I was like, *Here we go, I'm going to get it about not embracing Monsanto or something*. The guy starts pointing his finger even before they can get the mic in his face, and he says: 'No matter what you do, whatever research you do, we expect your work to remain in the public interest. We don't want you commercializing anything.' All the other farmers in the room nodded. I started to laugh, you know? It was crazy. I'd been fighting for that very thing for twenty years, and I'd been losing the fight. Here I was, *ordered* to do it— the demand was that I do the right thing. I mean, I wanted to just throw my arms around the guy and hug him."

BEYOND HEIRLOOMS

On our way to the greenhouse to see the Aragon 03 cross, Steve took me through the eight-acre research field just steps from his office. Agriculturally, it was the most arresting display of diversity I'd ever seen. The field was divided into an orderly grid of four-by-twelve-foot rows, but the overall effect was dazzling in its variation. Each row held a unique kind of wheat, creating a kaleidoscope of colors and shapes. Even Glenn's fields looked bland by comparison.

Steve stopped and turned to me. "Whenever I'm out here, I'm standing in history," he said, sounding just like Klaas on my first visit to his farm. Since we were, technically, standing in the middle of a field trial comprised of largely untested varieties, I asked him what he meant. "All these varieties have genes that can be traced to antiquity. So I start to think about the farmer that saved them, or the community that preserved them. Stand in a tomato field and it's like, 'Okay, nice red fruit' or whatever. But you're really not going to capture any depth."

I asked how many potential varieties he was tracking; there looked to be a couple thousand. "It's more like forty thousand," Steve said.

"You keep track of forty thousand different varieties?" I asked.

"Technically speaking, these are experimental lines, not varieties—not yet anyway," he said. "But the math is just beautiful with this stuff." The genetic variation in the field, and the potential for every plant in front of me to be bred again for even greater variety, was limitless, a theme park for extreme diversity.

Steve's mastery of it all is less conspicuous than the sheer delight he takes in playing the role of a head coach overseeing his young recruits. "Will you look at that right there?" he said, cradling a not-yet-mature seed head from a line he'd been working on for years. Steve is reflexively physical, accentuating his pronouncements with jabs, taps, and touches. I got the sense he viewed the wheats' successes and shortcomings as his own. "Look how pretty that one is—isn't that gorgeous? It's called Red Chief. Not great-quality wheat. But when it's like that we just keep it, 'cause it's so pretty. I mean, that's gorgeous—that yellowish red, that's just damn pretty."

Breeders write a kind of playbook for seeds. Different breeders have different playbooks. Steve's approach is mainly to create new varieties by marrying desired traits from various genetic lines. He is not controlling the seed's future as much as he is pushing it in a desired direction. And he can predict, with amazing accuracy, what the next generation will look like. Glenn's strategy, by comparison, is more freewheeling. He's not interested in crossbreeding varieties; if he's pushing anything, it's to let nature run its course and determine where it wants to go. "Sports" and outcrosses are celebrated. You could say Norman Borlaug followed an entirely different playbook, one rooted in an extreme command and control—to improve yield and efficiency at any cost.

So who has the most winning strategy? "Glenn's work sounds great," Steve told me when I pushed him to weigh in on the subject. "I love old ways and old things, but it is a good idea to realize that not all old varieties are good, and that ancient wheat landraces were, and are, highly adapted to their original environments. We use them and recommend them to some degree, but they are agronomically risky. Glenn is adapting them to his environment, but there aren't many Glenn Robertses in the world."

It was a point I'd never considered: when chefs advocate for older varieties, the assumption is that we're advocating for the farmer, too. But Steve was saying *Not so*. Unless the farmer takes the time to adapt the variety to his own environment, there's often a substantial risk in the form of low yields or poor disease resistance.

When I asked Steve if he saw his own strategy as more in line with Glenn than with Norman Borlaug, he hesitated. "I think I'm broader, actually," he said. The landraces Glenn works with don't make yield a priority, he explained. "I don't apologize for breeding varieties that have good yield, in the same way that I don't apologize for looking back at older genes. Disease resistance, flavor, functionality—we look at all of it. The idea that you can't have one without sacrificing another is preposterous."

"Don't lower yields mean better flavor?" I asked.

"I know," Steve said. "Chefs are pretty convinced of that."

"It's true of heirloom tomatoes," I said.

"With wheat, yield and flavor are not in inverse proportion. There's plenty of room for flavor." The trade-off between yield and flavor, he explained, happens mainly in crops that, in their domesticated form, contain a lot of water. "A true wild tomato is smaller than a cherry," he said, offering a pinch of his index finger. Wild foods tend to be more flavorful because they're not "washed out by water."

Steve pointed to a section of nice-looking wheat, an old French variety that he used as part of a cross. "That yielded 170 bushels of gorgeous, really flavorful, high-protein wheat last year," he said. "That's five tons per acre. Spectacular, right?" In Kansas, he pointed out, the average yield is 1.5 tons per acre.

"When it comes to wheat, the beauty is that you can have your cake and eat it, too," he said. "And there isn't one flavor gene—flavor is about an interaction and combination of genes. Thank God for that, right? Otherwise we'd be selecting for just one gene, which would probably spell disaster at some point."

I asked Steve if nutrition in wheat worked the same way, and he said it did. However, his research has found that older varieties contained more micronutrients than newer breeds that have come along since green-revolution dwarfs were introduced. The older wheats, he discovered, had as much as 50 percent more calcium, iron, and zinc. "You don't eat wheat by the acre; you eat it by the slice, and you'd have to eat a whole loaf of bread made from modern varieties of wheat to get the equivalent nutrition in just half a loaf made from the older varieties," Steve said.

He stopped to talk with one of his graduate students about a new trial. I kept walking. Every few feet revealed wildly different clusters of wheat plants—some dark and nearly mature, and others lightly colored, faintly white, even, with just a hint of a seed head developing. Some were so tall they towered over me like a canopy. The last time I had seen wheat this tall, I was with Glenn, in his test field at Clemson University.

Since Steve was occupied with his student, I phoned Glenn on a whim and described the scene. He was in perfect form. "Hell, yeah, wheat is tall. But you're focused on the wrong end up," he said. "It's the root system that really blows the mind."

I asked Glenn what he thought about wheat that had been crossbred in a laboratory, rather than in a landrace system like his. "How do I feel about it? I feel great about it. I'd say any way we can bring back nutrition and flavor back into wheat—which is the same damn thing, as every chef knows—any way we can do it, by landraces like I do or by inventing new varieties in a laboratory, like the ones you're looking at, I'm all for it. Because wheat is fundamental. It's everywhere in our culture. And if it's not correct, the culture starts crashing down."

He paused for a moment. "People always ask me, *Hey, Glenn, how are you going to feed the world with a landrace, mixed-agriculture system?* I tell them I have no idea how to feed the world. No clue. But what you're standing in over there in Skagit is perhaps one answer to our future."

When Steve returned to continue our walk, I told him I was starting to re-think my obsession with the wheats of antiquity as the only way to ensure better flavor. In a field of forty thousand exciting new possibilities, how could I not?

"Heirlooms are fine," he said. "We don't have to be antagonistic towards them, but we can move beyond them. If you look at an heirloom anything, it's stopped, genetically; someone took a moment in time and froze it. But that's not how they were developed in the first place—whether it's a tomato or a grain or an apple. They were constantly improved." Steve said he tries to continue that improvement by crossing landrace and heirloom wheats with modern, regional varieties.

"Like the Aragon 03," I said.

"Right. The question was: Can we improve the flavor of Aragon 03? Can we improve disease resistance for the farmer? Can we improve the nutritional value by, say, upping the iron and zinc? We looked at it, and we doubled it for free just by selecting for it."

"For free?" I asked.

"'For free' means it doesn't affect the yield," Steve explained. "And you don't have to go and pull it out of a jellyfish and put it into wheat. There's a tremendous amount of variation in the wheat already; some lines have more nutrition, others have less. It's all about capturing the characteristics you want. This is really no different than what has been practiced for ten thousand years in wheat, adapting lines to your own environment. It's what everyone did."

Steve said this kind of work is critical to the renaissance in local grain. "One hundred years ago, Maine had about thirty thousand acres of wheat, all milled and consumed locally. Today there's basically zero wheat grown in Maine. Zero wheat grown in Vermont, zero wheat grown in New Hampshire. And yet there are suddenly farmers starting to plant small amounts of wheat as part of their rotations—likely because a local market of passionate chefs and bakers are interested in flavorful whole grains."

The problem, he said, is that these farmers are often, like Klaas, planting very old varieties with low yield—the problem with heirloom anything—or they're planting conventional varieties with no flavor. "Without a breeder to support the continual betterment of the plant, an alternative to conventional wheat will never establish itself."

When I later told Klaas what Steve had said, expecting an elegant counterargument, he instead heartily agreed. "That's just it," he said. "There aren't a lot of choices out there. You can't just call up a seed company and buy old seed—heirloom, landrace, or whatever. Not if you want to plant two hundred acres. Not even if you want to plant fifty acres."

Klaas explained that, while his and Mary-Howell's seed business has helped in some ways to fill that niche, they didn't have the time or the resources for experimental breeding. I asked him if the answer was to have a Steve Jones–like breeder in his corner of New York state.

"What you need," he said, "is a Steve Jones in every corner of every state." Neither of us acknowledged the irony that, when Congress created the land-grant system 150 years ago, it provided the means to do exactly that.

BARBER WHEAT

Steve's greenhouse ran along one side of the eight-acre research field. It had a hospital-like feel—orderly, scrubbed clean, and hushed except for the whirring sound of overhead fans circulating fresh air. Hundreds of newly crossed varieties lined the rows like newborns. Clear baggies covered the seed heads.

We walked through several nursery rooms, each as immaculate as the last, until Steve stopped at a section and pointed to the Aragon 03 cross. It was a beauty. My newborn was surrounded by dozens of other crosses, but I have to say it stood out from the others—tall and well proportioned, with a slight curve at the top of the stalk.

"Barber wheat," I said to Steve.

"Okay, sure," he said, smiling. "I was also thinking Jones-Barber, to be honest."

I thought back to a time, twenty years earlier, when, at the end of a long and brutal internship at a restaurant in the South of France, I visited the local farmers' market for the first time. I roamed the stalls until I stumbled upon some red-spotted apricots. Stunned by their perfection (and, after sampling one, their incredible flavor), I turned to the elderly French farmer propri-etress and blurted out, "*Où sont-ils nés?*" ("Where were these born?") Here, in Steve's greenhouse, staring down at Barber (or Jones-Barber) wheat in its incubator, I felt equally moved.

"How was this created?" I asked.

Fairly easily, I found out. It's surprising how accessible the mechanics of breeding are—even to a chef whose education in biology began and ended in the ninth grade. Steve refers to wheat as the "perfect flower," which isn't so much an opinion as a botanical designation. Because it's self-pollinating, wheat has both male and female parts—anthers, which contain the pollen sacs, and the stigma, respectively. If the first step had been to arrange a marriage between Aragon 03 and Jones Fife, the next step was to designate a male and a female. Steve chose to emasculate the Aragon 03 by removing the anthers, then took the head of the Jones Fife (whose anthers were still intact) and held it in place above the Aragon 03. That may sound antiquated and perhaps sexist, the male lording over the female, but there's a functional necessity to it: the pollen must fall directly onto the stigma. Steve wrapped a baggie (called a dialysis sleeve, which looks like plastic but breathes) around the two wheats to secure their positioning. The next day, he flicked the baggie "like you would a cigarette butt," agitating the pollen off the Jones Fife where it fell onto the stigma of the Aragon 03—consummation. Five weeks later, Barber wheat was born, the seed ready to be planted.

"How do you make sure the offspring thrive?" I asked, my paternal instincts kicking in.

"Well, you plant it. The first harvest is your F_1, a hybrid that produces 50 percent of each parent," Steve explained. "Then you plant it again and you get an F_2, which is the generation with all the variation. That's when you can really get into it and know what you're working with."

"Which is when you select for the traits you want," I said.

"First we plant that generation and we grow it out," he said.

"And *then* you select the one you want," I guessed.

"We'll pick nice-looking plants, absolutely, but then we plant them back again."

"And then what?"

"And then we repeat."

"Repeat?"

"Right, you keep going. It can get pretty drawn out."

After several generations of selection, the resulting wheat might be absolutely uniform, a completely "pure" line. But unlike most breeders, Steve keeps some populations intentionally diverse—making selections based on groups of plants in order to retain some natural variety. His reasoning is similar to Glenn's—the in-built genetic diversity provides an insurance plan for the wheat and allows it to adapt more easily to its environment. It's a particular concern for Steve, as many of his varieties are intended for other parts of the country.

I asked him how these populations differed from a landrace system.

"Well, for one thing, we're actively making the crosses," he told me, whereas landrace systems leave more up to nature. And he's often doing it with tools that have come around only in the past half-century.

We tend to think of breeders as either old-school guys, who patiently make crosses and wait a lifetime for the new variety they want, or as Monsanto maniacs, who press a button to sequence new plant genes. Steve is somewhere in the middle. He embraces classical breeding but marries it with genome mapping, marker-assisted selection, chromosome painting, and other technologies that enable him to "look inside the guts of the plant" and understand how traits will express themselves, without waiting several generations to see it play out in the field.

I was already imagining a revolutionary new variety of wheat for the following spring. "There has to be a way to get this into my kitchen more quickly," I said. I told Steve I didn't want to spare any expense.

"Of course you don't," he said, indulging me. "It's easy to get wrapped up in all the gee-whiz shit, but in the end it all comes down to what goes on out there . . ." He pointed to the field we had just walked through. "Your wheat has to make it in the field, in the elements, through the winter and spring, the droughts and the floods. That takes time. People don't like to hear that, necessarily, I know. They want and expect technology to cut through reality, but it can't. Them's the breaks."

We stepped out of the greenhouse and back into the eight-acre experimental field. I worried that my namesake, so new to this world, would soon be forced to find its way among all those forty thousand experimental lines. It seemed a fate worse than competitive summer camp.

"There's always a paternal thing that happens with varieties you develop and put a name to," Steve said, picking up on my concern. As I collected my things to head back to the airport, he assured me that he would keep an eye out for Barber wheat.

CHAPTER 31

Wᴵᵀᴴ ʟɪᴛᴛʟᴇ Bᴀʀʙᴇʀ wheat set to fulfill, if not surpass, my high expectations, I began to wonder what might prevent all bakeries from using it, or any of the other new varieties Steve was shepherding in his field.

Access is one stumbling block, I discovered. There's only one breeder in the country like Steve Jones. (Really. Just one, which is incredible when you remember that wheat is grown on over fifty-six million acres in the United States.) And farmers like Klaas are hard to find. So there's still a marginal supply, and it's expensive.

Paula Oland, the founder of Balthazar Bakery, in New York City, told me price was the reason she ignores small farmers like Klaas. "People will only pay so much for bread," she said. "We're at the high end already. The margins are too tight." To grow the market and bring down the price, more farmers like Klaas are needed, but such farmers will emerge only if they know they have a market to sell into. Which, of course, brings things back to the baker.

Beyond price (assuming bakers can persuade customers to pay more for better-tasting bread), there is the problem of consistency. Like any crop harvested in nature, wheat is unpredictable, influenced by the weather and the local soil conditions. Inconsistency is guaranteed. When Klaas delivered wheat to Blue Hill, we had come to expect the flour to vary from harvest to harvest. Adjusting our recipes to the character of the wheat was manageable for the twenty loaves we baked every day. At that level of production, we'd even learned to appreciate the variances. But if we were baking two thousand loaves and the flour acted differently with each delivery, we wouldn't be in business for long.

Which is why almost all bakers buy from large millers. These millers are able to process fifty deliveries of wheat, from fifty different farms, and combine them (a higher-protein batch is mixed with another with lower protein, for instance) to derive highly consistent flour.

Nancy Silverton, the former owner of La Brea Bakery, in Los Angeles, learned to rely on this system because customers returned to her bakery expecting the same loaf every time. "When you buy flour from a good miller, they spec a certain flour for you, with only small variances," she explained. "They make a flour cocktail that will give you a consistent bread."

The corner bakery wants consistent bread as much as the industrial behemoth does, and buys flour from the same source. The prerequisite? Flour that acts the same each time it's mixed with water.

"Look, as a baker, I loved those days back when we never really knew what would come out of the oven," Nancy told me. "I mean, that is what bread baking was all about for me. But while people understand the change in seasons when it comes to the availability of fruits and vegetables, or the different cheeses at different times of the year, they see bread as more than a staple. People view bread as stability itself. You really can't mess with that."

Jim Lahey, of Sullivan Street Bakery, believes the culture of bread bakers—and bread consumers—needs to change. "I think that the bread community doesn't participate in the pride and occasion that you see in so many parts of restaurants, where they talk about the sourcing of the ingredients, and the chef goes in the garden and picks the herbs and the people come in and order nice wines from the sommelier," he once said.

If grapes are soaked with rain one year, the wine tastes different, but people don't reject it for being different, he told me. "We don't give that kind of slack to bread. People romanticize bread bakers, but at the end of the day we're laborers. We're just above the guys that come out to dry off your car after the wash. We don't dictate the rules. We obey them."

THE BREAD LAB

The rules are changing. Perhaps too slowly, but the world of wheat—which includes wheat itself, but also farmers, millers, bakers, and breeders—is digging itself out of a rut. At the moment, wheat is like the preacher in George Bernard Shaw's *Too True to Be Good*, standing "midway between youth and age like a man who has missed his train: too late for the last and too early for the next."

"We know the one-size-fits-all farm—the monoculture behemoth growing ten thousand acres of one variety of wheat—we know that's passé," Steve told me, adamant that the green-revolution way of thinking is on its way out. "The emerging paradigm, the future of this, is to grow wheat locally, all over the place, in ridiculously small plots. Hyper-local wheat. That's where this is headed. But we're in a strange space right now, because the entire production chain—the fucking entire food chain, really—is still in the old paradigm. There will still be people pushing more of the same. Monsanto isn't going anywhere soon, and I'm not fighting them."

Not directly, he's not. But Steve is fighting for the future of wheat, and he's doing it on the fringe. That crystallized for me one morning several months after my visit to Washington when I opened an e-mail that he had sent at 4 a.m.

I had come to expect and look forward to these early-morning notes. Steve may spend his days in the lab and the field, but at night—sometimes all night—he experiments with his wheats and bakes bread in his backyard wood-burning oven. He is the most complete insomniac I have ever known. Between the hours of 1:00 and 6:00 in the morning, I've received e-mails with subjects ranging from preparing his own malt extract ("Blows my mind, food for the Gods") to drying his bread starter—a mother dough used to increase flavor—into candy-size bars he rehydrates on road trips ("I get an itch to bake sometimes when I'm traveling and can't sleep"). He once sent a long, detailed note titled "Water Boarding My Wheat," in which he described drowning and starving his new varieties to stress them into sprouting, for an extra pop of flavor. ("I had to set the alarm every two hours—not real popular with the wife.")

But this particular morning's e-mail didn't have any of his usual wry humor. It was exuberant, announcing his decision to build a new laboratory for his life's work. Steve explained that when he was hired at Mount Vernon, the position included some startup money from the university. But he had asked them to hold the funds for a few years. "I didn't quite know what I wanted to do," he wrote. "Now I do."

Steve recognized that farmers would be persuaded to grow his wheat only if bakers were eager to bake with it. Most bakers were interested; few were eager. Bakers, as I'd already discovered, didn't want to take risks with flours they didn't know. And millers were just as conservative—they didn't seek out wheat that bakers wouldn't buy. Without demand, Steve could only get so far with farmers.

"How's that for circular?" he said when I called him later that morning. "Bakers ask for certain flours because they know what they're getting; millers know what bakers need because that's what they always ask for, and farmers—well, farmers just get told what to grow, based on demand. And around and around it goes." To break the cycle, Steve wanted to give everyone involved a better understanding of his wheats and their distinctive qualities.

"Selling flour to most bakers is like selling steel to Chevy and Ford," he said. "It doesn't change. Breeders are breeding for a specific type of flour— we throw everything else away. If it doesn't yield enough in the field, it's gone. If the volume is not quite right, or the texture is unpredictable, or any of these things are 'off,' it's tossed aside. If the flour comes out a little yellow? Throw it out! Well, I got tired of throwing it out. I've wondered—my whole career I've wondered—what was in that yellow that I just tossed? Because there was something there, for sure, probably something pretty damn delicious or terrific for our health. I got tired of doing that. I didn't want to make steel all my life."

Instead of asking the university to fund a laboratory filled with scientists analyzing traits and genes, Steve used the money to build a new kind of laboratory, unlike anything that exists in this country, if not the world—a public research space that brings together farmers, chefs, bakers, and breeders to freely experiment with new wheat varieties. He calls it the Bread Lab.

———

Several months later, I flew back out to Washington to check in on Barber wheat. Steve had just transplanted the first cross to the eight-acre experimentation field, but when I arrived, he didn't seem interested in baby Barber wheat or any of the forty thousand other lines. He wanted only to show me the Bread Lab, and very nearly ran through the center's corridors to tour me through the newly completed space.

There were mixers and a large deck oven, just as you might find in a small neighborhood bakery. And then, awkwardly juxtaposed, there were some machines I didn't recognize, like an alveograph. Steve demonstrated its use, placing a few grams of bread dough into the center of its small, concave dome. Air pressure from the machine inflated the dough into a softball-size bubble, simulating the carbon dioxide gas released during baking. This enables Steve to measure the wheat's "extensibility"—its ability to hold air bubbles and produce the kind of open, airy texture we expect in a loaf of great artisanal bread. He suddenly reached over and popped the balloon, a large, satisfied smile crossing his face.

"Without this you're back in the Middle Ages," he said (a little grandly, but Steve was feeling a little grand).

He gestured toward the next machine. "This one tests for falling number." Falling number is a measurement of sprouting caused by dampness or rain just before harvest. The sprouting—the beginnings of germination—causes an increase in alpha-amylase, an enzyme that breaks starches into sugars. To get the falling number, Steve filled a large test tube with a fixed measurement of flour and water, shaking it to form a thick slurry that smelled slightly of stone-milled oatmeal. He inserted a plunger into the tube, then placed the whole thing into a hot-water bath.

"If the starch is present, if it hasn't been destroyed by sprouting, this plunger thing is going to fall slowly," he said. Indeed, when he dropped the plunger, it slowly, almost lazily, made its way down the slurry, taking several minutes to arrive at the bottom of the tube. "The starch is adding viscosity, meaning the wheat will make for a nice, sturdy dough," he told me.

Had the slurry been made with sprouted wheat, the starches would have converted to sugars, and the plunger would have dropped quickly through a thinner liquid. The resulting bread would be sticky and doughy, with poor structure. A low falling number is a large problem for industrial bakeries—if the dough is sticky, mechanical slicers won't slice the loaves evenly—but it's also a problem for artisanal bakers. If you've ever tried a bread that tastes sweet but has a gummy structure, it's likely because the flour lost its starch from excessive sprouting. Low falling number is also the enemy of al dente pasta, ensuring a limp noodle no matter how quickly you've cooked it.

I wondered about its relevance for a breeder. Wasn't falling number simply measuring rainfall before harvest? And wasn't the level of sprouting in nature's hands?

"Not necessarily," Steve said. "If we grow a thousand different kinds of wheat side by side and we get a few days of rainfall, we'll look at how each of the varieties sprouted." Some sprout more easily in damp conditions. Steve referred to these as precocious. "You just look at them funny and they sprout," he said. "Others aren't so easy. There's a large range."

But then, if genetics determine sprouting, why not select for wheat that doesn't sprout, and ensure better bread?

"It's complicated," he said. "By selecting just for that, you're getting rid of a lot of other traits, too. Remember: it's never one trait. It's the interaction of traits that's important. The key is balance," he said. "You want balance." That kind of balance is not a priority for industrial bakeries, which have pushed breeders to select for high-gluten flour geared toward assembly line production. Steve calls it the "Wonder Bread–ification" of bread.

"Basically, fifty or sixty million acres' worth of wheat has been bred, grown, and sold for high-speed mixing," he told me. The system is efficient, well established, and built around what Americans are willing to pay for bread.

Steve pointed to still another machine, this one called the NIR, which measures for protein content. Like the alveograph and the falling-number machine, the NIR looked awkwardly out of place, its design perhaps inspired by a low-budget 1960s sci-fi film. "It's measuring light absorption," Steve explained. "The higher the absorption, the more protein present in the wheat."

He poured a handful of Bauermeister, a wheat he developed nearly ten years ago, into a funnel. With the press of a button, it was released into the machine. There was a muffled whooshing sound, and sixteen seconds later Steve read the measurement.

"Ten point five percent. Good number, if you ask me. Not to mention Bauermeister is a sexy wheat—chocolate-like flavor is what they tell me. But bring this to an industrial bread-baking company and they'll laugh in your face. Ten point five percent is wimpy wheat. Fourteen percent is manly-man wheat. They'd either make this part of a blend or feed it to pigs."

I told Steve I once read that in France, bread wheat measuring *over* 11.5 percent is considered junk wheat. "That's right," he said. "And yet the French aren't maligned for wimpy flour. I mean, everyone likes to dump on the French, but have you ever heard 'Man, those French people, they don't know how to bake a good loaf of bread'?"

The problem, as Steve described it, is that industrial bread bakers have the most muscle. Their interests determine not only what breeders select for, but also how the wheat is grown. "To get these whopping percentages, you have to fertilize your soil. There's a direct correlation between healthy soil and good protein content, but farmers are forced to do it artificially to get these numbers every time. *Every* time," Steve said. "Nature doesn't work like that."

In this way, he said, wheat farmers—and, when you start to follow the logic, bread bakers and eaters—are doing their part to poison the environment. Many farmers deliver an extra shot of synthetic nitrogen just before harvest to pump up the protein percentage. "They're overfertilizing sixty million acres," Steve said. "And for what? So the industry can bake more loaves per hour, and so they can charge you more to 'enrich' and 'fortify' it—which just means they can add tons of shit into the bread and it won't collapse."

He tapped his fingers on his Bauermeister wheat a few times, acknowledging its presence. "Yeah, sure," he went on, to no one in particular, "I wish everyone grew wheat organically. But when the industry is demanding 14 percent protein, forget it. The end. It's just not possible without nitrogen."

This is why Steve built the lab to appeal to the people who weren't as obsessed with protein percentage, or falling number, but with the quality of the loaf. Craft bakers, as Steve likes to distinguish them, many of whom come out of the French tradition of bread baking, rarely add sugar, and the best of them don't use yeast or excessive amounts of salt. The natural yeast and bacteria present in the flour are activated by fermentation (mix flour and water, and wait—the *wait* being the crucial ingredient). The starches break down, and the sugars help develop a depth of flavor and complexity that doesn't exist in industrial breads.

"Granted," Steve said, "it's a day and a half to make a loaf, versus twenty minutes, but you develop texture, flavor, and most likely a lot better nutrition."

Like Glenn Roberts, Steve has become the neighbor, and, to a certain extent, he's become the community organizer, too. He got here by default. Steve recognized that to breed distinctive wheat, he would have to become much more than a breeder.

His work required broadening the conversation, in concentric circles. The inner circle is the farmers—after all, they have to be persuaded to grow his varieties in the first place. But in order to persuade the farmers, he had to convince the bakers they needed better wheat. He didn't do that the old-fashioned way—presenting the wheats to the bakers and telling them to figure out how to use them. Instead Steve turned the equation on its head. He asked bakers, *What do you like?* and figured out how to make it work agronomically.

"That's what's needed," he said. "Because if you do not actively breed for a trait, for a flavor or a unique functionality or whatever, odds are damn close to 100 percent that you will not find them."

Steve soon realized that bread bakers, like farmers, were only the beginning. He introduced me to Brook Brouwer, a graduate student who had recently been recruited to work on barley for malting. Barley is a hardy, disease-resistant crop, and already a valuable one for animal feed. But Steve was betting he could make it even more valuable, because Washington state microbreweries—which use over 25,000 tons of malt per year, none of it from local sources—were beginning to demand distinctive malts.

"These small breweries are just like the artisan bread bakers: they want flavor, and they want local," Steve said. "There's a huge opportunity to return some of the big, bold flavors we had before Prohibition." But the malsters are more interested in selling to the big brewers, he said—"the Budweiser boys and the like." For the past several decades, farmers around the country have grown barley with little distinction because, in yet another example of circular logic, that's all the malsters buy—and that's all the malsters buy because that's all the commercial beer industry orders.

Steve was trying something new. "We're breeding varieties of barley with flavors that will blow your mind," he told me.

Steve had recently decided to broaden the community even further by inviting millers to the Bread Lab. The local flour mill is as much of a relic as the local slaughterhouse, he explained. "The mill closest to here was built in the '20s," he said. "Today it's an Italian restaurant. The old mill in Jefferson County, next door to us? An Italian restaurant. The mills near Seattle? Torn down to make kitchen cabinets and old wood floors in Seattle and Portland."

The small mills that remain all operate the same way: take a single harvest of wheat from a single farm and a single year, mill it, and sell it to a few local bakers. They don't blend flours like the larger mills, which means the quality of what they produce is always in flux.

"Buying your wheat from one farmer, however romantic, is not sustainable," he told me.

I remember thinking, *It isn't?* Wasn't I doing that with Klaas—just doing the milling ourselves?

"You're not a bread bakery," Steve said. "Even the bakers we work directly with can only afford so much funkiness. The reality is, the bread better taste good, and it better not be crumbly. It has to have the right texture and color and flavor, every time." I thought of Nancy Silverton: *People view bread as stability itself. You really can't mess with that.*

The local grains network won't work, Steve said, until the baker can request a particular kind of flour, the way the big bakers do, and explain to the millers what traits he's looking for. "That's going to take blending," he said. "And that's what the Bread Lab is about—to help those millers learn to do that. That will be a beautiful maturation of the whole system."

He showed me a corner of the room devoted entirely to milling. There was a small stone mill, a steel-plated mill, and a lab-hammer mill. "This puppy runs at nineteen thousand rpms," he said, pointing to the hammer mill. "It just pulverizes the wheat nice and fine."

But in his zeal to improve the consistency of flour, Steve isn't ignoring the opportunity to change the culture of bread. He also sees the Bread Lab as a haven for experimentation—a place where chefs and bakers can learn to embrace the inconsistencies and complexities of wheat.

"When bakers spec flour, they know what they're going to get. They don't know what they *could* get," he said. "What other kinds of complexities could you add in? It takes chefs and the bakers to push that, to offer these things. But their margins are so tight, they can't afford to mess with it. They can't afford to make mistakes. Here we welcome mistakes. We encourage them."

———

I saw Steve's new paradigm in action when we visited a local bakery called Breadfarm, in the tiny town of Edison (population 133), a few miles from Steve's research center. We parked down the street. A field had been mowed nearby, and the smell of freshly baked bread mingled stubbornly with the fragrances of wet earth and cut grass. Passing a poster with the image of Che Guevara, Steve turned to me. "Welcome to the revolution," he said.

In a funny way, Steve's small satellite research center is bringing the local food revolution straight to the people. He's following through on the promise of the land-grant university system in the truest sense. Only, instead of merely visiting farmers in the field, he reaches out to bakeries like this one, where, as Steve described it, "the rubber meets the road."

Inside the bakery, Steve approached the bread counter and suddenly seemed transformed. A friend of the French writer Honoré de Balzac once described the famed gourmand at mealtimes: "[His] lips quivered, his eyes lit up with delight, his hands shook with pleasure on seeing a pyramid of pears or beautiful peaches . . . tie whipped off, shirt open, knife in hand . . . [he] laughed explosively, like a bomb. . . . He melted for joy."

Balzac came to mind as Steve feasted on the free samples, providing a play-by-play of his favorites. "Samish River Potato Bread. That's a natural ferment, local potatoes. Real nice. Not too sour. And look here," he said, waving his hand. "Stone Ground Wheat Miche. Local wheat in there—some of the ones I bred. Makes a hearty loaf. It's big and serious-looking, don't you think?"

Having surveyed the selection, he started to shake his head. The look on his face—steady, sensible, resigned, yet essentially disappointed—almost spoke for him: *Why aren't more people enjoying great bread like this?*

"I get asked, why the Bread Lab and not the Wheat Lab, or something more indicative of what I'm supposed to be doing?" Steve paused to sample a sourdough, ignoring a small bowl of olive oil meant for dipping. "The problem is land-grant universities all have laboratories devoted to a particular industry. You want a pig that doesn't produce soft meat under stressful conditions? Or a cheese stick that has a shelf life of ten thousand years? A potato chip that stays crisp in high humidity? We can breed for that! It's supposedly in service to agriculture, but really it's in service to big food." The Bread Lab does not work off of an industry standard, he said. "We're creating our own."

Scott Mangold, the owner and baker, appeared, disheveled, covered in flour, and clearly interested in getting back to his doughs. He gave us a quick tour and answered a few questions. Steve thanked him for his time. As we gathered some bread before leaving, I asked Scott about the different flavors he's discovered with new varieties of wheat.

He said that at first he was skeptical that wheat even *could* have undiscovered flavors—at least on the level that Steve described. "When I first moved to Washington State, I heard crazy crap like 'The wheat from around here can express apricot or chocolate overtones.' You know, really pretentious shit."

Even after he began experimenting with Steve's grains, Scott was still focused on functionality, the technical side of the equation. But one day he decided to set up a blind tasting with his staff.

"We're all just standing there," he said, "and it gets real quiet or whatever. I mean, there was a long, uncomfortable silence, and finally I say, "*I'm tasting chocolate.*" And everyone's eyes just lit the hell up, like, *Yeah, that's what that is. Chocolate!* We all started high-fiving each other. It was pretty awesome."

I asked Scott if he remembered the variety of wheat. "Do I remember? I sure do. It was one of Steve's wheats. Bauermeister."

Klaas was right. We've lost the taste of wheat. And yet here it was, in the tiny town of Edison. Not just the flavors of the past, but of the future, too.

———

Before I left for the airport, Steve again walked me through his eight-acre test plot, with its forty thousand experimental lines. It was as extraordinary as I had remembered it.

The kaleidoscope of varieties made a mockery of the idea that wheat is generic and anonymous, that it doesn't have as much diversity or potential for flavor as a tomato. The possibilities seemed endless, which I suppose is the point of keeping all these potential lines going. More diversity, more opportunity. It was, when I stopped to think about it, the aboveground manifestation of what Jack argued should be happening belowground. In healthy soil, the diversity of microorganisms helps ensure healthy plants—but exactly how that happens is a bit of a mystery. Looking out at these wheat varieties, I felt that same sense of mystery underlying all the possibility.

My hope that Barber wheat would become the most desired in the world suddenly felt naive. In the best scenario, Barber wheat would do what all of the wheat I was looking at had the potential to do: thrive in a specific environment and remain open for selection on the most local of terms. It was Borlaug in reverse, really—breeding for greater distinction rather than dumbing down wheat in an effort to make it grow everywhere.

Steve's field—and the community of farmers, bakers, chefs, millers, brewers, and breeders that orbits around it—is a new kind of botanical ark, one that preserves the past by allowing those genes to thrive in the future. It dares us to re-create our relationship to wheat, to do what we all really want for our children, which is to provide them with the genetics to succeed while at the same time celebrating their distinctiveness.

I wanted a bird's-eye view of the diversity, just as, a few years earlier, I'd gotten a look at the *dehesa* from a rooftop. Back then, I'd seen how farming could sculpt a landscape—the pigs leading the dance of grazing cattle and sheep, the enormous oaks interspersed among the famously lush grasses, and the community of homes and churches dotting the landscape. From that perch, I'd seen how interconnected it all was.

Steve's experimental plot didn't require a rooftop for me to see that the real connective power wasn't in the field. It was in the way he'd fashioned the Bread Lab. In opting out of building what he *should* have built—a laboratory focused only on creating new wheat varieties—Steve enlarged the vision. He complicated the picture. The Bread Lab functions like those rich striations of *jamón ibérico* fat Eduardo held up to the light. It keeps all the disparate parts together. It also forces them to mingle.

John Muir, who described how everything in nature is "hitched to everything else in the Universe," would not have said the same about our current food system. Because our food system is disconnected. It operates in silos: vegetables here, animals there, grains somewhere else—each component part separate from the others and unhitched to any kind of culture.

Steve devised the Bread Lab to do the work that wheat, a social crop, always used to do: create community. He forces the socialization a little, much as Klaas and Mary-Howell's grain mill forces the community of Penn Yan farmers to exchange ideas. He clears the way for connections to happen. He helps the hitching.

Steve's work is not merely the difference between the tabletop monocultures of Kansas and a field like this one. It is the difference between an antiquated system of agriculture and the possibility for a future of delicious food.

EPILOGUE

Ever since I started cooking professionally, I have been sensitive to the evening hours between 5:30 and 7:00, from the late spring through the first weeks of fall. Photographers and film directors call this the golden hour, the magic time in the day when the sun is low in the sky and the light turns soft, diffusing over everything it touches.

Chefs miss out on the golden hour because most professional kitchens don't have windows. But I've never heard a chef grumble about missing this time of day. I've never heard one even mention it, though I think about it all the time. No one warned me that my career, which is to say most of my life, would require that I not only miss the most magical time of the day but also that I feel the *most stressed* at the most magical time of the day. The first orders of the evening arrive at the pass just after 5:30 p.m., as the cooks finish their preparations and launch into dinner service. Our pressurized little world gets much more pressurized during those golden hours we cannot see.

The first time I visited Stone Barns, four years before the center opened to the public, it was late in the day. I walked into the former milking area, which was already slated to become Blue Hills's kitchen. It wasn't the size of the space or the plans for its transformation that gripped me. It was the south wall, with its five sun-drenched windows. They looked out on a courtyard surrounded by the collection of Norman-style stone barns. *Windows*, I said to myself, as if this were all that mattered. This kitchen would have windows. I felt like the luckiest chef in the world.

Imagine my surprise when, after just a week of watching that golden-hour light bouncing off the old barns, I found myself turning away from the scene. I realized the windows only made me more miserable: I could see what I was missing.

I was thinking about those missed hours one summer evening not long ago while having dinner in Klaas and Mary-Howell's kitchen. It was my first visit to their farm in two years. Looking out the back window, I watched the sun splash golden light across a field of emmer wheat.

Chefs, in my experience, seem especially captivated by farms scenes like this. That's not because our success depends on the quality of what those fields produce (though of course it does) but because in our windowless, turbulent little world, a farm seems like a rare dose of what endures. Watching the wheat field as it played tricks with the fading sun, I felt a sense of serenity and stasis.

But for Klaas, who dines with Mary-Howell in the same seat at the kitchen table every evening, the view elicited an entirely different reaction. For many years, he told me, he'd looked out over his fields with a deep restlessness. He longed for a different view, one that included animals grazing the pastures, which is precisely what came into focus as I stretched for a slice of Mary-Howell's homemade spelt bread. There were dairy cows foraging one of the fields, and just below a dilapidated barn I could make out pigs rooting in straw. Mary-Howell served the bread with butter made from the cows, alongside three hulking pork chops.

—

During the previous year, Klaas had expanded the farm yet again by bringing in livestock, something he once had not wanted to do. Blue Hill was slated to buy most of what he raised, so I came to learn about how the animals added to the already robust farm and to understand why, in his late fifties—an age when most farmers scale back if not retire—Klaas decided to "complicate things just a bit" and engage in a part of farming he knew very little about.

Over dinner, I reminded him that he was already doing groundbreaking, experimental work, not only by managing complex rotations for his soil and propagating important older varieties of grain but also by inspiring an entire community to farm without chemicals and, along with Mary-Howell, providing those farmers with an infrastructure to make it happen. How much better could he get?

He spooned a helping of butter so large onto his bread that I thought he was joking. "I met you ten years ago," he said matter-of-factly. "Our sound bite then was: Go organic, change the world. It was sort of said tongue-in-cheek, but since then what's evolved is a better understanding of what's actually needed to change the world."

Even if farmers are rotating their crops to support soil health and even if they are preserving vast landscapes for food production, they're not doing enough unless they're also feeding the maximum number of people directly, Klaas told me.

"If they're not doing that, they have more work to do," he said. He and Mary-Howell had always made a conscious effort to grow as much as they could for human consumption. That's why they started growing bread wheat in the first place and it's why they've continued to grow crops like kidney beans, even though they haven't been particularly profitable.

"We wanted to balance the profits with the ethics," he said. "And now that the bills are paid and the children have grown, we're able to think more this way. So we looked at our farm more critically and went to work trying to figure out the right balance—feed people, feed the soil, and make a profit. The answer was staring me in the face. I just didn't see it."

It wasn't until Klaas and Mary-Howell traveled to Europe the previous summer that he was able to see. "Everywhere we went we passed animals, cows especially, and then walked over incredibly rich soil—the richest soil I'd ever seen."

Klaas took Mary-Howell to Laverstoke, where nearly a decade before I had first met him and the Fertile Dozen of visionary farmers. The former race-car driver turned über-farmer Jody Scheckter now had a thriving farm that fed thousands of people. From Jody's dining room table, Klaas and Mary-Howell could see grain fields and pastures for cows, chickens, and pigs. There was even a small herd of buffalo. The soil was dark and rich. They ate lunch with everything sourced from the farm.

Jody had apparently listened carefully to the Fertile Dozen's advice. Klaas was struck by how in so short a time he had "closed the nutrient cycle," creating a self-sustaining farm wherein the waste from one part became food for another. The crop rotations were finely tuned with the animal rotations. The whole farm revolved around supporting the biological health of the soil.

Nothing is more typical of Klaas, or more estimable, than this: in that moment of humbling irony—the former student surpassing the teacher—he wanted only to return to his farm, because he realized there was *more work to do.*

"All during lunch I kept saying to myself, *This is the view I need to have from my kitchen window. This is what's been missing.*"

———

Klaas and Mary-Howell started small, introducing a few dairy cows to the farm. Their daughter, Elizabeth, was especially drawn to cows and wanted at one point to become a vet. Pretty soon, more cows. Klaas grazed the animals on the cover crops that were meant to benefit the soil, like clover. The cows ate the tops—what Klaas calls "rocket fuel for ruminants"—and the rest of the plant was plowed in, along with the cows' manure, to feed the soil's microorganisms.

I was surprised to learn that Klaas's brothers were also beneficiaries of the new system. Profits on their dairy farm had been declining for years as feed costs continued to rise. Finally, they decided to transition to organic in order to sell their milk at much higher prices.

I wondered if Klaas felt vindicated. After all, his brothers had once scorned his decision to go organic. When I suggested that a "Hey, bro, you blew this one" wouldn't have been uncalled for, he merely shrugged. "It's turned out to be a very good development for all of us."

Klaas harvests some of his cover crops and sells them to his brothers for high-quality feed. And his brothers, who are required by organic regulations to stop milking part of their herd for a certain period, can send these "dried-out" cows over to Klaas to graze with his daughter's cows—adding even more manure to the soil.

"We're in a symbiotic relationship now," he said, smiling. "As Joel Salatin said at Laverstoke, the goal is 'more than the sum of its parts.' I subscribe to that. I see the wisdom of that now in ways that I didn't see ten years ago."

After dinner, Klaas and Mary-Howell laid out several large pieces of cheese that had arrived earlier in the day from their eldest son, Peter, who was interning at an organic dairy in Germany. Klaas praised the farm for its work in renewable energy and raw-milk cheese.

"Cheese is a wonderful value-added product," he said, hinting that Peter might one day further expand the dairy.

I got the sense that Klaas was looking beyond just Peter. Bringing in livestock, creating an infrastructure for cheese-making, and reimagining how the farm draws its energy are projects for multiple generations. It was Mennonite-like thinking (*When do you start raising a child?*), preparing for a farm that will sustain the family one hundred years from now.

Over the previous few years, I had seen how the very best farming systems—whether the oak-lined pastures of the *dehesa* or the intricate canals of Veta la Palma—are constantly in flux, adapting and readapting to balance the needs of a healthy ecology with the imperative to feed people. Sitting at the kitchen table, staring out at the view of Klaas's fields, I could see evidence of that same evolution. It was (for me, anyway) a peaceful scene, but there wasn't stasis.

In just two decades, Klaas and Mary-Howell have gone from harvesting a few organic grains to complex rotations that include heirloom wheat, vegetables, and legumes—many of them farmed on newly leased land. They've added seed production to the mix, and a seed distribution company to supplement the thriving mill and grain distribution business. And now they've created a small dairy.

One hundred years from now, I suspect that Klaas's heirs will look out on a similar view. If I'm right, it won't be because anyone fought against change. It will be because each generation embraced it.

Perhaps the most dynamic transformations are yet to come. Klaas no longer thinks of the farm as an entity all to itself (not that he ever really did) but as an interlocking piece of a larger community.

"I'm recognizing more and more that not every farm needs to, or should, do every operation," he said, a large bite of the cheese punctuating his thoughts. "Which is really the point. Can you build a community where different enterprises fit into each other and make better use of resources? That's the challenge." By "challenge," he meant *his* new challenge for the future— changing the farm, forever striving to make it more than the sum of its parts.

━◦━

Of all the insights and observations I've gained from farmers, breeders, and chefs during the research for this book, I can't help dialing back, again and again, to the one that sits with me most heavily.

It was after I told Wes Jackson about Klaas's farm, arguing that it was a good example of sustainability. Wes didn't buy it. "It won't last," he said. And just like that, he rejected not only Klaas's work but also a generation of farmers looking to transition their farms in similar ways. As much as I longed to dismiss him as an old crank, I had the nagging suspicion that he might be right. History shows that at some point, good farming unravels with just a few shortsighted decisions.

The vagaries of our country's food preferences don't help. Even with the farm-to-table movement running high at the moment, we're still guilty of reducing sustainability down to what we buy for dinner. Rarely do we imagine the whole picture, which means that rarely are we forced to realize that a truly sustainable food system is not simple. It is not built on one or two principles of farming, and it does not produce merely one or two good things to eat.

That whole picture might look like my rooftop view of the *dehesa*. With its two-thousand-year history of diverse farming and its carefully cultivated landscape, even Wes acknowledges that the *dehesa* has lasted, and actually thrived, over generations.

The Skagit Valley is perhaps another exception, if Steve Jones continues to have something to do with it. His work with farmers and his creation of the Bread Lab is about building a community around the right kind of farming and baking.

I bet his vision will endure. But then in a funny sort of way, the time I spent with Steve only underscored Wes's argument. Working together, farmers, chefs, and breeders can become part of a complex web of relationships that supports the health of the land. And yet, as Steve understands (and the *dehesa* got me to see), those relationships don't last without a permanent food culture to sustain them. Few farmers have a Steve Jones to connect the pieces.

Let me put a finer point on this, closer to home. Klaas has the opportunity to create something just as important and as lasting as what Steve created. But missing from his story are crops ingrained into people's culture through good cuisine. That's something only I, along with other chefs and home cooks, can provide.

And it was clear I wasn't doing a very good job. If the chef's role is like that of a musical conductor, our goal is to create harmony—to avoid amplifying one section of the orchestra at the expense of others. As successful and enlightening as my wheat experiments had been, they were still too single-minded, too focused on promoting only one product of Klaas's farm. I had yet to address the countless "bycatch" crops—the millet, flax, soy, buckwheat, rye, and dozens of other grains and legumes—that made his whole wheat so delicious. And, of course, now there were more additions to consider, such as dairy. Working with these crops seemed like an opportunity—and, the more I thought about it, an obligation—to support the land's long-term ecological health.

The same was true of my relationship to Stone Barns, and to the countless other farms that supplied Blue Hill. Like any farm-to-table chef, I supported these systems by purchasing the daily harvests. But by privileging only the ingredients I wanted to cook instead of championing a whole class of integral yet uncelebrated crops and cuts of meat, I had ignored what was really required to produce the most delicious food.

In order for these farms to last, to be truly sustainable, I needed to learn to cook with the whole farm.

What does whole farm cooking look like?

The more I thought about it, the more I realized that whole farm cooking is what peasants around the world figured out thousands of years ago. They did not choose their dietary preferences by sticking a wet finger up to the prevailing wind, as we do today. They never had that kind of freedom. Instead, they developed cuisines that adhered to what the landscape provided.

Take the cooking of Extremadura, with its regional variations of *migas*— a traditional dish of fried old bread that might include a lowly cut of braised rib meat from their famous pigs. It's a plate as economical as it is delicious. Here in the United States, Hoppin' John, Lowcountry cuisine's combination of rice and field peas with a brassica like collard greens (and a small taste of pork) is based on the same logic. The dish is an ode to soil fertility: the cowpeas provided the soil with enough nitrogen to grow the rice, and the collards usually took up whatever salt was left over from the seawater that flooded into the basin. There are too many other culinary examples to count, but all of them took their shape and form from what the local landscape could offer.

Not long ago, I sketched out a vision for a Third Plate with a similar ethic in mind: a "carrot steak," flattened and roasted to resemble a juicy sirloin, with braised beef shank (an underutilized part of the animal) playing a supporting role as a sauce. I meant to invalidate our Westernized, meat-centric conception of a plate of food. As a first stab at the future of good cooking, it wasn't bad. But it was merely one entry in a possible menu—a hit single without an album to sustain it. *What would a meal look like?*

In a nod to the Mennonite belief that you begin raising a child long before it's born, I set out to create a menu that Blue Hill will serve a generation from now. I wanted it to be in the spirit of my rooftop view of the *dehesa*, built around the sum of what a farm, or network of farms, can offer. It was a playbook for a new cuisine, one designed to create demand for soil-improving crops and enlarge our sense of what is delicious.

I was picturing specific plates of food, yes, but beyond that, it was an exercise in imagining how the view outside my kitchen window would change as these new ideas took root on our menu.

It will look something like this:

A MENU FOR 2050

MILKY OAT TEA AND CATTAIL SNACKS

How do you begin a meal?

Ángel León doesn't start with fish. He starts with bread, infusing it with a homemade brew of plankton. It's a first bite with a larger idea: without plankton, there won't be any fish left in the sea.

Like Ángel, I'll begin with a larger idea—two of them, actually. The first will take the form of tea made from an infusion of milky oats. Milky oats are baby oats, very nearly mature but still soft and sweet. Klaas, like many farmers, grows oats as a cover crop, mowing them down before maturity so they can enrich the soil and become fertility for the next crop.

Without restoring fertility to the soil, delicious food is not possible. Which is really the message of the milky oats. I'll cook with just the tip of the plant, the immature oat, full of sweet oak milk that makes an aromatic infusion. The rest will remain in Klaas's field to profit the soil.

It's a "take half, leave half" equation, just as with Eduardo's geese. Eduardo explained how the birds eat half of his olives and figs and leave half for the harvest. "The geese are always quite fair," he told me. We will be, too.

If this works—which means if the tea is delicious and memorable—we may well create a market for cover crops, incentivizing more growers to incorporate them into their farms. But more important, we'll create a consciousness about feeding the soil that feeds us.

The second idea will take the form of something wild. It came to me when a forager showed up at the kitchen door with young cattails—native plants that grow next to ponds and lakes. Cattails are a filtering plant, which is why they're so important next to water sources. They act like a sponge, absorbing chemical run-off from the soil and reducing water contamination. You don't want to eat cattails originating from polluted places in the same way you wouldn't want to eat mullet feeding from a polluted pond. Their flavor reflects the health of the environment.

We'll scrape the cattails and sauté their mossy skins in butter and lemon juice. Like scrambled eggs—runny, rich, uncomplicated, perfect—they're a nice way to start any meal. They say: relax, you're about to eat food that's been grown in healthy soil.

Milky oats are agriculture's improvement crop; cattails are nature's wild equivalent. Creating something delicious out of both makes food the measure through which we better understand nature. It defines cuisine around cooking with the whole farm.

FIRST COURSE: *Whole Wheat Blue Brioche with Blue Hill Farm Single Udder Butter*

The meal will formally begin with a slice of our whole wheat brioche, which will taste even better than it does now.

How is that possible? If you're thinking that the superior bread will be about improved versions of Barber wheat, you're correct—sort of. Because by 2050 we'll be baking our bread with Blue wheat, a delicious, nutrient-rich, disease-resistant variety developed for Blue Hill.

Here's what happened: Barber wheat matured and went through several years of selection—all under the watchful eye of Steve Jones in his Bread Lab—and eventually became a much better version of itself by marrying with a wild wheat relative (which happened to have an attractive blue beard).

"A hundred years ago, breeders never stopped innovating," Steve told me. "We shouldn't either." More lab work, more selection, and the resulting Blue wheat tastes like roasted nuts, with a bright, grassy finish.

Steve sketched out his vision for the wheat in 2013 when he came to visit Stone Barns. I confess I had a larger plan in mind when I persuaded Steve to fly to New York. I wanted to grow wheat at Stone Barns, closer to home, instead of leaving it entirely in Klaas's hands. It was a crackpot idea; cultivating wheat in Jack's eight-acre vegetable field is like trying to fit a car-manufacturing plant into a tortilla factory. Then again, I knew from my time with Steve that our conception of wheat as a monoculture crop, much like our expectation of beef as a seven-ounce portion of steak, urgently needed to be turned on its head. What better place to do it than in Pocantico Hills, New York, thirty miles from midtown Manhattan?

But during Steve's tour, Jack had a more practical concern: Where exactly do you plant the wheat?

Steve had the answer. Several, even. After just a few yards, he pointed to an empty patch of land just opposite the greenhouse, suggesting it as a good spot for wheat. Passing a grassy area in front of the restaurant, he bent his large frame and broke a momentary silence with, "What about wheat over there?" A few feet later, another spot. "Or here?" And then he'd tap my shoulder, point again, and murmur under his breath, "Wheat would look pretty damn good there, too." After wrapping my head around the image of wheat planted in rows like zucchini, I looked at Jack, who takes suggestions about what to plant about as well as a host receives decorating advice from an overnight guest. He seemed skeptical. But then Steve removed a plastic vial from his pocket, filled with dark bluish gray seeds, and handed it to Jack. He apologized for the muddy color.

"We'll fix that. Pretty soon, blue like this," Steve said, pointing to Jack's shirt. "Flavor is great. And the antioxidants are off the charts. Literally never seen anything like it." He added that the wheat wasn't genetically engineered or patented by the university, and would be open to public use.

And that was that. Jack planted Blue wheat in the fall. Our brioche has been blue ever since.

———◆———

Will Blue wheat one day make its way to Blue Hill Farm in Massachusetts? Maybe. But in the meantime our Blue Brioche will get buttered with Blue Hill Farm's "single udder" butter, which we already use today.

The idea for single udder butter started with the first delivery of grass-fed milk from our family's newly converted farm. But you could argue it really began a few years earlier, when my brother David and I began to notice the forest encroaching on Blue Hill's pastureland.

It was two decades after my grandmother Ann had passed away. The beef cattle were gone, and that vibrant edge of semiwilderness I'd studied as a child from my perch on the tractor looked increasingly like plain old wilderness. More bramble, thicker ferns. Our grandmother's cherished fields were slowly shrinking. Which is why, in 2006, David and I decided to resurrect the farm as an all-grass dairy. The grazing cattle preserved the integrity of the open space; the dairy began supplying the restaurants with grass-fed milk.

Soon after the first batch arrived, the farmer we'd hired to oversee the land posed a question: Did I know that cows with names produce more milk than cows without names? Apparently he knew as much, so he hung signs on every stall: Annabelle, Daffodil, Jillian, Sunshine, and twenty others. Different names, but also different breeds, and mixes of breeds—a United Nations of bovine diversity.

For me, it brought to mind a slightly different question: do cows with names produce more *distinctive* milk than cows without names? If the answer was yes—and it wasn't hard to imagine, watching the parade of colors and sizes on display as the herd filed into the barn twice a day, that it *had* to be yes—why blend all that distinctiveness into one vat?

As soon as we began separating the cows' milks, their differences became clear. Annabelle, a Dutch Belted cow, is a catholic grazer, producing butter that is bright yellow. Sunshine, a cross of Kerry and Shorthorn, is more discerning. Her butter is ivory white, except in the heat of summer when it turns golden and tastes like pound cake.

Connecting the butter to the personality of the breeds is one thing; connecting it to the condition of the grass is another. Without the flavor-flattening effect of grain feed, the character of the butter changes according to the quality of the grass. We already taste the differences from season to season. A generation from now, we'll go deeper. A recent visit to the restaurant from an older French chef got me to see how.

The chef tasted Annabelle's butter. Mildly disappointed, he declared it a bit bland. He offered rain as a possible cause. (It had, in fact, rained earlier in the week at Blue Hill Farm.) Then he asked where the cows were grazing—close to the barn, or in a field farther away? He guessed farther away. When I called the farmer to check, he confirmed the cows had been grazing in the field farthest from the barn.

How did the chef know? As Eliot Coleman once taught me, in an all-grass dairy, the most distant pasture is usually the least grazed. The soil in these fields is sparsely fertilized as a result, and the grass is almost always less lush and nutritious. The chef, apparently, could taste the difference.

When Klaas and I walked his fields, he schooled me in the mechanics of identifying a healthy pasture. "The trick," he told me, "is to learn the language of the soil."

Single udder butter will help with the translation. Grass-fed cows produce butter that's varied not only by season or by week but also by field and breed. The next generation will know this for sure because they'll be able to taste the difference.

SECOND COURSE: *Rotation Risotto, 898 Squash*

The second course is really an ode to the first. The secret to great wheat, I had learned, is that it's not about the wheat. Klaas's complex rotations of small grains and legumes create much of the flavor of what he harvests, and they all require gastronomic support. I had already imagined a role for milky oats on my menu. Could I create a dish honoring more of these lesser-known crops?

I did it just recently, serving them in the style of a risotto, with the grains replacing the rice. There was rye (Klaas's go-to crop to build carbon in the soil), barley (weed suppression), buckwheat (the cleanser—ridding the soil of any buildup of toxins), and millet (important for drier weather conditions). And I included legumes, like red kidney beans and soybeans (nitrogen fixers).

Without the high starch of risotto rice, the grains lacked a creamy consistency. So I stirred in a puree of brassicas—kale, broccoli, and cabbage—representing the mustard plants (also nitrogen boosters) in Klaas's rotations. The porridge took on the uniformity of a risotto and, with its dozen or so grains and beans, was deeply flavorful.

We called it Rotation Risotto, and it's been on Blue Hill's menu in the city ever since. One night I walked through the dining room and heard a diner ask, "What's rotation risotto?" The waiter fielded the question with a fine explanation, and then he added, "It's the plant equivalent of 'nose-to-tail' eating." The analogy struck me as exactly right. Nose-to-tail means cooking with the entire animal, not merely the most coveted parts; Rotation Risotto means cooking with the whole farm.

I can be certain that a version of Rotation Risotto will appear in the future, and that's not only because I feel confident about Klaas's rotations continuing, and continuing to evolve through the work of his son Peter, but also because Blue wheat has already introduced a whole new system of rotations to Stone Barns.

Having managed to persuade Craig to give up a full acre of pasture for the project, Jack mapped out a series of crops that would follow the wheat. He was thinking about the soil, of course, and how best to enrich it. I was thinking about my menu, and how many other new grains I could add into my Rotation Risotto.

"You might get a few, we'll see," Jack told me when I asked what to expect. "But you've got your head stuck in Klaas's system. He's a grain farm. We're not." Instead, Jack described a very different set of rotations. "After the wheat is harvested, we'll plant rye and a cover crop to rest the soil," he said. "Some may go to you, some may go to feed the pigs; since we took this land out of Craig's pasture rotations for a while, he'll have to benefit somehow. After that we'll follow with Mazourek's winter squash."

Jack was referring to the Cornell University breeder Michael Mazourek, who years earlier had admitted to me that no one ever asks him to breed for flavor. Upon our request, Michael had developed a super-sweet variety of winter squash, experimental line number 898, which Jack had been trialing for a few years. It looked like a shrunken butternut squash, with an intense sweetness and an almost pudding-like consistency.

I couldn't believe my luck—a full acre of it to work with. Grains are terrific, but this squash was like nothing I'd ever tasted, and from Jack's point of view, it would make a lot more money than grain. Even better, Michael assured us it would yield only slightly less per acre than conventional varieties of butternut squash. That's because this new variety, like those Mountain Magic tomatoes, was designed to produce good yields for the farmer.

One certainty for the menu of the future: chefs won't just celebrate heirlooms. In a half-century, as long as we embrace like-minded breeders, the idea that feeding a growing world requires sacrificing deliciousness for yield will be considered a silly, antiquated idea. So will the one-size-fits-all mentality of seed breeding. Working together, chefs and breeders can help ensure the production of flavorful and nutritious crops ideally suited to their localities.

Dishes like Rotation Risotto, with a puree of 898 squash swirled in, will help broadcast the message.

Third Course: *Grilled Crossabaw, Blood Sausage, and Pig-Bone Charcoal*

The future of nose-to-tail eating? Bone to blood. And a few decades from now you will have it at Blue Hill, because I'm predicting Stone Barns will one day raise more pigs.

That's an odd conviction in light of what Steve Jones said later during his visit to Stone Barns, and yet it's one I'm prepared to stand by. We had stopped at a group of Berkshire pigs hustling around the grain bin for their lunch. I mentioned that they'd provide enough manure to fertilize the land where we'd be growing wheat.

"A lot of growers think, *Well, I've got manure from the animals, so I've got plenty of nitrogen and phosphorus to fertilize my soil,*" Steve said, measuring his words carefully as he stood next to Jack. "But where's the manure coming from? It's from grain."

As he pointed out, we were still importing nutrients extracted from another farm in the form of grain feed in order to get the manure (and therefore the fertility) for our soil. "You're a bit like a mining operation," Steve said cheerfully. His point was an unconscious nod to Wes—how long could a system like this last?

But then, over the next several months, I started to see a different system take shape, and the culinary results were as exciting as the ecological impact. Craig, who may or may not have overheard Steve calling out our Achilles' heel, began removing the dead and unproductive trees from a small area of forest just behind the greenhouse. He used the pigs to clear the underbrush, rotating them around the land and planting grass seeds in their wake.

I happened to pass the area one afternoon the following spring, and what I saw amazed me. I swear it looked like a replica of the *dehesa*. There were no oaks—or rather, there were a lot of oaks, just not the manicured and prolific acorn-producing ones of Extremadura—but there was an abundance of grass and roots to forage, and the trees created the same savanna effect.

Craig's plan, unannounced but long in the making, was to replicate this *dehesa*-like system around the entire perimeter of the pasture. The goal was to reduce grain imports and ramp up the potential for homegrown forage.

"Sustainability by way of disturbance," Jill Isenbarger, the director of Stone Barns Center, told me one day about the project. Disturb the land by sculpting it to be more productive but also more in balance with what the environment could provide.

I thought back to Miguel's description of Veta la Palma as "a healthy artificial system." ("Yes, artificial," he said. "But what's natural anymore?") As he showed me, the best ecosystems do not preclude human intervention—in fact, they often depend on it, as long as people operate in service to the ecology. It seemed as though Jill had the right idea.

To this Craig added two more ideas, which is where the culinary benefits began to align with the environmental ones. Perhaps inspired by his mini-*dehesa*, or by the breeding genius of Steve, or both, Craig purchased an Ossabaw boar and had it sire two of his Berkshire sows. Direct descendants of the Iberian pigs of the *dehesa*, Ossabaws have the same barrel-like design, muscular legs, and long, pointed snouts, perfect for sniffing out acorns.

The resulting offspring—we nicknamed them "Crossabaws"—were, by far, the best pork I'd ever cooked, with a layer of outstanding fat that was so integrated throughout the muscle of the animal that it was hard not to think of Eduardo proudly holding that slice of *jamón* up to the light. It was also hard not to think of how many more Crossabaws Craig might raise for the restaurant in the next several decades, their flavor improving (to be honest, that *is* hard to imagine) as the land continues to thrive.

The second culinary benefit was a kind of gravy—not necessary, really, but I feel very lucky to have it as part of the pork plate. It started when Craig created a program to convert those dead and unproductive trees into charcoal, a nice benefit for Blue Hill's grilling station, which up until that point was using store-bought charcoal.

Inspired by the range of homemade wood charcoals, each one offering its own aroma, we got to thinking about carbonizing other things as well. It was the leftover bones from the pigs that first caught our attention. We'd always used the bones for broths and sauces at the restaurant, extracting the flavor and then tossing them in the trash. What if they were carbonized instead, just like wood? And what if, just as with wood, we could retain some of the flavor when we grilled? We succeeded in doing both. The grilled Crossabaw is seasoned in the cooking process by the pig-bone charcoal, and it's outstanding.

And the pigs' blood? Adam Kaye, our charcuterie maker, will use this other overlooked part of the animal for his version of a *boudin noir*, a traditional French blood sausage that often incorporates grains or discarded meat scraps. Adam's version, marrying superior technique and the miracles of coagulation, is pure blood. It's intense, a sausage with an attitude. And the next generation of eaters will be ready for it. Celebrating the whole animal will mean celebrating *every* part of the animal, blood and bones included.

FOURTH COURSE: *Trout with Phytoplankton*

Before writing this book, I would have predicted that a generation from now there would be only a sad selection of fish left for our menus. Having met Ángel, and seen his approach to the oceans' offerings, I'm more optimistic. His imaginative sourcing and cooking will inspire other chefs (who will inspire home cooks) to seek out and celebrate less well-known—and lower trophic level—alternatives.

One such alternative appeared last year, when Jack was exploring the forest acreage around Stone Barns. He stumbled upon an old trout ladder, a simple apparatus that uses the natural current to raise the fish upstream. The structure had been built in the 1940s to raise fish for the Rockefeller family.

Right now, almost all farm-raised trout in the United States comes from enormous aquaculture ponds, where the conditions are as questionable as the quality of the fish. But trout, fed the right diet and farmed in conditions that mimic nature, can be incredibly tasty.

Down the road, Jack hopes to supply Blue Hill with brook and rainbow trout, which will feed on compost worms that he is cultivating near the greenhouse. The worms will break down waste from the restaurant's vegetable scraps, providing a nice little sanitation service for the kitchen, and, at the end of the day, some farm-raised fillets of trout.

I imagine we'll serve the trout the same way Ángel might: in a pool of phytoplankton sauce. As luck would have it, a biology graduate student at Dartmouth recently heard of my interest in phytoplankton and offered to sell us his home brew. The first attempts were fine, nothing approaching what Ángel had introduced me to, but I have no doubt that in the next few decades, Ángel's new restaurant—with its fish farm and phytoplankton facility attached—will ignite even more worldwide interest in protecting and promoting "the origin of life."

It will be part of this menu of the future because, speaking to Ángel's point, without it there will be little else to serve.

FIFTH COURSE: *Parsnip Steak, Grass-Fed Beef*

We don't end tasting menus these days with a vegetable course, but a generation from now we probably will, or we'll come close, just as I predicted with the carrot steak.

As with everything on this menu of the future, the reason to turn away from a meat-centric plating strategy will be based on the demands of the ecology. If our menus are going to work in partnership with what the land can provide, vegetables and grains will inevitably take center stage.

My vision for a vegetable steak came to life recently when Jack harvested a crop of parsnips in the dead of winter. They'd been in the ground for nearly a year, about five months longer than required. The parsnips were as big as hulking T-Bones and, in the cold conditions, had converted nearly all their starches into sugars.

They were sweet enough to pass as dessert, but we showed off the bravura roots by roasting them like steaks and carving them for our diners tableside. In a few decades, we'll still be serving something like it—only this time with poached marrow, a bit of braised beef shank, and richly flavored Bordelaise sauce made with the bones.

The addition of the beef will be based on Craig's decision to raise beef cattle at Stone Barns. As the health of the system improves—as the grass becomes even more abundant because of the complexity of the rotations—there will be an opportunity to expand it even further by adding a new herbivore to follow the sheep.

But that expansion won't happen without the right kind of cooking to support it. So the lowly shank, a tough cut (to eat and to sell), will act as a kind of sauce for the parsnip. Sauce-making, like the kind Jean-Louis Palladin practiced, will have returned to prominence by then, a perfect way to create a deeply flavored, resonant dish out of these disparate and lowly cuts.

DESSERT: *Rice Pudding, Beer Ice Cream*

The other day I looked out the kitchen window and spotted Glenn Roberts standing in the courtyard. He was dressed in his usual white Polo and khakis, staring up at one of the stone buildings and smiling to himself.

It was a Sunday in late spring, and he and his wife, Kay, had spent the morning walking around Stone Barns. I got the sense that he was a little embarrassed not to be on his own farm, planting. When the conversation turned to Steve Jones and his new variety of Blue wheat, Glenn's eyes nearly popped out of his head.

Not to be outdone, he casually offered something himself. "How about getting Jack to plant some black rice?" he asked. Kay, knowing her husband, left us to talk.

"Rice?" I said. "I don't remember Jack having access to a paddy field."

"This is for dry rice farming," he said. I told him I didn't know there was such a thing. "Oh, hell yeah. You think they have flooded fields in the Himalayas?"

I almost told him I didn't know they *grew* rice in the Himalayas, but luckily Jack happened to pass by. Sensing an opportunity, I corralled them both into the restaurant's bar and asked Glenn to explain.

It turns out he had his hands on a rare class of black rice, an aromatic short grain that grows like a row crop in regular topsoil—no flooding required. He offered Jack the opportunity to be the first farmer in America to trial the seed, just as he had done with the Eight Row Flint corn after we first opened.

Jack agreed, but not before hesitating. He may have been hoping that

Glenn would offer a couple of thousand dollars to plant it, as he had done for the corn. Finally, he stuck out his hand. "I'm in," he said, and they shook on the deal.

Three weeks later I visited Jack in the greenhouse. The sprouted rice was already in seeding trays, awaiting transport to the soil. Jack was planning to plant it in the farthest corner of the vegetable field, in the same spot where Eliot Coleman had turned and raised his arm to the setting sun on that frigid November afternoon more than ten years ago.

In the midst of forecasting the future, it's worth noting that the present would have been difficult to envision even a decade ago. Would Eliot have predicted that this small farm could support not only the diversity of new breeds of animals and vegetables but also three of the world's major commodity grain crops—corn, wheat, and rice?

And yet these grains, and others, have become integral additions to the landscape. In the months after Jack decided to plant wheat and rice, we received three calls, in quick succession. One was from Philipsburg Manor, an eighteenth-century grain mill turned museum five miles from Stone Barns, offering to grind our wheat with their newly refurbished stone mill.

Another call came from a new malting company, requesting barley to malt for beer. Microbreweries are exploding in the Northeast—many of them driven by young entrepreneurs motivated to improve the quality of what we drink—and there is a severe local barley shortage. They wondered if Jack might consider adding barley into his wheat rotations.

And then a local microbrewery called to inquire about Jack selling some of his wheat for a new beer. They asked whether Craig would have interest in feeding his pigs spent wheat from the beer-making process.

I'm as confident about these opportunities materializing in one form or another as I am about Glenn's black rice finding itself in useful partnership with the grain rotations around the farm.

That's the day we'll marry the beer and the rice together in a dessert, if for no other reason than to celebrate these late-inning additions to our menu and the revitalized community of producers and distributors they've helped to build.

—⁓—

Sitting at the bar with Glenn and Jack that afternoon, I asked Glenn if he saw rice playing the role of an adopted Fourth Sister on the farm. Native American farmers had seen the wisdom in the companion planting of corn, squash, and beans. Could rice join into this symbiotic relationship?

"Oh hell, I don't know," he said, throwing up his arms. "But since there was never a Three Sisters, I doubt it." I figured he was joking, but when I looked over at Jack, he signaled agreement.

"What do you mean?" I asked, in a tone that must have sounded as if I had just been told Santa Claus doesn't exist.

"The idea for Three Sisters has been nice and distilled over time, passed down in a way that's easy for everyone to understand. It's fundamental, the architecture is there," Glenn said, "but it's so totally simplistic."

Jack punctuated Glenn's "so totally simplistic" with vigorous head nods. Looking at me he said, "What did you think? That Native Americans were growing things in row crops?" He shook his head. "No way. They were growing things everywhere. Nothing was disconnected from anything else. It was all one big farm."

—⁓—

In many ways the research for this book began with the opposite idea: take one great ingredient, and discover how it was grown or raised. Learn the recipe, I figured, and help make the harvests from the farmers that surround me more delicious and ecological. But the greatest lesson came with the realization that good food cannot be reduced to single ingredients. It requires a web of relationships to support it.

Aldo Leopold once wrote that the right kind of farming doesn't discard any of nature's component parts. "To keep every cog and wheel is the first precaution of intelligent tinkering." The right kind of cooking does the same thing: it promotes the vibrant communities, above and below ground, that make food delicious in the first place.

It's not unlike what I hoped we, as chefs, and as eaters, could do in imagining a Third Plate. The idea is to think not only about ingredient combinations but about *how* they're combined and how that reflects the larger picture.

Those connections begin before the plate and extend beyond it. They are more than the sum of their parts. They have the power to change culture and to shape landscapes.

Maybe what Jack said that day at the bar with Glenn is exactly the right way to think about a Third Plate for the future. Rooted in the natural world, it becomes a blueprint for *one big farm*—forever in flux, connected to a larger community, narrated by a cook through his food.

ACKNOWLEDGMENTS

This book began as much in the pasture fields of Blue Hill Farm as it did in the restaurant kitchens of Blue Hill. I have my grandmother, Ann Marlowe Strauss, to thank for that view, and for bringing the farm into existence for our family.

The other view of the farm was from the dinner table, and this book began there, too. It's where my aunt Tobe introduced me to French cooking—to scrambled eggs cooked over a double boiler—and where my uncle Steve, a gourmet and gourmand, shared the same fiery enthusiasm for a just-picked tomato as the Hellmann's mayonnaise he paired it with. He, above all, left me with the greatest gift a chef or a writer could ever own: an insatiable curiosity.

Along the way, a great many English teachers encouraged me to follow that curiosity and write. Two in particular inspired me to read. Thank you, Andrew Glassman, for making American nature writers like Thoreau and Emerson meaningful and accessible, and thank you, Sol Gittleman, for helping me see the genius of Philip Roth. (And, what the hell? Thank *you*, Philip Roth.)

If a writer needs something to write about, a chef needs someplace to cook. That is why the origins of this book can also be traced to the day my brother, David, insisted that my vague notion for a restaurant become a reality. He backed that up by becoming my business partner. Young cooks often ask for advice on starting their own restaurant. I tell them that as long as someone as brilliant as David is counting the coins, you'll be fine. I can practically guarantee success if you add a talent like my sister-in-law, Laureen. She is the eye of—and designer for—everything Blue Hill. For added insurance, my advice is to get a lawyer as loyal as my sister, Carolyn—who, with furrowed brow, has given me counsel, insight, and comfort my whole life.

I advise luck, too. In my case, there's been a lot of it. James Ford, tasked with finding a restaurant tenant for Stone Barns Center, introduced us to

David Rockefeller, Sr., and his daughter Peggy Dulany. They blessed the project and gave me a reason to research this book. In the years since, I've become indebted to the farmers at Stone Barns, who constantly rescue me from ignorance, and the entire staff, led by Jill Isenbarger, who motivate the Blue Hill team in more ways than I can name.

This book more or less officially began ten years ago, when Amanda Hesser suggested I write a monthly column in a magazine. To get the column approved, she asked for eight of the first twelve essays. Had her editor, the great Gerry Marzorati, not had the good judgment to reject those pieces (and the column itself), this book would not have happened.

Through all the morass of those early essays, David Black generously saw the potential for a book. D. Black is an agent, but he's also a rabbi, huggable henchmen, cheerleader, and coach. You want him in your dugout.

Ann Godoff took a risk and signed me on to her team. Ann: 127 years later, thank you. I'm grateful for your patience and straight talk, and for your keeping your hand on the wheel whenever I tried to steer the other way. And to the rest of the Penguin team, especially Ben Platt, Tracy Locke, and Sarah Hutson—thank you.

I will never fully understand why the farmers in this book were so generous with their time and so patient in explaining why they do what they do. One way to thank them, along with the chefs, ecologists, breeders, and scientists mentioned in these pages, is to honor their work through the writing of this book. I hope I have succeeded.

Behind the scenes, a great many people educated me along the way. Among the most influential have been Fred Kirschenmann, a farmer, scholar, and good friend, and Michael Pollan, who sets the standards for journalism and generosity, as well as many others, including: Ingrid Bengis, Bob Cannard, Betty Fussell, Thomas Harttung, Sam Kass, Ridge Shinn, Gary Nabhan, Marion Nestle, Bill Niman, Fred Magdoff, Autar Mattoo, Bill McKibben, Kathleen Merrigan, John Mishanec, Frank Morton, Joel Salatin, Eric Schlosser, Rick Schnieders, Gus Schumacher, and Sean Stanton.

If Lisa Abend didn't exist, I would have had to invent her. Much of this book takes place on two farms in Spain. I could never have gotten my foot inside the gates without her persistence. The endless hours of translation and historical background only add to my deepest sense of gratitude.

Charlotte Douglas has worked with me since I began writing this book. Her ear is unerring, her judgment unfailing, and her skill at politely amputating whole paragraphs—and her steadfast silence as I hurled obscenities and flailed about—is a kind of art form. She's read these pages as many times as I have. Every sentence is better because of her support and criticism.

I'm grateful to Sarah Bowlin, Mary Duenwald, Michael Gitter, Carol Hamburger, Liz Schaldenbrand, and Wendy Silbert for their critical reads along the way, and to David Shipley and Jacob Weisberg, two wise men who offered thoughtful and critical advice. Thank you.

Running a kitchen is like quarterbacking a football game, every night. No chef can survive, let alone prosper, without a phalanx of devoted cooks. They somehow push me over the goal line even when I'm more than ready to walk away. Thanks especially to chefs Adam Kaye, Trevor Kunk, Joel de la Cruz, and Michael Gallina—beside me at the stove for a combined forty-five years. Their care and attention are in everything good that I cook.

My colleagues at team Blue Hill include Franco Serafin, Philippe Gouze, Christine Langelier, Katie Bell, Michelle Biscieglia, Charles Puglia, Charlie Berg, Danielle Harrity, Peter Bradley, and John Jennings. These are, in restaurant terms, the "front of house," and they are the most generous people—and the finest professionals—I could ever know. Irene Hamburger (still with us sixteen years after signing on to help out for "a couple of weeks") deserves a lifetime of gratitude for her sustained support, not to mention the indefatigable attention to a thousand details.

When I first called my father after graduating from college to inform him of my decision to cook in a restaurant kitchen, he asked why. I told him I loved food. He said, "I love books, but I don't *read* for a living." He did, however, live to read, and while he died before I could finish this book—but not before encouraging my career at every turn—his passion for reading motivated me to write.

I've been told that my mother, who died when I was four, also loved books. I didn't learn until recently that she had once longed to be a writer. Is this book as much for her as anyone else? The answer is not in the book. It is the book.

A chef's life is not easy, undoubtedly most challenging for those living with the chef. Likewise, I've heard it's no cakewalk for those who live with writers. For not only enduring this double whammy but for being the infallible and loving editor of all that I cook and write (including these acknowledgments, which she made me redo), I am luckiest of all to have my wife, Aria, by my side. All that is joyful and lovely in my life is because of her, including our daughter, Edith.

NOTES

INTRODUCTION

10 **"die of its own too much"**: Aldo Leopold, "Wilderness" (1935), in *The River of the Mother of God and Other Essays*, ed. Susan L. Flader and J. Baird Callicott (Madison: University of Wisconsin Press, 1992), 228–9. Leopold borrowed the phrase from Shakespeare's *Hamlet*.

11 **"inescapably an agricultural act"**: Wendell Berry, "The Pleasures of Eating," in *What Are People For?* (New York: North Point Press, 1990), 149.

16 **took root in the philosophy of extraction**: For more on colonial American agriculture, see: Willard W. Cochrane, *The Development of American Agriculture: A Historical Analysis* (Minneapolis: University of Minnesota Press, 1979); Arnon Gutfeld, *American Exceptionalism: The Effects of Plenty on the American Experience* (Brighton, UK: Sussex Academic Press, 2002); and Steven Stoll, *Larding the Lean Earth: Soil and Society in Nineteenth Century America* (New York: Macmillan, 2003).

16 **American cooking was characterized, from the beginning**: See Harvey A. Levenstein, *Revolution at the Table: The Transformation of the American Diet* (New York: Oxford University Press, 1988), 7. Levenstein writes, "To nineteenth-century observers, the major differences between American and British diets could usually be summed up in one word: abundance. Virtually every foreign visitor who wrote about American eating habits expressed amazement, shock and even disgust at the quantity of food consumed." For more on early American cooking, see: James McWilliams, *A Revolution in Eating: How the Quest for Food Shaped America* (New York: Columbia University Press, 2005); Trudy Eden, *The Early American Table: Food and Society in the New World* (Dekalb: Northern Illinois University Press, 2008); Jennifer Wallach, *How America Eats: A Social History of U.S. Food and Culture* (Plymouth, UK: Rowman, 2013).

16 **"so much wasted from sheer ignorance, and spoiled by bad cooking"**: Juliet Corson, *The Cooking Manual of Practical Directions for Economical Every-Day Cookery* (New York: Dodd, Mead & Company, 1877), 5.

18 **"the attitude of the farmer"**: Lady Eve Balfour, quoted in Eliot Coleman, *Winter Harvest Handbook: Year-Round Vegetable Production Using Deep-Organic Techniques and Unheated Greenhouses* (White River Junction, VT: Chelsea Green Publishing, 2009), 204–5.

20 **"hitched to everything else"**: John Muir, *My First Summer in the Sierra* (1911; repr., Mineola, NY: Dover Publications, 2004), 87.

20 **the "culture" in agriculture**: See Wendell Berry, *The Unsettling of America: Culture & Agriculture* (1977; rev. ed., San Francisco: Sierra Club Books, 1996).

PART I: SOIL

27 **pounds of grain to produce one pound of beef:** See Erik Marcus, *Meat Market: Animals, Ethics, and Money* (Ithaca, NY: Brio Press, 2005), 187–8.

34 **"whereas the foundations provided":** Peter Thompson, *Seeds, Sex & Civilization: How the Hidden Life of Plants Has Shaped Our World* (London: Thames and Hudson, 2010), 31.

35 **we eat more wheat:** See "Wheat: Background," US Dept. of Agriculture, Economic Research Service, March 2009 briefing; and USDA, Office of Communications, *Agricultural Fact Book 2001–2002* (Washington, DC: Government Printing Office, 2003).

38 **"the conjugation of seemingly unrelated events":** Karen Hess, "A Century of Change in the American Loaf: Or, Where Are the Breads of Yesteryear" (keynote address at the History of American Bread symposium, Smithsonian Institution, Washington, DC, April 1994).

38 **The Spanish were the first to bring wheat:** See Charles Mann, *1493: Uncovering the New World Columbus Created* (New York: Knopf, 2011).

38 **one for every seven hundred Americans in 1840:** See Dean Herrin, *America Transformed: Engineering and Technology in the Nineteenth Century: Selections from the Historic American Engineering Record, National Park Service* (Reston, VA: American Society of Civil Engineers, 2002), 18.

39 **grown in every county in New York:** See Jared van Wagenen Jr., *The Golden Age of Homespun* (Ithaca, NY: Cornell University Press, 1953), 66; and Tracy Frisch, "A Short History of Wheat," *The Valley Table*, December 2008.

39 **Massachusetts "Red Lammas":** For more information on heritage New England wheats, see Eli Rogosa, "Restoring Our Heritage of Wheat" (working paper, Maine Organic Farmers and Gardeners Association, 2009).

40 **nutritional benefits of whole grains:** See David R. Jacobs and Lyn M. Steffen, "Nutrients, Foods, and Dietary Patterns as Exposures in Research: A Framework for Food Synergy," *American Journal of Clinical Nutrition* 78, no. 3 (September 2003): 508S–513S; and David R. Jacobs et al., "Food Synergy: An Operational Concept for Understanding Nutrition," *American Journal of Clinical Nutrition* 89, no. 5 (2009): 1543S–1548S.

41 **"native to its place":** See Wes Jackson, *Becoming Native to This Place* (Washington, DC: Counterpoint, 1994).

43 **Great American Desert:** See Walter Prescott Webb, *The Great Plains* (Waltham, MA: Ginn and Co., 1931), 152; and Henry Nash Smith, *Virgin Land: The American West as Symbol and Myth* (Cambridge, MA: Harvard University Press, 1950). In chapter 16, Smith has a good discussion of the prairie as both garden and desert in the American imagination.

45 **"Mistaking wisdom for backwardness":** Janine M. Benyus, *Biomimicry: Innovation Inspired by Nature* (New York: HarperCollins, 2009), 16.

45 **"the failure of success":** Jackson mentions this idea in his book *New Roots for Agriculture* (Lincoln: University of Nebraska Press, 1980).

45 **wave of settlement became a tsunami:** See Timothy Egan, *The Worst Hard Time: The Untold Story of Those Who Survived the Great American Dust Bowl* (New York: Houghton Mifflin Harcourt, 2006), 43.

46 **"The War integrated the plains farmers":** Donald Worster, *Under Western Skies: Nature and History in the American West* (Oxford, UK: Oxford University Press, 1992), 99.

46 **The soil . . . turned to dust:** For more on the Dust Bowl, see Donald E. Worster, *Dust Bowl: The Southern Plains in the 1930s* (New York: Oxford University Press, 1979); and Egan, *The Worst Hard Time.*

47 **"A cloud ten thousand feet high":** Egan, *The Worst Hard Time,* 113.

47 **"This, gentlemen, is what I'm talking about":** Egan, *The Worst Hard Time,* 227–8. See also Wellington Brink, *Big Hugh: The Father of Soil Conservation* (New York: Macmillan, 1951).

48 **"We came with visions, but not with sight":** Wendell Berry, "The Native Grasses and What They Mean," in *The Gift of Good Land: Further Essays Cultural and Agricultural* (New York: North Point Press, 1981), 82.

48 **"have disregarded every means":** George Washington, President of the United States, to Arthur Young, Esq., November 18, 1791, in *Letters on Agriculture from His Excellency, George Washington, President of the United States, to Arthur Young, Esq., F.R.S., and Sir John Sinclair, Bart., M.P.: With Statistical Tables and Remarks, by Thomas Jefferson, Richard Peters, and Other Gentlemen, on the Economy and Management of Farms in the United States,* ed. Franklin Knight (Washington, DC: Franklin Knight, 1847), 49–50.

49 **"presented the scares of fierce extraction":** Stoll, *Larding the Lean Earth,* 19.

49 **"a spanning of the scale of genetic possibilities from A to B":** Richard Manning, *Grassland: The History, Biology, Politics and Promise of the American Prairie* (New York: Penguin, 1997), 160.

50 **Wheat Belt is emptying out:** Wil S. Hylton, "Broken Heartland: The Looming Collapse of Agriculture on the Great Plains," *Harper's,* July 2012.

50 **enabling fewer farmers to farm even more land:** See William Lin et al., "U.S. Farm Numbers, Sizes, and Related Structural Dimensions: Projections to Year 2000," US Dept. of Agriculture, Economic Research Service, Technical Bulletin No. 1625 (1980).

51 **"what nature has made of us and we have made of nature":** Verlyn Klinkenborg, "Linking Twin Extinctions of Species and Languages," *Yale Environment 360,* July 17, 2012.

52 **Leopold asked the same question:** Aldo Leopold, "What Is a Weed?" (1943), in *River of the Mother of God: and Other Essays by Aldo Leopold,* 306–9.

54 **pests overtake its natural defenses:** See Philip S. Callahan, *Tuning in to Nature: Infrared Radiation and the Insect Communication System,* 2nd rev. ed. (Austin, TX: Acres U.S. A., 2001).

55 **"The organic farmer would look for the cause":** Eliot Coleman, "Can Organics Save the Family Farm?" *The Rake,* September 2004.

61 **supposedly "dumb beasts":** William A. Albrecht, *The Albrecht Papers, Volume I:*

Foundation Concepts, ed. Charles Walters, Jr. (Metairie, LA: Acres U.S.A., 1996), 279, 282.

61 **walked past what was commonly considered "good grass":** Charles Walters, Jr., "Foreword," *The Albrecht Papers, Volume I: Foundation Concepts*, x.

61 **"The cow is not classifying":** Albrecht, *The Albrecht Papers, Volume I: Foundation Concepts*, 170.

65 **compared a good organic farmer to a skilled rock climber:** See Coleman, *Winter Harvest Handbook*, 202.

69 **Soil is alive:** For more on the life of soil, see William Bryant Logan, *Dirt: The Ecstatic Skin of the Earth* (New York: W. W. Norton Limited, 2007); Fred Magdoff and Harold van Es, *Building Soils for Better Crops*, 3rd ed. (Waldorf, MD: SARE Outreach Publications, 2010); David Montgomery, *Dirt: The Erosion of Civilizations* (Berkeley: University of California Press, 2007); and David W. Wolfe, *Tales from the Underground: A Natural History of Subterranean Life* (New York: Basic Books, 2002).

69 **"not a thing but a performance":** Colin Tudge, *The Tree: A Natural History of What Trees Are, How They Live, and Why They Matter* (New York: Crown Publishers, 2006), 252.

70 **the Law of Return:** Sir Albert Howard, *The Soil and Health: A Study of Organic Agriculture* (1947; repr., Lexington: The University Press of Kentucky, 2007), 31.

71 **soil its "constitution":** William A. Albrecht, *The Albrecht Papers, Volume II: Soil Fertility and Animal Health*, ed. Charles Walters, Jr. (Kansas City, MO: Acres U.S.A., 1975), 101.

71 **soils stopped producing:** See Evan D. G. Fraser and Andrew Rimas, *Empires of Food: Feast, Famine, and the Rise and Fall of Civilizations* (New York: Atria Books, 2010).

72 **dowry measured by the amount of manure:** Logan, *Dirt: The Ecstatic Skin*, 38.

72 **"Now a farmer just had to mix the right chemicals into the dirt":** David Montgomery, *Dirt: The Erosion of Civilizations*, 184–5.

74 **"original sin" of agriculture:** Michael Pollan, *The Omnivore's Dilemma: A Natural History of Four Meals* (New York: Penguin Press, 2006), 258.

74 **In 1900, diversification:** See Bill Ganzel, "Shrinking Farm Numbers," in *Farming in the 1950s & 60s* (Wessels Loving History Farm, York, Nebraska, 2007), www.livin ghistoryfarm.org/farminginthe50s/life_11.html.

75 **three billion people depend on synthetic nitrogen:** Fred Pearce, "The Nitrogen Fix: Breaking a Costly Addiction," *Yale Environment 360*, November 5, 2009.

76 **"treating the whole problem":** Howard, *The Soil and Health*, 11.

76 **"learning more and more about less and less":** Howard, *The Soil and Health*, 250.

76 **"professors of agriculture":** Howard, *The Soil and Health*, 111.

76 **"tough, leathery and fibrous":** Sir Albert Howard, *An Agricultural Testament* (1940; repr., London: Benediction Classics, 2010), 82.

76 **"The maintenance of soil fertility is the real basis of health":** Howard, *Agricultural Testament*, 39.

77 **Artificial manures . . . "lead inevitably to artificial nutrition":** Howard, *Agricultural Testament*, 37.

81 **"innate tendency to focus on life and lifelike processes":** E. O. Wilson, *Biophilia* (Cambridge, MA: Harvard University Press, 1984), 1.

87 **more antioxidants and other defense-related compounds:** See Brian Halweil, "Still No Free Lunch: Nutrient Levels in U.S. Food Supply Eroded by Pursuit of High Yields" (Washington, DC: The Organic Center, September 2007), 33; and Charles M. Benbrook, "Elevating Antioxidant Levels in Food Through Organic Farming and Food Processing," Organic Center, State of Science Review, January 2005.

88 **"more complex than we can think":** Frank Egler, *The Nature of Vegetation: Its Management and Mismanagement* (Norfolk, CT: Aton Forest Publishers, 1977), 2.

89 **"Imagine a wonderfully balanced Italian main course":** Thomas Harttung, "Sustainable Food Systems for the 21st Century" (Agrarian Studies Lecture, Yale University, New Haven, CT, October 2006).

90 **"substitute a few soluble elements":** Coleman, *Winter Harvest Handbook*, 197.

92 **mycorrhizal fungus:** See David Wolfe, *Tales from the Underground*, and Albert Bernhard Frank, "On the Nutritional Dependence of Certain Trees on Root Symbiosis with Belowground Fungi (An English Translation of A. B. Frank's Classic Paper of 1885)," *Mycorrhiza* 15 (2005): 267–75.

92 **"subterranean-impaired":** Wolfe, *Tales from the Underground*, 6.

93 **"industrial organic . . . shallow organic":** Pollan, *Omnivore's Dilemma*, 130; Coleman, *Winter Harvest Handbook*, 205–7.

94 **"To be well fed is to be healthy":** Albrecht, *Soil Fertility and Animal Health*, 45.

94 **"You are what you eat eats, too":** Pollan, *Omnivore's Dilemma*, 84.

94 **correlation between recruits . . . and soils:** See Steve Soloman, *Gardening When It Counts: Growing Food in Hard Times* (Gabriola Island, BC: New Society Publishers, 2005), 19.

95 **nutrient declines . . . "biomass dilution":** See Donald R. Davis, "Declining Fruit and Vegetable Nutrient Composition: What Is the Evidence?" *HortScience* 12, no. 1 (February 2009); Donald R. Davis, "Trade-offs in Agriculture and Nutrition," *Food Technology* 59 (2005); Donald R. Davis et al., "Changes in the USDA Food Composition Data for 43 Garden Crops, 1950–1999," *Journal of the American College of Nutrition* 23, no. 6 (2004), 669–82; and David Thomas, "The Mineral Depletion of Foods Available to Us as a Nation (1940–2002)," *Nutrition and Health* 19 (2007): 21–55 (Thomas traces similar trends in the UK).

96 **840 million people suffer from chronic hunger:** "The State of Food Insecurity in the World 2013: The Multiple Dimensions of Food Security," Food and Agriculture Organization of the United Nations, Rome, 2013.

97 **"If we humans have this same basic tendency":** John Ikerd, "Healthy Soils, Healthy People: The Legacy of William Albrecht" (The William A. Albrecht Lecture, University of Missouri, Columbia, MO, April 25, 2011).

97 **"The sedentary lifestyles of many Americans":** Ikerd, "Healthy Soils, Healthy People."

99 **"to be well fed is to ward off the insects":** Albrecht, *The Albrecht Papers, Volume I: Foundation Concepts*, 304.

PART II: LAND

105 **equivalent to eating about forty-four pounds of pasta:** Lee Klein, "Foie Wars," *Miami New Times*, July 13, 2006.

108 **Palladin was smuggling:** See Stewart Lee Allen, *In the Devil's Garden: A Sinful History of Forbidden Food* (New York: Ballantine Books, 2002), 236.

109 **The French tradition of foie gras:** For more on the history of foie gras, see Mark Caro, *The Foie Gras Wars: How a 5,000-Year-Old Delicacy Inspired the World's Fiercest Food Fight* (New York: Simon & Schuster, 2009); Michael Ginor with Mitchell A. Davis, *Foie Gras: A Passion* (Hoboken, NJ: John Wiley & Sons, 1999); and Maguelonne Toussaint-Samat, *A History of Food* (Hoboken, NJ: John Wiley & Sons, 2009), 385–94.

110 **"The goose is nothing":** Charles Gérard, *L'Ancienne Alsace à Table: Étude Historique et Archéologique Sur L'Alimentation, Les Mœurs et Les Usages Épulaires De L'Ancienne Province D'Alsace*, 2nd ed. (Paris: Berger-Levrault et Cie, 1877). Quoted in Caro, *The Foie Gras Wars*, 35–6.

111 **thirty-five million Moulard ducks . . . eight hundred thousand geese:** Caro, *The Foie Gras Wars*, 33.

111 **"This cannot be called foie gras":** Graham Keeley, "French Are in a Flap as Spanish Force the Issue over Foie Gras," *The Guardian*, January 2, 2007.

113 **"Nothing is more stupid than a cow":** Alan Richman describes a similar diatribe in his article "A Very Unlikely Fish Story: Brother and Sister from Brittany Open Restaurant, Hook New York," *People*, August 4, 1986.

113 **"take half, leave half" rule of grazing:** Grass farmer Joel Salatin refers to this as the "law of the second bite." See Pollan, *Omnivore's Dilemma*, 189. In chapter 10, "Grass: Thirteen Ways of Looking at a Pasture," Pollan has an excellent discussion of grass farming and its history.

114 **"taste the misery":** Garrison Keillor, "Chicken," *Leaving Home: A Collection of Lake Wobegon Stories* (New York: Penguin Books, 1990), 45.

115 **"The challenge of cooking in America":** Eric Asimov, "Jean-Louis Palladin, 55, a French Chef with Verve, Dies," *New York Times*, November 26, 2001.

115 **"it trickles down to everybody":** Thomas Keller, quoted in Dorothy Gaiter and John Brecher, "The Genius That Was Jean-Louis," *France Magazine*, Winter 2011–12. For more on Jean-Louis Palladin's influence, see Justin Kennedy, "Raising the Stakes: Jean-Louis Palladin Pioneered Fine Dining in D.C.," *Edible DC*, Summer 2012.

116 **mimic what bison herds had been doing:** See Allan Savory with Jody Butterfield, *Holistic Management: A New Framework for Decision Making*, 2nd ed. (Washington, DC: Island Press, 1999).

137 **Restaurants, after all, are named:** Rebecca L. Spang, *The Invention of the Restaurant: Paris and Modern Gastronomic Culture* (Cambridge, MA: Harvard University Press, 2000), 1. For more on the history of restaurants, see Elliott Shore, "Dining Out: The Development of the Restaurant," in *Food: The History of Taste*, ed. Paul Freedman (Berkeley: University of California Press, 2007); and Adam Gopnik, *The Table Comes First: Family, France, and the Meaning of Food* (New York: Knopf, 2011).

137 **"locked up, cloistered in his smoke-filled basement"**: Paul Bocuse, quoted in Nicolas Chatenier, ed., *Mémoires de Chefs* (Paris: Textuel, 2012), 21 (translated from French).

138 **"a stifling, low-ceilinged inferno of a cellar"**: George Orwell, *Down and Out in Paris and London* (1933; rev. ed., New York: Mariner Books, 1972), 57.

138 *la nouvelle cuisine française*: For more on nouvelle cuisine, see Chatenier, ed., *Mémoires de Chefs*; Alain Drouard, "Chefs, Gourmets and Gourmands: French Cuisine in the 19th and 20th Centuries," in *Food: The History of Taste*, 301–31; and David Kamp, *The United States of Arugula: How We Became a Gourmet Nation* (New York: Broadway Books, 2006).

139 **"Down with the old-fashioned picture of the typical bon vivant"**: Julia Child, "La Nouvelle Cuisine: A Skeptic's View," *New York*, July 4, 1977.

141 **"extends on either side of the borders of simplicity and artifice"**: Paul Freedman, "Introduction: A New History of Cuisine," in *Food: The History of Taste*, 29.

145 **"the bigness of modern agriculture"**: Berry, *The Unsettling of America*, 61.

146 **ancient Egyptians observed how wild geese**: See Caro, *The Foie Gras Wars*, 24–7.

146 **The story of the chicken in this country**: See Steve Striffler, *Chicken: The Dangerous Transformation of America's Favorite Food* (New Haven, CT: Yale Agrarian Studies Series, 2007); Donald Stull and Michael Broadway, *Slaughterhouse Blues: The Meat and Poultry Industry in North America* (Stamford, CT: Cengage Learning, 2003); Roger Horowitz, *Putting Meat on the American Table: Taste, Technology, Transformation* (Baltimore: Johns Hopkins University Press, 2005); and Janet Raloff, "Dying Breeds," *Science News* 152, no. 14 (Oct. 4, 1997), 216–8.

147 **Mrs. Cecile Steele**: Donald Stull and Michael Broadway, *Slaughterhouse Blues*, 38.

148 **Arthur Perdue went into the poultry business**: See Melaine Warner, "Frank Perdue, 84, Chicken Merchant, Dies," *New York Times*, April 2, 2005.

148 **"The barnyard chicken was made over"**: Striffler, *Chicken: The Dangerous Transformation*, 46.

149 **first poultry company to differentiate its product with a label**: See Stull and Broadway, *Slaughterhouse Blues*, 47.

150 **"He had a weird authenticity"**: "Frank Perdue: 1920–2005," *People*, April 18, 2005.

150 **sales of chicken rose by nearly 50 percent**: See U.S. Environmental Protection Agency, "Poultry Production," *Ag 101*, www.epa.gov/agriculture/ag101/printpoultry.html.

150 **cost-effective way to feed the troops**: Horowitz, *Putting Meat on the American Table*, 119.

151 **about $18,500 per year**: See Jill Richardson, "How the Chicken Gets to Your Plate," La Vida Locavore, April 17, 2009, www.lavidalocavore.org/showDiary.do?dia ryId=1479.

153 **"protein paradox"**: Paul Roberts, *The End of Food* (New York: Houghton Mifflin Harcourt, 2009), 208–12.

153 **thirty-three minutes a day preparing food**: Karen Hamrick et al., "How Much Time Do Americans Spend on Food?" Economic Information Bulletin no. 86 (Washington, DC:

US Dept. of Agriculture, November 2011). See also: Michael Pollan, "Out of the Kitchen, Onto the Couch," *New York Times Magazine*, August 2, 2009.

154 **By the end of the 1990s those numbers had completely reversed:** See Striffler, *Chicken: The Dangerous Transformation*, 19.

155 **"It's easy to cook a filet mignon":** Thomas Keller, "The Importance of Offal," *The French Laundry Cookbook* (New York: Artisan Books, 1999), 209.

155 **tripled its production of chickens:** See Roberts, *End of Food*, 71.

156 **falsely low prices:** See "China Launches Anti-Dumping Probe into US Chicken Parts," *China Daily*, September 27, 2009; and Guy Chazan, "Russia, U.S. Are in a Chicken Fight, the First Round of New Trade War," *The Wall Street Journal*, March 4, 2002.

156 **Jalisco's poultry workers:** See Peter S. Goodman, "In Mexico, 'People Do Really Want to Stay,'" *Washington Post*, January 7, 2007.

163 *Jamón ibérico*'s **significance:** For this point and more on *jamón ibérico*, see Peter Kaminsky, *Pig Perfect: Encounters with Remarkable Swine and Some Great Ways to Cook Them* (New York: Hyperion, 2005), 66.

165 **The *dehesa* system originated:** See Vincent Clément, "Spanish Wood Pasture: Origin and Durability of an Historical Wooded Landscape in Mediterranean Europe," *Environment and History* 14, no. 1 (February 2008): 67–87.

166 **"Any person caught chopping down":** Quoted in Clément, "Spanish Wood Pasture."

175 **"immeasurable gift":** Wendell Berry, "The Agrarian Standard," *Orion Magazine*, Summer 2002.

178 **"The bottom layer is the soil":** Aldo Leopold, "The Land Ethic" (1948), in *A Sand County Almanac and Sketches Here and There, 2nd ed.* (New York: Oxford University Press, 1968), 215.

178 **"an extension of ethics":** Aldo Leopold, Foreword, *A Sand County Almanac and Sketches Here and There*, viii–ix.

179 **Extremaduran food is unadorned and simple:** For more on this cuisine, see Turespaña, "The Cuisine of Extremadura," www.spain.info/en_US/que-quieres/gastronomia/cocina-regional/extremadura/extremadura.html.

186 **"a fountain of energy flowing":** Aldo Leopold, "The Land Ethic," *A Sand County Almanac and Sketches Here and There*, 216.

PART III: SEA

208 **"like having a second tongue in my mouth":** Jeffrey Steingarten, *It Must Have Been Something I Ate* (2002; repr., New York: Vintage, 2003), 11–12.

210 **Atlantic tuna populations dropped by up to 90 percent:** See Carl Safina, *Song for the Blue Ocean: Encounters Along the World's Coasts and Beneath the Seas* (New York: Henry Holt and Co., 1998), 8.

211 **The article ran that fall:** Caroline Bates, "Sea Change," *Gourmet*, December 2005.

212 **"sea ethic":** Safina, *Song for the Blue Ocean*, 440.

212 **"honest inquiry into the reality of nature":** Ibid.

212 **"soft vessels of seawater":** Ibid., 435.

213 **"nothing we do seriously affects the number of fish":** Thomas Henry Huxley, "Inaugural Address" (Fisheries Exhibition, London, 1883).

213 **taking too many fish from the sea:** For information and statistics on the decline of world fish stocks, see FAO Fisheries and Aquaculture Department, *The State of World Fisheries and Aquaculture 2012* (Rome: Food and Agriculture Organization of the United Nations, 2012); and Wilf Swartz et al., "The Spatial Expansion and Ecological Footprint of Fisheries (1950 to Present)," *PLOS ONE* 5, no. 12 (December 2010).

214 **depleting certain populations of fish for ages:** See W. Jeffrey Bolster, *The Mortal Sea: Fishing the Atlantic in the Age of Sail* (Cambridge, MA: Belknap Press, 2012). Bolster identifies the degradation of some fisheries as far back as the Middle Ages.

214 **"too far" and "too deep":** Carl Safina and Carrie Brownstein, "Fish or Cut Bait: Solutions for Our Seas," in *Food and Fuel: Solutions for the Future*, ed. Andrew Heintzman and Evan Solomon (Toronto: House of Anansi Press, 2009), 75.

215 **"dragging a huge iron bar across the savannah":** Charles Clover, *The End of the Line: How Overfishing Is Changing the World and What We Eat* (Berkeley: University of California Press, 2008), 1.

215 **Estimates of "bycatch":** See Dayton L. Alverson et al., *A Global Assessment of Fisheries Bycatch and Discards*, FAO Fisheries Technical Paper no. 339 (Rome: Food and Agriculture Organization of the United Nations, 1994).

216 **dead zones worldwide:** See R. J. Diaz and R. Rosenberg, "Spreading Dead Zones and Consequences for Marine Ecosystems," *Science* 321, no. 5891 (August 15, 2008): 926–9.

216 **"decomposing bodies lying in sediment":** Nancy Rabalais, quoted in Allison Aubrey, "Troubled Seas: Farm Belt Runoff Prime Source of Ocean Pollution," *Morning Edition*, National Public Radio, January 15, 2002.

217 **food web begins with phytoplankton:** For more on the role of phytoplankton, see Sanjida O'Connell, "The Science Behind That Fresh Seaside Smell," *The Telegraph*, August 18, 2009; I. Emma Huertas et al., "Warming Will Affect Phytoplankton Differently: Evidence Through a Mechanistic Approach," *Proceedings of the Royal Society B—Biological Sciences* 278, no. 1724 (2011): 3534–43; and John Roach, "Source of Half Earth's Oxygen Gets Little Credit," *National Geographic News*, June 7, 2004, http://news.nationalgeographic.com/news/2004/06/0607_040607_phytoplankton.html.

217 **decline in phytoplankton:** See Daniel G. Boyce, Marlon R. Lewis, and Boris Worm, "Global Phytoplankton Decline over the Past Century," *Nature* 466, no. 7306 (July 29, 2010): 591–6.

217 **El Niño climate cycles:** See Mike Bettwy, "El Niño and La Niña Mix Up Plankton Populations," NASA, June 22, 2005, www.nasa.gov/vision/earth/lookingatearth/plankton_elnino.html.

218 **one-third of the seafood . . . ordered in restaurants:** See *The Marketplace for Sustainable*

Seafood: Growing Appetites and Shrinking Seas (Washington, DC: Seafood Choices Alliance, 2003), 9.

218 **Rising CO2 levels:** Bärbel Hönisch et al., "The Geological Record of Ocean Acidification," *Science* 335, no. 6072 (March 2012): 1058–63.

219 **rise in the trophic levels of the fish used in recipes:** See Phillip S. Levin and Aaron Dufault, "Eating up the Food Web," *Fish and Fisheries* 11, issue 3 (September 2010): 307–12.

219 **"Does this matter?":** Clover, *End of the Line*, 189.

235 **The business of fish farming:** For statistics on aquaculture, see FAO Fisheries and Aquaculture Department, *The State of World Fisheries and Aquaculture 2012*; and R. L. Naylor et al., "Effects of Aquaculture on World Fish Supplies," *Nature* 405 (2000): 1017–24.

236 **substituting grains and oilseeds:** See Emiko Terazono, "Salmon Farmers Go for Veggie Option," *Financial Times*, January 21, 2013.

238 **Veta la Palma was born:** See J. Miguel Medialdea, "A New Approach to Ecological Sustainability Through Extensive Aquaculture: The Model of Veta la Palma," Proceedings of the 2008 TIES Workshop, Madison, Wisconsin; and J. Miguel Medialdea, "A New Approach to Sustainable Aquaculture," *The Solutions Journal*, June 2010.

239 **"the primeval meeting place":** Rachel Carson, *The Edge of the Sea* (1955; repr., New York: Mariner Books, 1998), xiii.

240 **writing "the wrong kind of book":** Sue Hubbell, introduction to *The Edge of the Sea*, xvi–xviii.

245 **most important private estate for aquatic birds in all of Europe:** Carlos Otero and Tony Bailey, *Europe's Natural and Cultural Heritage: The European Estate* (Brussels: Friends of the Countryside, 2003), 701.

247 **"A gastronome who is not an environmentalist is stupid":** Carlo Petrini, Report from the European Conference on Local and Regional Food, Lerum, Sweden, September 2005.

248 **bird populations . . . have decreased:** See Robin McKie, "How EU Farming Policies Led to a Collapse in Europe's Bird Population," *The Observer*, May 26, 2012.

248 **seabird populations:** See Jeremy Hance, "Easing the Collateral Damage That Fisheries Inflict on Seabirds," *Yale Environment 360*, August 9, 2012.

248 **long before industrialized agriculture:** See Christopher Cokinos, *Hope Is the Thing with Feathers: A Personal Chronicle of Vanished Birds* (New York: Penguin 2009), 53. As Cokinos observes, "Prehistoric islanders in the Pacific killed off some 2,000 bird species, diminishing by one-fifth the global number through a variety of activities, including habitat destruction."

248 **"the world has lost at least eighty species":** Colin Tudge, *The Bird: A Natural History of Who Birds Are, Where They Came From, and How They Live* (New York: Random House, 2010), 400.

249 **"a knock-on effect":** Alasdair Fotheringham, "Is This the End of Migration?" *The Independent*, April 18, 2010.

249 **"The birds of Walden":** Jonathan Rosen, *The Life of the Skies: Birding at the End of Nature* (New York: Picador, 2008), 94.

253 **fifteen pounds per person:** Alan Lowther, ed., *Fisheries of the United States 2011* (Silver Spring, MD: National Oceanic and Atmospheric Administration, 2012).

265 **"Gravity is the sea's enemy":** Carl Safina, "Cry of the Ancient Mariner: Even in the Middle of the Deep Blue Sea, the Albatross Feels the Hard Hand of Humanity," *Time*, April 26, 2000.

266 **longtime advocate of traditional diets:** Sally Fallon Morell, "Very Small Is Beautiful" (lecture, Twenty-eighth Annual E. F. Schumacher Lectures, New Economics Institute, Stockbridge, MA, October 2008).

267 **Cowan spent twenty years contemplating the question:** See Thomas Cowan, *The Fourfold Path to Healing: Working with the Laws of Nutrition, Therapeutics, Movement and Meditation in the Art of Medicine* (Washington, DC: Newtrends Publishing, 2004). Cowan discusses Steiner's understanding of the heart in chapter 3.

267 **"the heart as a pump":** Rudolf Steiner, "Organic Processes and Soul Life" (1921), in *Freud, Jung, and Spiritual Psychology*, 3rd ed. (Great Barrington, MA: Anthroposophic Press, 2001), 124–5.

267 **"It is the blood that drives the heart":** Rudolf Steiner, "The Question of Food" (1913), in *The Effects of Esoteric Development: Lecures by Rudolf Steiner* (Hudson, NY: Anthroposophic Press, 1997), 56.

268 **scientific revolution . . . masters and possessors of nature:** Frederick Kirschenmann, "Spirituality in Agriculture" (academic paper, Concord School of Philosophy, Concord, MA, October 8, 2005).

272 **"fishing down the food chain":** For more on this idea, see Taras Grescoe, *Bottomfeeder: How to Eat Ethically in a World of Vanishing Seafood* (New York: Bloomsbury, 2008).

296 **"International Conspiracy to Catch All Tuna":** Safina, *Song for the Blue Ocean*, 13.

307 **"reminding one of barn-yard fowls feeding from a dish":** Alan Davidson, *North Atlantic Seafood: A Comprehensive Guide with Recipes* (New York: Ten Speed Press, 2003), 115.

PART IV: SEED

326 **"a town with a bombed out center":** Jim Hinch, "Medium-Size Me," *Gastronomica: The Journal of Food and Culture* 8, no. 4 (Fall 2008), 72.

330 **in another decade most of them will be gone:** For more on midsize farms, see Fred Kirschenmann et al., "Why Worry About the Agriculture of the Middle? A White Paper for the Agriculture of the Middle Project" (n.d.), http://grist.files.wordpress.com/2011/03/whitepaper2.pdf.

339 **nothing "intrinsically sweet" about sugar:** Daniel C. Dennett, *Breaking the Spell: Religion as a Natural Phenomenon* (New York: Viking, 2006), 59.

339 **refined wheat in sociocultural terms:** For more on the sociocultural history of white

bread, see Aaron Bobrow-Strain, *White Bread: A Social History of the Store-Bought Loaf* (Boston: Beacon Press, 2012); H. E. Jacob and Peter Reinhart, *Six Thousand Years of Bread* (New York: Skyhorse Publishing, 2007); Steven Laurence Kaplan, *Good Bread Is Back: A Contemporary History of French Bread, the Way It Is Made, and the People Who Make It* (Durham, NC: Duke University Press, 2006); Harold McGee, *On Food and Cooking: The Science and Lore of the Kitchen* (New York: Scribner, 2004); Michael Pollan, *Cooked: The Natural History of Transformation* (New York: The Penguin Press, 2013); and William Rubel, *Bread: A Global History* (London: Reaktion Books, 2011).

340 **"The bread is as soft as floss":** Theodore Roszak, *The Making of a Counter Culture: Reflections on the Technocratic Society and Its Youthful Opposition* (Berkeley: University of California Press, 1969), 13.

340 **The countercuisine movement:** See Warren J. Belasco, *Appetite for Change: How the Counterculture Took on the Food Industry* (Ithaca, NY: Cornell University Press, 2006), 46–50.

341 **"The worst loaf of bread":** Jeffrey Steingarten, "The Whole Truth: Jeffrey Steingarten Searches for Grains That Taste as Good as They Are Good for You," *Vogue*, November 2005.

345 **rice kitchen:** See Karen Hess, *The Carolina Rice Kitchen: The African Connection* (Columbia: University of South Carolina Press, 1992), 3.

346 **agriculture in the South became largely experimental:** For information on the South's age of experimental agriculture, see Burkhard Bilger, "True Grits," *The New Yorker*, October 31, 2011, 40–53; Interview with Glenn Roberts, "Old School," *Common-place* 11, no. 3 (April 2011); David Shields, "The Roots of Taste," *Common-place* 11, no. 3 (April 2011); and David Shields, ed. *The Golden Seed: Writings on the History and Culture of Carolina Gold Rice* (Charleston: The Carolina Gold Rice Foundation, 2010).

352 **Carolina Gold was exported:** See Hess, *The Carolina Rice Kitchen*, 20; and Richard Schulze, *Carolina Gold Rice: The Ebb and Flow History of a Lowcountry Cash Crop* (Charleston, SC: The History Press, 2005).

360 **plant breeders discovered a way to farm more efficiently:** For more on the history of plant breeding, see Noel Kingsbury, *Hybrid: The History and Science of Plant Breeding* (Chicago: University of Chicago Press, 2009); Jonathan Silvertown, *An Orchard Invisible: A Natural History of Seeds* (Chicago: University of Chicago Press, 2009); and Jack R. Kloppenburg, *First the Seed: The Political Economy of Biotechnology*, 2nd ed. (Madison: The University of Wisconsin Press, 2004).

362 **the Green Revolution:** See Susan Dworkin, *The Viking in the Wheat Field: A Scientist's Struggle to Preserve the World's Harvest* (New York: Walker & Company, 2009); Cary Fowler and Patrick Mooney, *Shattering: Food, Politics, and the Loss of Genetic Diversity* (Tucson: University of Arizona Press, 1990); Richard Manning, *Against the Grain: How Agriculture Has Hijacked Civilization* (New York: North Point Press, 2004); Peter Thompson, *Seeds, Sex & Civilization*; and Roberts, *The End of Food*.

363 **Borlaug began growing new semidwarf crosses:** See Gregg Easterbrook, "Forgotten

Benefactor of Humanity," *The Atlantic Monthly*, January 1, 1997; and Henry W. Kindall and David Pimentel, "Constraints on the Expansion of the Global Food Supply," *Ambio* 23, no. 3 (May 1994).

363 **Bourlag next sent his dwarf wheat to India:** Roberts, *The End of Food*, 148–9.

364 **From 1950 to 1992, harvests increased:** Easterbrook, "Forgotten Benefactor of Humanity."

364 **more than 70 percent of the wheat grown in the developing world:** See Maximina A. Lantican et al., "Impacts of International Wheat Breeding Research in the Developing World, 1988–2002," Impact Studies 7654 (Mexico City: International Maize and Wheat Improvement Center [CIMMYT], 2005), 30.

364 **global increase in diet-related diseases:** See Knut Schroeder et al., *Sustainable Healthcare* (Chichester, West Sussex: John Wiley & Sons, 2013).

365 **From 1950 to 2000, the amount of irrigated farmland tripled:** See Lester Brown, *Plan B 4.0: Mobilizing to Save Civilization* (New York: W. W. Norton & Company, 2009); and Sandra Postel, *Pillar of Sand: Can the Irrigation Miracle Last?* (New York: W. W. Norton & Company, 1999).

365 **"Although high-yielding varieties":** Vandana Shiva, "The Green Revolution in the Punjab," *The Ecologist* 21, no. 2 (March–April 1991).

365 **"akin to the relationship of the chicken and the egg":** Fowler and Mooney, *Shattering*, 60.

365 **synthetic fertilizers . . . not exactly green:** See Donald L. Plucknett, "Saving Lives Through Agricultural Research," Issues in Agriculture no. 1 (Washington, DC: Consultative Group on International Agricultural Research, May 1991).

365 **more chemicals are needed to get the same kick:** Stuart Laidlaw, "Saving Agriculture from Itself," in *Food and Fuel: Solutions for the Future*, 10–11. Laidlaw writes, "Decades of monoculture had robbed the soil of its nutrients so that it now needed regular nitrogen applications to keep productive. Nitrogen also increases soil acidity, which slows biologic activity, hurting the soil's ability to produce food on its own, so even more nitrogen must yet again be applied. The land, in short, is addicted to nitrogen."

366 **"They're looking at the swollen belly":** Interview with Susan Dworkin, *Acres U.S.A.*, February 2010.

367 **"so-called miracle varieties":** Vandana Shiva, *Stolen Harvest: The Hijacking of the Global Food Supply* (Cambridge, MA: South End Press, 2000), 12.

368 **achieved with old-world farming techniques:** See Colin Tudge, *Feeding People Is Easy* (Grosseto, Italy: Pari Publishing, 2007), 75–6.

368 **"the result of intelligent, innovative minds":** Fowler and Mooney, *Shattering*, 139. For more on Vavilov, see Gary Paul Nabhan, *Where Our Food Comes From: Retracing Nikolay Vavilov's Quest to End Famine* (Washington, DC: Island Press, 2009).

368 **peasant farmers working with nature:** See Shiva, *Stolen Harvest*, 79. Shiva cites a few remarkable examples: "Indian farmers have evolved thousands of varieties of rice.

Andean farmers have bred more than 3,000 varieties of potatoes. In Papua New Guinea, more than 5,000 varieties of sweet potatoes are cultivated."

384 **developed and trialed by land-grant university plant breeders:** The Mountain Magic tomato was developed by Dr. Randy Gardner at North Carolina State University's Mountain Horticultural Crops Research and Extension Center (hence the "Mountain" in its name).

385 **Genetically modified foods:** For more on the controversy surrounding genetically modified foods, see Daniel Charles, *Lords of the Harvest: Biotech, Big Money, and the Future of Food* (Cambridge, MA: Perseus Publishing, 2001); Brian J. Ford, *The Future of Food: Prospects for Tomorrow* (London: Thames & Hudson, 2000); Craig Holdrege and Steve Talbott, *Beyond Biotechnology: The Barren Promise of Genetic Engineering* (Lexington: University Press of Kentucky, 2008); Peter Pringle, *Food, Inc.: Mendel to Monsanto—The Promises and Perils of the Biotech Harvest* (New York: Simon & Schuster, 2003); Pamela C. Ronald and Raoul W. Adamchak, *Tomorrow's Table: Organic Farming, Genetics, and the Future of Food* (New York: Oxford University Press, 2008); and Josh Schonwald, *The Taste of Tomorrow: Dispatches from the Future of Food* (New York: HarperCollins, 2012).

386 **land-grant colleges:** For more on land-grant institutions, see Jim Hightower, *Hard Tomatoes, Hard Times* (Cambridge, MA: Schenkman Publishing Company, 1973); George R. McDowell, *Land-Grant Universities and Extension into the 21st Century: Renegotiating or Abandoning a Social Contract* (Ames: Iowa State Press, 2001); and Roger L. Geiger and Nathan M. Sorber, eds., *The Land-Grant Colleges and the Reshaping of American Higher Education* (New Brunswick, NJ: Transaction Publishers, 2013).

396 **"prefer to talk in terms of a 'division of labor'":** Fowler and Mooney, *Shattering*, 138.

396 **funding of agricultural research:** See Food and Water Watch, "Public Research, Private Gain: Corporate Influence on University Agricultural Research" (Washington, DC: Food and Water Watch, April 2012); P. W. Heisey et al., *Public Sector Plant Breeding in a Privatizing World* (Washington, DC: US Dept. of Agriculture, Economic Research Service, 2001); and Jorge Fernandez-Cornejo, "The Seed Industry in U.S. Agriculture: An Exploration of Data and Information on Crop Seed Markets, Regulation, Industry Structure, and Research and Development," US Dept. of Agriculture, Economic Research Service, Agriculture Information Bulletin No. 786 (2004).

403 **older varieties contained more micronutrients than newer breeds:** See Kevin M. Murphy, Philip G. Reeves, and Stephen S. Jones, "Relationship Between Yield and Mineral Nutrient Content in Historical and Modern Spring Wheat Cultivars," *Euphytica* 163, issue 3 (October 2008): 381–90.

411 **"I think that the bread community":** Gabe Ulla, "Pizzaiolo Jim Lahey on Fire, Craft, and Tactile Pleasure," Eater Online, May 8, 2012, http://eater.com/archives/2012/05/08/pizzaiolo-jim-lahey-on-fire-craft-and-tactile-pleasure.php#more.

412 **"midway between youth and age":** George Bernard Shaw, *Too True to Be Good* (New York: Samuel French Inc., 1956), 118.

420 **"his eyes lit up with delight"**: Anka Muhlstein, *Balzac's Omelette: A Delicious Tour of French Food and Culture with Honoré de Balzac* (New York: Other Press, 2011), 7.

EPILOGUE

447 **"keep every cog and wheel"**: Aldo Leopold, "Conservation," in *Round River: From the Journals of Aldo Leopold* (1953; repr., New York: Oxford University Press, 1993), 147.

FURTHER READING

SOIL

Ausubel, Kenny, with J. P. Harpignies, ed., *Nature's Operating Instructions: The True Biotechnologies* (San Francisco, Sierra Club Books, 2004).

Balfour, Lady Eve, *The Living Soil* (London: Faber and Faber, 1943).

Buhner, Stephen Harrod, *The Lost Language of Plants: The Ecological Importance of Plant Medicines for Life on Earth* (White River Junction, VT: Chelsea Green Publishing, 2002).

Carson, Rachel, *Silent Spring* (New York: Houghton Mifflin, 1962).

Coleman, Eliot, *The New Organic Grower: A Master's Manual of Tools and Techniques for the Home and Market Gardener* (White River Junction, VT: Chelsea Green, 1989).

Fromartz, Samuel, *Organic, Inc.: Natural Foods and How They Grew* (Orlando, FL: Harcourt, 2006).

Gershuny, Grace, and Joseph Smillie, *The Soul of the Soil: A Guide to Ecological Soil Management*, 3rd ed. (Davis, CA: agAccess, 1995).

Graham, Michael, *Soil and Sense* (London: Faber & Faber, 1941).

Holthaus, Gary, *From the Farm to the Table: What All Americans Need to Know About Agriculture* (Lexington, KY: University Press of Kentucky, 2006).

Jackson, Wes, *Consulting the Genius of the Place: An Ecological Approach to a New Agriculture* (Berkeley: Counterpoint, 2010).

Mabey, Richard, *Weeds: In Defense of Nature's Most Unloved Plants* (New York: HarperCollins, 2010).

Morton, Oliver, *Eating the Sun: How Plants Power the Planet* (New York: HarperCollins, 2009).

Robinson, Raoul A., *Return to Resistance: Breeding Crops to Reduce Pesticide Dependence* (Davis, CA: agAccess, 1996).

Stoll, Steven, *The Fruits of Natural Advantage: Making the Industrial Countryside in California* (Berkeley: University of California Press, 1998).

Sykes, Friend, *Food, Farming and the Future* (Emmaus, PA: Rodale, 1951).

Tompkins, Peter, and Christopher Bird, *The Secret Life of Plants: A Fascinating Account of the Physical, Emotional, and Spiritual Relations Between Plants and Man* (1973; repr., New York: Harper Perennial, 1989).

Voisin, André, *Soil, Grass, and Cancer: The Link Between Human and Animal Health and the Mineral Balance of the Soil* (New York: Philosophical Library, 1959).

———, *Grass Productivity* (New York: Philosophical Library, 1959).

Walters, Charles, *Weeds: Control Without Poisons* (Kansas City: Acres U.S.A., 1991).

Wedin, Walter F., and Steven L. Fales, *Grassland: Quietness and Strength for a New American Agriculture* (Madison, WI: American Society of Agronomy, 2009).

Willis, Harold, *Foundations of Natural Farming: Understanding Core Concepts of Ecological Agriculture* (Austin, TX: Acres USA, 2007).

LAND

Fussell, Betty, *Raising Steaks: The Life and Times of American Beef* (Orlando, FL: Harcourt, 2008).

Imhoff, Daniel, ed., *The CAFO Reader: The Tragedy of Industrial Animal Factories* (Berkeley: Watershed Media, 2010).

Lappé, Frances Moore, *Diet for a Small Planet* (1971; repr., New York: Ballantine Books, 1991).

Nierenberg, Danielle, *Happier Meals: Rethinking the Global Meat Industry* (Washington, D.C.: Worldwatch Institute, 2005).

Robinson, Jo, *Pasture Perfect: How You Can Benefit from Choosing Meat, Eggs, and Dairy Products from Grass-Fed Animals* (Vashon, WA: Vashon Island Press, 2004).

Schlosser, Eric, *Fast Food Nation* (Boston: Houghton Mifflin, 2001).

Sinclair, Upton, *The Jungle* (1906; repr., London: Penguin, 1985).

SEA

Bowermaster, Jon, ed., *Oceans: The Threats to Our Seas and What You Can Do to Turn the Tide* (New York: PublicAffairs, 2010).

Carson, Rachel, *The Sea Around Us* (New York: Oxford University Press, 1951).

Danson, Ted, *Oceana: Our Endangered Oceans and What We Can Do to Save Them* (Emmaus, PA: Rodale Books, 2011).

Ellis, Richard, *The Empty Ocean* (Washington, D.C.: Island Press, 2003).

Greenberg, Paul, *Four Fish: The Future of the Last Wild Food* (New York: Penguin Press, 2010).

Jacobsen, Rowan, *The Living Shore: Rediscovering a Lost World* (New York: Bloomsbury, 2009).

Molyneaux, Paul, *Swimming in Circles: Aquaculture and the End of Wild Oceans* (New York: Thunder's Mouth Press, 2007).

Whitty, Julia, *The Fragile Edge: Diving and Other Adventures in the South Pacific* (New York: Houghton Mifflin, 2007).

SEED

Brown, Lester, *Full Planets, Empty Plates: The New Geopolitics of Food Scarcity* (New York: W. W. Norton, 2012).

Conway, Gordon, *The Doubly Green Revolution: Food for All in the Twenty-first Century* (London: Penguin Books, 1997).

Cribb, Julian, *The Coming Famine: The Global Food Crisis and What We Can Do to Avoid It* (Berkeley: University of California Press, 2010).

Eldredge, Niles, *Life in the Balance: Humanity and the Biodiversity Crisis* (Princeton, NJ: Princeton University Press, 1998).

Kunstler, James Howard, *The Long Emergency: Surviving the End of Oil, Climate Change, and Other Converging Catastrophes of the Twenty-first Century* (New York: Atlantic Monthly Press, 2005).

Manning, Richard, *Food's Frontier: The Next Green Revolution* (New York: North Point Press, 2000).

Nabhan, Gary Paul, *Coming Home to Eat: The Pleasures and Politics of Local Food* (New York: W. W. Norton, 2002).

———, *Why Some Like It Hot: Food, Genes, and Cultural Diversity* (Washington, D.C.: Island Press, 2006).

Pfeiffer, Dale Allen, *Eating Fossil Fuels: Oil, Food and the Coming Crisis in Agriculture* (Gabriola Island, BC: New Society Publishers, 2006).

Ruffin, Edmund, *Nature's Management: Writings on Landscape and Reform, 1822–1859*, Jack Temple Kirby, ed. (Athens, GA: University of Georgia Press, 2000).

Solbrig, Otto, and Dorothy Solbrig, *So Shall You Reap: Farming and Crops in Human Affairs* (Washington, DC: Island Press, 1994).

GENERAL FURTHER READING

Ackerman-Leist, Philip, *Rebuilding the Foodshed: How to Create Local, Sustainable, and Secure Food Systems* (White River Junction, VT: Chelsea Green Publishing, 2013).

Berry, Wendell, *The Gift of Good Land: Further Essays Cultural and Agricultural* (San Francisco: North Point Press, 1981).

Capra, Fritjof, *The Hidden Connections: Integrating the Biological, Cognitive, and Social Dimensions of Life into a Science of Substainability* (New York: Doubleday, 2002).

———, *The Web of Life* (New York: Anchor Books, 1996).

Diamond, Jared, *Guns, Germs, and Steel: The Fates of Human Societies* (New York: W. W. Norton, 1997).

Dubos, René, *The Wooing of the Earth: New Perspectives on Man's Use of Nature* (New York: Charles Scribner's Sons, 1980).

Dumanoski, Dianne, *The End of the Long Summer: Why We Must Remake Our Civilization to Survive on a Volatile Earth* (New York: Three River Press, 2009).

Fraser, Caroline, *Rewilding the World: Dispatches from the Conservation Revolution* (New York: Metropolitan Books, 2009).

Freidberg, Susanne, *Fresh: A Perishable History* (Cambridge, MA: Harvard University Press, 2009).

Goleman, Daniel, *Ecological Intelligence: How Knowing the Hidden Impacts of What We Buy Can Change Everything* (New York: Broadway Books, 2009).

Halweil, Brian, *Eat Here: Reclaiming Homegrown Pleasures in a Global Supermarket* (New York: W. W. Norton, 2004).

Jackson, Dana L., and Laura L. Jackson, ed., *The Farm as Natural Habitat Reconnecting Food Systems with Ecosystems* (Washington, D.C.: Island Press, 2002).

Jackson, Louise E., ed., *Ecology in Agriculture* (San Diego: Academic Press, 1997).

Kirschenmann, Frederick, *Cultivating an Ecological Conscience: Essays from a Farmer Philosopher* (Lexington, KY: University Press of Kentucky, 2010).

Lopez, Barry, ed., *The Future of Nature: Writing on a Human Ecology from Orion Magazine* (Minneapolis, MN: Milkweed Editions, 2007).

McKibben, Bill, *The End of Nature* (New York: Anchor Books, 1989).

McNeely, Jeffrey A., and Sara J. Scherr, *Ecoagriculture: Strategies to Feed the World and Save Wild Biodiversity* (Washington, D.C.: Island Press, 2003).

Meine, Curt, *Aldo Leopold: His Life and Work* (Madison, WI: University of Wisconsin Press, 1988).

Patel, Raj, *Stuffed and Starved: The Hidden Battle for the World Food System* (London: Portobello Books, 2007).

Smith, J. Russell, *Tree Crops: A Permanent Agriculture* (New York: Harcourt, Brace and Company, 1929).

Sokolov, Raymond, *Why We Eat What We Eat: How Columbus Changed the Way the World Eats* (New York: Touchstone, 1991).

Soule, Judith, and Jon Piper, *Farming in Nature's Image: An Ecological Approach to Agriculture* (Washington, D.C.: Island Press, 2009).

Stuart, Tristram, *Waste: Uncovering the Global Food Scandal* (New York: W. W. Norton, 2009).

Tannahill, Reay, *Food in History* (New York: Stein and Day Publishers, 1973).

Taubes, Gary, *Good Calories, Bad Calories: Fats, Carbs, and the Controversial Science of Diet and Health* (New York: Anchor Books, 2007).

Tudge, Colin, *So Shall We Reap: What's Gone Wrong with the World's Food—and How to Fix It* (London: Allen Lane, 2003).

Wilson, Edward O., *The Future of Life* (New York: Vintage Books, 2002).

Wirzba, Norman, ed., *The Essential Agrarian Reader: The Future of Culture, Community, and the Land* (Lexington, KY: University Press of Kentucky, 2003).

INDEX